Dynamic Response of Advanced Ceramics

Dynamic Response of Advanced Ceramics

Ghatu Subhash
University of Florida
Florida, USA

Amnaya Awasthi
University of Florida
Florida, USA

Dipankar Ghosh
Old Dominion University
Virginia, USA

The American Ceramic Society

WILEY

Published by John Wiley & Sons, Inc., Hoboken, New Jersey.
Published simultaneously in Canada.

For general information on our other products and services or for technical support, please contact our Customer Care Department within the United States at (800) 762-2974, outside the United States at (317) 572-3993 or fax (317) 572-4002.

Wiley also publishes its books in a variety of electronic formats. Some content that appears in print may not be available in electronic formats. For more information about Wiley products, visit our web site at www.wiley.com.

Library of Congress Cataloging-in-Publication Data Applied for:

ISBN: 9781119599777

Cover Design: Wiley
Cover Image: © sensationaldesign/iStock/Getty Images; Courtesy of Ghatu Subhash, Amnaya Awasthi

Set in 9.5/12.5pt STIXTwoText by SPi Global, Chennai, India

In loving memory of our parents whose sacrifices have shaped our lives...

Contents

Preface

This book is a compilation of recent advances in the field of dynamic behavior of brittle materials with emphasis on advanced structural ceramics and transparent materials. As this book covers basic concepts and few advanced topics in the field, it is intended as an introductory guide as well as a comprehensive review for students, beginning researchers, and practicing engineers. The few books written on the dynamic behavior of materials mainly cover metals, and there is a lack of compiled knowledge on the response of nonmetallic materials, especially brittle ceramics and transparent materials with emphasis on their impact behavior. In the last few decades, these materials have become workhorses for applications against threats from high velocity impacts. Significant progress has been made in understanding the mechanisms of deformation under such conditions, and these advances have motivated the authors to create this monograph focused on the current research on the dynamic deformation of ceramics and transparent materials.

The book begins with a brief history of ceramics and then presents a few common examples of dynamic deformation, explaining real-life scenarios and providing an introductory rationale for some key observations. The importance of understanding fundamentals of mechanical wave propagation principles is highlighted, followed by a description of various experimental methods to achieve dynamic deformation. The book often refers to the impact of a projectile on a confined ceramic target, which serves as an overarching example of dynamic loading and the complexities of material response ranging from cracking to phase transformations to spalling. Models for a variety of material deformation modes are presented to help understand the complex mechanisms in play during dynamic loading.

Realizing the limitations of experimental methods to identify material responses under high-strain-rate loading, we have emphasized state-of-the-art atomistic simulations to unearth phenomena which simply cannot be observed experimentally. These simulations have resolved two longstanding concepts, amorphization of boron carbide and if failure waves exist in glasses. The monograph explains amorphization behavior in boron carbide, a mystery that had remained unresolved for almost a quarter century. Using atomistic simulations, the book describes the thermodynamics approach central to solving this mystery. Similarly, the existence of failure waves in glasses upon shock loading was critically examined and it was concluded that the damage propagation does not follow wave characteristics, indicating that there is no failure wave propagation in glasses. Thus, the book is unique in its selection of fundamental concepts.

The authors acknowledge that there is certain overemphasis on their own research topics, but effort has been made to include a broad range of topics of general interest. Not just a repository of recent knowledge, the contents covered in this book are well suited for a semester of graduate level course or can be condensed for a two-day short course for practicing engineers and scientists in industry and defense laboratories. The first author actively covers these contents in advanced graduate level course at the University of Florida. The core content was also adapted for a two-day technical course taught at large companies, conferences, and universities: General Motors (March 1997 – Toluca, Mexico; February 1998 – Vandalia, OH; and September 1998 – Milford, MI); Creative Composites (April 2011 – Crystal River, MI); Combined Annual Meeting of the American Ceramic Society and MS&T (October 2009 – Pittsburgh, PA); and the Hopkins Extreme Materials Institute (HEMI) at Johns Hopkins University (JHU) (June 2014 and March 2017 – Baltimore, MD).

Finally, we hope that this book will inspire the next generation of young researchers to pursue careers in the dynamic field of material behavior under high-strain-rate loading. The need for new ways of thinking is fundamental in breaking traditional boundaries and making leaps in any scientific field. By summarizing recent developments on broad range of experimental, computational and theoretical concepts in this field, we made this knowledge accessible to both novices and experts in the field. Our sincere hope is that readers will not only gain new insights into dynamic behavior of ceramics and glasses but also be inspired to move the field forward in exciting new directions.

Acknowledgments

I was exposed to the field of dynamic behavior of materials by two eminent scholars: Professor Sia Nemat-Nasser*, my graduate advisor at the University of California San Diego and Professor G. Ravichandran, my postdoctoral advisor at California Institute of Technology. Since then, I have continued to learn much from fellow researchers at various universities. However, most of my excitement for knowledge and insights have come from interactions with my exceptionally talented graduate students and postdoctoral researchers. They have been instrumental in forcing me to think beyond the obvious and fueled me with their innovative approaches to the challenges in this field. Their meticulous efforts and insightful arguments have pushed scientific boundaries and yielded new theories in this field. It has been my privilege to have advised these brilliant students, now working in academia, government labs, and private industry all over the world.

I started the journey of writing this book almost nine years ago with my then graduate student, Dipankar Ghosh, who is now a professor at Old Dominion University and also a coauthor. Demands of academic life did not deter our efforts, and we continued adding relevant contents bit by bit. Then, Dr. Amnaya Awasthi joined my research group from the University of Illinois Urbana-Champaign (UIUC) as a post-doctoral research fellow. His insightful work on atomistic simulations, and creative new thinking accelerated our efforts to decipher the mystery behind the amorphization behavior of icosahedral boron-rich ceramics. Amnaya and my former student Cody Kunka developed the computational framework for Raman spectra of boron carbide and then Amnaya extended the analysis to determine spectral shifts due to applied compression. He also guided my students in MD simulations of shock response of boron carbide that opened an entirely new direction for research of this enigmatic material. Due to his

* While the proofs of this book were being finalized, we heard the disheartening news of Prof. Nemat-Nasser's demise. The mechanics community has lost an eminent scholar. We dedicate this book to him in remembrance of his numerous noteworthy contributions in research, education, mentorship, and philanthropy.

exceptional knowledge in this area, I invited him to coauthor this book. His intrepid approach to confronting complex issues and talent to forge new theories have inspired many of my students and resulted in innovative approaches to decipher the mystery of amorphization behavior of boron carbide. I am grateful to my coauthors to have kept the flame alive while accompanying me on this long trek.

Once all the contents were compiled, this book was eye-hawked at the right time by one of my former students, Dr. Alison Trachet, whom I extend a special thanks to. Alison made a deep-dive into reviewing the substance and provided the much-needed organizational layout which the coauthors had overlooked. Incidentally, Alison was also instrumental in providing suggestions to several earlier drafts of this book, many years ago. I am extremely thankful to her critical review, meticulous attention to detail, and suggestions for improvement of this book, and quite serendipitously, over the timeline. This book would have hardly witnessed completion if not for the dedicated efforts of another former student, Dr. Salil Bavdekar, who edited many drafts. In addition, I have been blessed to have advised many talented graduate students who have directly or indirectly contributed to this effort: Drs. Brian Koeppel, Josh Loukus, Wei Zhang, Adam Loukus, Hong Wang, Chengyi Huang, Edward Haney, Qunli Liu, Hongwen Zhang, Timothy Walters, Jiwoon Kwon, Michael Klecka, Phillip Jannotti, John Pittari, Abir Bhattacharyya, Shafiq Muhammed, Canchi Saranya, Bryan Allison, Gregory Parsard, Cody Kunka, Zhichao Chen, Mathew DeVries, Amith Adoor Cheenady, James Nance, Hemanth Thandaga Nagaraj, and Kshitiz Upadhyay. Their creative arguments and outstanding research efforts have made worthy entry into many sections of this book. I also want to acknowledge the assistance of Kimia Ghaffari, who as an undergraduate research assistant contributed to figures and edits.

Efforts of this magnitude would not be successful if not for numerous interactions with prominent scholars in the field. I am thankful to Professor K.T. Ramesh of Johns Hopkins University and Professor K. Ravi-Chandar of University of Texas, Austin for their enthusiastic support. I was also inspired by many innovative works of Professor Marc Meyers of the University of California San Diego. My early collaborations during my tenure at Michigan Technological University with Professor Xin-Lin Gao, currently at Southern Methodist University, have been instrumental in developing innovative concepts regarding ceramic behavior. Finally, I would like to acknowledge my colleague, Professor Douglas Spearot at University of Florida, for many illuminating discussions on the topic of molecular dynamics simulations and for allowing me to sit in his lectures and learn Fundamentals of Quantum Mechanics.

Over the last 30 years, my research in the field of the dynamic behavior of ceramics has been funded by the Army Research Office, Army Research

Laboratory, National Science Foundation, and the Department of the Army. It has been my privilege to have had this support and thank all my program managers for endorsing my research efforts. I specifically want to acknowledge my technical interactions with numerous ARL scholars who have led this field with distinction and challenged me to explore uncharted paths to solve complex problems.

Finally, no words can express my gratitude to my loving family who have been the foundational pillars of my exciting journey. I have been blessed with unlimited support and love from my wife Shobha who has been vital in my efforts and accomplishments to date. She is my best friend, a worthy critic and wholesome adviser in all aspects of my work and life. During the last five months before the completion of this book, our daughter Sanika and son Sohan returned home from their colleges due to the COVID-19 shutdown. It was a blessing I never expected to receive in my life after they left for college. Their presence for 24hours a day over several months at home was invigorating and allowed me to work tirelessly to complete this book. In addition to the scintillating discussions around the dinner table each day, Sanika's incredible international cuisine kept my spirits up and gave me the needed impetus to keep working. With no demands and only love, my family has graciously allowed me to work diligently on this book, and I am forever grateful. This effort would not have been completed without the unwavering support of my sisters, brothers, and their families in India who have been taking care of my ailing mother. I am forever indebted to them for their sacrifices and support.

Ghatu Subhash

1

A Brief History of Ceramic Materials and Introduction to Their Dynamic Behavior

1.1 Introduction

The word "ceramic" has its origins from Greek word "Keramos" which refers to "pottery" or "potter's clay" (Rice 2015). The art of pottery making may have originated independently in various locations during prehistoric times, but the first use of clay to sculpt figures was traced to over 25 000 years ago (Vandiver et al. 1989). Making ceramic figurines or pottery involves mixing moist clay with temper (e.g. sand, small rocks, or crushed pottery), forming the desired shape, and firing the mixture over an open fire or in a kiln (Rice 2015). In ancient times, ceramics were used for storage (e.g. Greek and Roman storage vessels, called amphorae), as building materials (e.g. Roman concrete and clay bricks), and for decoration (e.g. beads, porcelain, and salt-glazed tiles). Ceramics were perhaps the earliest human-made materials from processed raw materials. Today, the word "ceramics" encompasses many materials beyond pottery. Ceramics are nonmetallic, inorganic solids with ionic and covalent bonds (Callister 2013). Most ceramics are metallic oxides or nonoxide compounds. They have high melting point, high hardness, and superior strength compared to metals. A ubiquitous, naturally occurring ceramic is sand, which is composed of silicon dioxide (SiO_2) or silica, and it is the basis for glass, an amorphous solid that has many features common to a ceramic. Diamond, the hardest material known to humankind, is another naturally occurring, highly desired ceramic composed entirely of carbon. Cubic zirconia (zirconium oxide or ZrO_2), an inexpensive substitute for diamond, is also a ceramic. The high melting point of some ceramics such as aluminum oxide (Al_2O_3 or alumina) allows them to be used in kilns and high temperature furnaces. Transitional metal diborides such as zirconium diboride (ZrB_2) and hafnium diboride (HfB_2) are used as ultrahigh temperature ceramics for applications in leading edges and external tiles for hypersonic vehicles. High hardness and wear resistance allow ceramics such as tungsten carbide and silicon carbide to be used as cutting tools, grinding wheels, abrasives, and in brake pads.

Dynamic Response of Advanced Ceramics, First Edition.
Ghatu Subhash, Amnaya Awasthi, Dipankar Ghosh.

The high hardness (more than two to five times that of hardened steel) of alumina, boron carbide (B_4C), and silicon carbide (SiC) makes them preferred candidates for armor applications. The same properties also make silicon nitride (Si_3N_4) a good candidate for aerospace bearings while the high resistance to radiation and temperature of uranium oxide (UO_2) allows it to be used as a nuclear fuel.

Compared to metals, most structural ceramics also typically exhibit excellent chemical resistance, high elastic moduli, high hardness, but lower densities, as can be seen in Figure 1.1. Ceramics also tend to have high compressive strength, but low tensile strength and low fracture toughness. Upon application of load, ceramics mostly undergo elastic deformation followed by brittle fracture. Although dislocations have been observed in alumina (Staehler et al. 2000), zirconium diboride (Ghosh et al. 2009, 2010), and boron carbide (Reddy et al. 2013), brittle fracture is still the dominant macroscopic deformation mechanism when the applied stress exceeds the ultimate strength of the material. Other inelastic deformation mechanisms are also observed such as transformation toughening in partially stabilized zirconia (Subhash and Nemat-Nasser 1993) and phase transformation in aluminum nitride (AlN) (Grady 1998; Heard and Cline

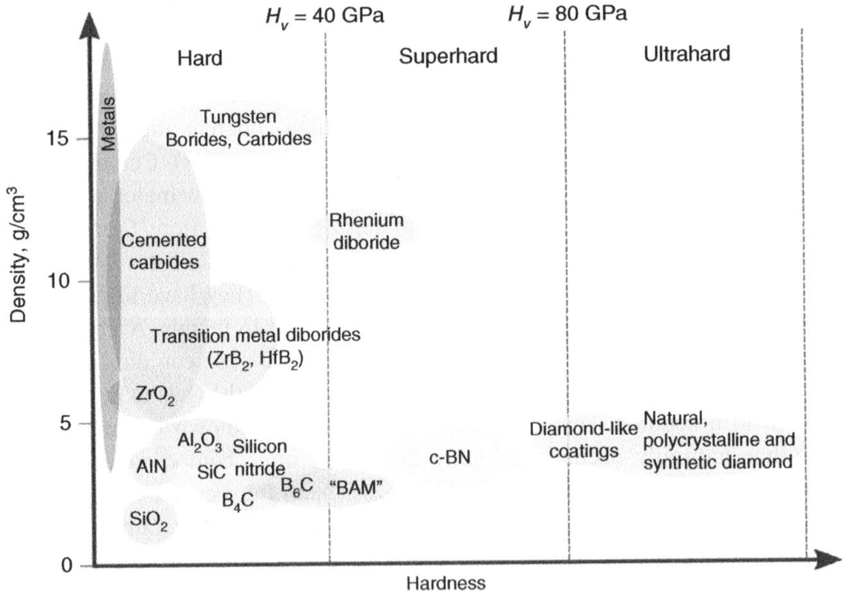

Figure 1.1 Distribution of density versus hardness for various structural ceramics. Note low hardness and high density of metals compared to ceramics. Source: Reproduced with permission from Awasthi and Subhash (2020). © 2020, Elsevier.

1980). Though most ceramics exhibit poor electrical and thermal conductivities, several others also serve as semiconductors (e.g. SiC and B_4C), and a few (e.g. ZrB_2) have thermal conductivities approaching those of metals (Callister 2013). Diamond even has a thermal conductivity greater than copper (Callister 2013). Several ceramics such as ZrB_2 have excellent resistance to thermal shock, in addition to exhibiting thermal conductivity close to that of copper (Ghosh et al. 2009, 2010; Subhash et al. 2008). Depending on the properties of interest, ceramics are attractive candidates for a variety of applications including structural uses (construction materials such as bricks, roof and floor tiles, cookware, and pottery), aesthetic purposes (gemstones such as sapphires and rubies and paint pigments), industrial components (toll bits, high-temperature crucibles, tiles for space shuttles, nozzles, disk brakes, biomedical implants, nuclear fuel and cladding, and bearings), and military applications (body armor, ceramic bullets, nose cones for missiles, and vehicle windows).

While the use of ceramics dates to prehistoric times, efforts have only recently been underway to better understand their behavior, particularly for complex, modern applications such as machine tools, nuclear cladding, and protective armor. Early studies from the 1950s to the 1980s have shown that ceramics are highly sensitive to defects and susceptible to unexpected fracture (Griffith 1921; Irwin 1957; Zhou and Molinari 2004). Probabilistic approaches based on defect population and distribution (Weibull 1939) were developed for predicting ceramic strength and to establish the reliability of ceramics for industrial use. In recent years, advancements in processing technology have resulted in ceramics with fewer inhomogeneities and more uniformity in microstructure, properties, and composition. Most of the initial studies conducted on processed ceramics obtained a quasi-static mechanical response, where the load is applied slowly, over several seconds to hours. Such slow loading rates lead to a deformation rate in the range of $10^{-5}\ s^{-1}$ to $10^{-2}\ s^{-1}$. Note that the rate of deformation is defined as "strain rate," which is the magnitude of strain divided by duration of loading in seconds. However, ceramic materials are also used extensively in applications where dynamic loading conditions persist such as those encountered during projectile impact on an armor plate, grinding grit interaction with a ceramic workpiece, meteor impact on celestial objects, and explosive loading on structures. In these applications, the load is applied rapidly in only a few microseconds, resulting in a deformation (strain) rate well in excess of $10^2\ s^{-1}$. Thus, in the last few decades, research focus has shifted from determining quasistatic mechanical response to identifying dynamic deformation mechanisms, especially dynamic fracture and fragmentation characteristics of ceramics as well as understanding the influence of defects on dynamic behavior.

1.2 Examples of Material Response to High-strain-rate Loading

The rate of deformation is an important mechanistic measure that dictates the deformation response of materials. Many engineering materials exhibit strain-rate-dependent mechanical responses in their deformation and failure behaviors. In the following, the influence of high-strain-rate loading on material behavior is illustrated with several examples: Silly Putty™, a mixture of corn starch and water, sandbags as protection against bullets, cannon ball impact on masonry structures, fracture of glass due to impact of a sharp projectile, and spallation due to axial impact of a long cylindrical glass rod.

Example 1.1 Fun for children of all ages, Silly Putty™ is a silicone-based polymer that was accidentally invented by James Wright during World War II (Wright 1951). Interestingly, the material response changes with the rate of applied load; when pulled slowly, it elongates to a large strain in excess of several thousand percent, but when pulled quickly, it snaps, exhibiting a characteristic brittle fracture. These behaviors are evident in Figure 1.2. Silly Putty™ is thus strain-rate sensitive, where the rate of loading influences deformation and failure behavior. For this reason, the polymer can easily be molded into a pancake shape by slowly pressing with hand, but it will shatter into many fragments when struck with a hammer.

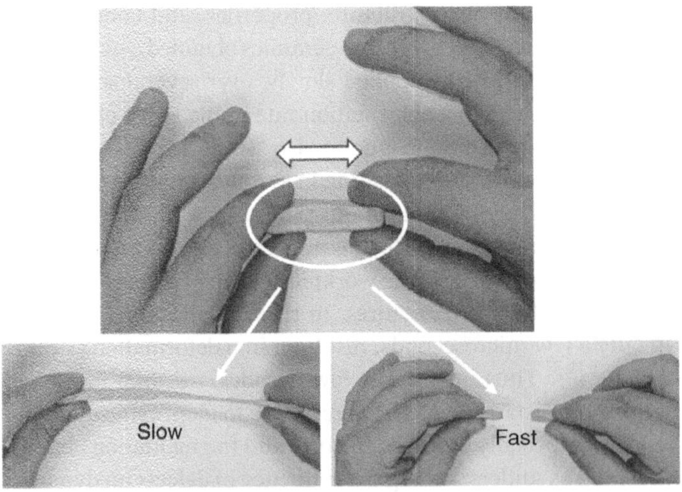

Figure 1.2 Deformation behavior of Silly Putty when subjected to different rates of loading. When pulled slowly, it elongates over a large strain (bottom left) and when pulled rapidly, it snaps (bottom right).

Thus, at higher loading rates the mechanism of deformation changes from ductile behavior to brittle fracture. Loss of ductility is one of the major consequences of a high rate of loading.

Example 1.2 A mixture of cornstarch and water exhibits a solid-like response when compressed quickly, but it behaves like a fluid upon release of pressure. Also, when cornstarch is stirred fast with a rod, it strongly resists the motion of the rod, but it is much easier to move the rod if stirred slowly. There are numerous videos that show a person running on the surface of a pool of cornstarch and water, but the person sinks if he or she tries to walk slowly. Although this behavior can be explained in terms of non-Newtonian fluid behavior, the emphasis here is the rate-dependent response of the material.

In both of the above examples, a high loading rate causes loss of ductility or an increased rigidity in a material. While the above two polymeric materials were chosen to illustrate an exaggerated form of rate-dependent behavior, many ceramics and metals also exhibit rate effects to a varying degree in their inelastic responses and failure mechanisms.

Example 1.3 Why do soldiers duck behind sandbags for protection against artillery fire, although the sandbag can be pierced by a sharp knife when pushed slowly? Why will a high-velocity bullet not fully penetrate a sandbag? In both cases, the sharp object must push the densely packed sand particles away from its path to achieve penetration. However, these particles are compacted by the surrounding sand particles that exert pressure. The bullet performs work against this sand mass to move particles from its path in a very short time in order to overcome the inertial resistance of the particles. Unlike the knife-edge that is slowly pushed through the sandbag, the bullet has mere microseconds to displace the sand particles to continue penetration. Thus, a compaction wave encompassing compressed particles and a fracture wave where sand particles are further pulverized, are propagated in the confined sand mass as shown in Figure 1.3. This process draws an enormous amount of energy from the bullet, as it involves friction between the sand particles, physical displacement of particles within a short time and friction between sand and the projectile surface. Faster displacement of sand particles is associated with higher inertial forces and hence draws more energy from the bullet. Thus, the kinetic energy (KE) of the bullet is quickly dissipated through these mechanisms, and the bullet stops just within a few inches of the initial penetration point into the sandbag. The inertial resistance offered by the confined sand is a protective mechanism against the KE of the bullets, and the use of sandbags makes perfect sense for protection against small arms projectiles. Faster bullets will increase the resistance from the sandbag. In summary, another characteristic of high rate loading, dominance of inertia, is illustrated.

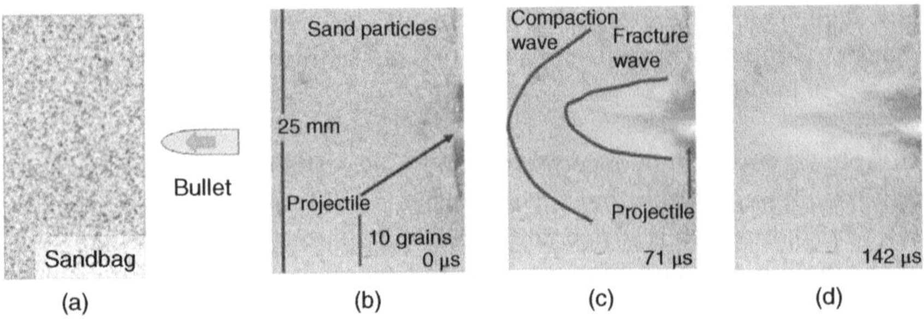

Figure 1.3 (a) Schematic illustration of a bullet (dart) impacting a confined sand mass. (b–d) Time resolved images of sand particles being pushed away to allow penetration. The speed of the bullet is 100 m/s. This process involves inertial movement of particles at high velocities, friction among sand particles and friction between the bullet and the particles, all of which dissipate KE rapidly. Source: Reproduced with permission from Van Vooren et al. (2013).

Example 1.4 One of the unique characteristics of rapid loading of ceramics is extreme fragmentation. Avid movie fans have likely observed the impact of a cannon ball or a projectile on a fort wall. Upon impact, fine debris and powdered material ejects off the impact site, leaving behind a crater and a heavily damaged (cracked) structure. The magnitude of KE in the incoming cannon ball is so great that it cannot be dissipated by the generation of a single crack in the fort wall. Multiple cracks nucleate simultaneously at the impact site and quickly coalesce to form fine debris or small fragments that eject off, taking away some of the incident KE. Away from the impact site, large cracks emanate from the boundary of the impact location and propagate long distances to form larger fragments when they merge. Excess energy is also dissipated rapidly through other deformation mechanisms such as plasticity, phase transformations, and heat generation in the target material. Similar scenarios and deformation modes can also be envisioned when an explosion occurs near a structure or a projectile impact on a brittle target.

Figure 1.4 illustrates the fracture pattern induced by a bullet penetrating a large transparent glass plate (McCauley et al. 2013). The region near the bullet hole appears misty due to a high concentration of cracks and numerous fine fragments. This misty region is surrounded by material with a few larger cracks radiating away from the point of impact. In contrast, if the same material is pushed slowly with a sharp object, no dense cracking or fine fragmentation will be seen; only a few large cracks extending farther into the glass will be noted. Thus, high-rate loading often causes severe fragmentation in brittle materials.

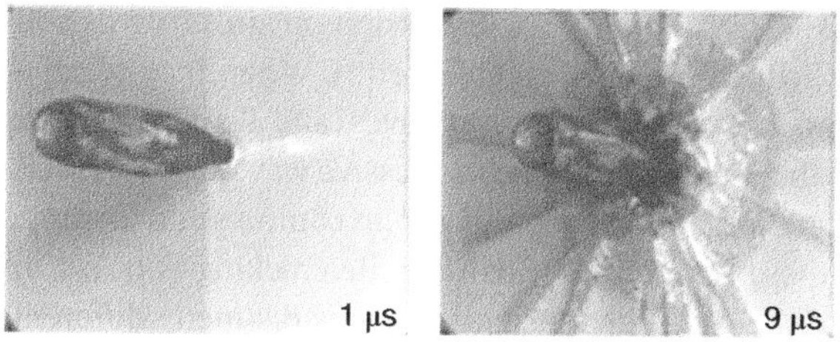

Figure 1.4 Incipient damage in a transparent ceramic (AlON) impacted by a high-velocity projectile. Note the high density of fragmentation close to the impact location and larger fragments and cracking farther away. Source: Reproduced with permission from McCauley et al. (2013).

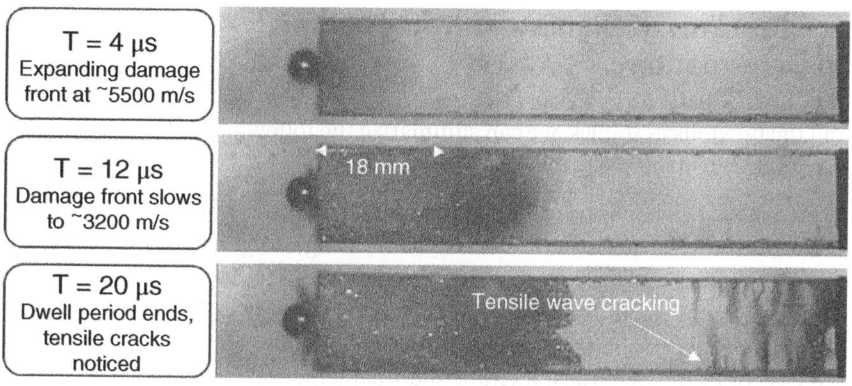

Figure 1.5 Evolution of damage in a transparent ceramic target due to impact by a steel ball. Fine fragmentation and ejecta formation are seen close to the impact location. A reflected tensile wave causes cracking at the free-end at a significantly later time. Source: Reproduced with permission from Haney and Subhash (2013).

Example 1.5 Figure 1.5 depicts another example of impact where a steel sphere is propelled at a glass rod. Upon impact, when the induced compressive stress amplitude exceeds the compressive strength, fragmentation of the material occurs. Small pieces of rubble and fine debris eject from the impact location. Farther away, grain-level cracking is noted (Haney and Subhash 2013). With an increase in impact velocity, the distance over which fine fragmentation occurs also increases. However, the curious feature to note is cracking at the opposite end of the bar seen in the last image.

Why is there delayed fracture at the opposite end of the glass rod, well away from the impact site? A characteristic feature of high-rate loading is stress wave propagation. High velocity impact by a projectile on any surface generates stress waves that carry the impact energy and travel farther away from the impact location. These waves are primarily compressive in nature (more discussion on stress waves and interactions with boundaries will be provided in Chapter 2). When they reach a boundary, they reflect back into the material as tensile waves (and also shear waves if the boundary is inclined to the direction of the compressive wave). As ceramics are more susceptible to cracking under tensile loading than compression loading, damage is initiated close to these boundaries. Such tensile cracking is noted in Figure 1.5 on the right end of the rod. The tensile damage is distinctly different from the compression-induced damage at the impact site. If the amplitude of the tensile stress wave is large enough, it is perceivable that the end of the rod will break apart. This mode of fracture is termed "spall" fracture and will be discussed in detail in subsequent chapters.

1.3 Features of High-strain-rate Loading and Deformation

From the previous examples, we can summarize the following features of ceramic behavior during dynamic loading:

1. The material response transitions from ductile to brittle when the loading rate changes from the quasi-static to the dynamic range. High-strain-rate deformation promotes brittleness (e.g. snapping of Silly Putty® when pulled fast).
2. Under dynamic loads, the material response is strongly influenced by inertial effects (e.g. effectiveness of sandbags against bullet threats).
3. In case of brittle materials, fracture and fragmentation characteristics change with the rate of loading. Higher loading rate often results in greater damage and smaller fragments (e.g. fine debris is generated close to the impact site).
4. Stress waves emanate in the target from an impact location and carry energy over long distances. Though the location of impact undergoes initial compressive loading and extensive damage, additional damage is experienced elsewhere in the component due to wave reflection from boundaries (e.g. tensile cracking in rods impacted by a ball at the other end of the rod).

These are only a few examples. More in-depth discussion and analysis are provided in the ensuing chapters. Suffice it to say, knowledge of wave propagation principles is fundamental to understanding the effects of dynamic loading and the associated fracture and fragmentation behavior of brittle materials.

Several monographs that focus on different aspects of dynamic behavior are currently available. Two comprehensive compilations are *Dynamic Behavior of Materials* by Meyers (1994), and *Physics of Shock and Impact* by Grady (2017a, 2017b). A specialized book by K. Ravi-Chander titled *Dynamic Fracture* (Ravi-Chandar 2004) deals with advanced concepts relevant to few topics addressed in this book. These contents will not be repeated, although a few overlaps are unavoidable. The current book on *Dynamic Response of Advanced Ceramics* differs in content by contrasting static and dynamic behaviors, discussing low-velocity impact damage, describing constitutive models for intermediate strain rate deformation, and discussing the latest research on molecular dynamics simulations. It merges the gap between fundamental theory and practical application and is better suited for students and engineers who have started to work in the field of dynamic behavior and intend to quickly familiarize themselves with both basic and advanced concepts.

1.4 Organization of the Book

In the last few decades, significant progress has been made in developing an in-depth understanding of the behavior of ceramics under high-strain-rate loading. The widespread use of ceramics for lightweight, protective armor applications has been the prime driver of this research endeavor. Impact of a projectile on a ceramic target activates numerous deformation and failure mechanisms in the ceramic to counter the incident KE. This monograph summarizes experimental, analytical, and computational efforts from a large body of literature concerning the mechanisms of deformation during dynamic behavior of ceramics and transparent materials. The emphasis is to provide a mechanistic understanding and description of the deformation processes and quantify the energy absorption. The goal is to summarize major findings and highlight fundamental concepts that characterize the deformation behavior of a large class of ceramics. The book is organized as follows:

Chapter 1 presents a brief history of ceramics and a few real-world examples to highlight the differences between static and dynamic deformation behavior in a variety of materials. It is then followed by a summary of features that govern the high strain rate deformation.

Chapter 2 focuses on experimental methods frequently deployed to assess high-strain-rate deformation and failure of brittle solids. Wave propagation in solids is explained, and key mechanistic concepts with regard to their application in experimental techniques are emphasized. In addition to the well-established uniaxial stress and uniaxial strain experimental methods, both of which probe the constitutive response of materials, new techniques such as dynamic hardness, dynamic fracture toughness, and edge-on-impact experiments are presented. The benefits of each technique in deciphering the underlying physics of brittle fracture are enumerated.

Chapter 3 describes the quintessential example of dynamic loading: a projectile impacting a confined ceramic target. The evolution of complex deformation modes and mechanisms such as shock wave propagation, dynamic fracture, fragmentation and comminution, phase transformation, and others are detailed. These mechanisms are central to understanding the dynamic behavior of ceramic materials. Both prepenetration as well as postpenetration phases are described from a mechanistic perspective, and relevant terms (e.g. dwell, interface defeat, terminal velocity, and V50) for a long-rod-projectile impact on a ceramic target are defined. Early modeling efforts that described the penetration process are also summarized. Finally, the relative importance of wave propagation, material bulk properties, fragmented material properties, and mechanisms of deformation for ballistic performance are highlighted.

Chapter 4 presents a broad review of experimental, analytical, and computational efforts currently available in the literature for describing the dynamic constitutive behavior of intact and damaged ceramics. The influence of defects, microstructure, rate of loading, pressure, stress state, and material condition (intact and damaged) on the dynamic response of ceramics is discussed. The limitations of phenomenological models including the Johnson–Holmquist models, Mohr–Coulomb models, and Drucker–Prager models in capturing the experimentally observed transition from static to dynamic regimes at high pressure are emphasized. The newly developed extended-Mohr–Coulomb model incorporated into dynamic expanding cavity models are presented to show how it better captures the pressure-dependent response as well as the physics of the penetration process. Starting from the wing-crack model, dynamic crack growth and dynamic damage progression in unconfined and confined ceramics are described with the help of a variety of micromechanical, energy-based, and stochastic models. Guided by the knowledge that severe damage evolves in ceramic targets ahead of the projectile during penetration, and given that fragmented material offers significant resistance due to its confined state, new approaches that describe the constitutive response of damaged ceramics are presented.

Chapter 5 presents the shock response of brittle materials at strain rates in the range of $10^5\ s^{-1}$ and beyond. This chapter covers shock wave anatomy, fundamentals of shock propagation in thin and thick ceramic plates, mechanisms of deformation under shock, equations of state, Hugoniot response, spall, strain-rate dependence of strength at ultra-high pressures, viscosity effects, and models for shock response. Of interest are the molecular dynamics simulations of the shock response of ceramics, which provide atomistic detail of deformation modes and thermal effects.

Chapter 6 describes the dynamic deformation behavior of a unique class of advanced structural ceramics with an icosahedral crystal structure. Examples of these ceramics include B_4C, B_6O, and boron–aluminum–magnesium (BAM) compounds. They possess high hardness, second only to diamond-like structural solids, and have emerged as promising light-weight materials for resistance against ballistic loads. However, they possess a unique deleterious high-pressure deformation mechanism called "amorphization." Recent efforts to mitigate amorphization through nano-twinning, enhancing boron content, nanograining, and doping with foreign atoms are presented. Because amorphization occurs at the nanoscale and under high pressures, quantum mechanical and molecular dynamics simulations have recently emerged as powerful tools to capture the underlying physics of amorphization and are highlighted here.

Chapter 7 narrates the dynamic behavior of a variety of transparent ceramics (e.g. single-crystal sapphire, spinels, and AlON) and glasses (e.g. leaded glass, borosilicate glass, soda-lime glass, and chemically strengthened glass). Stress wave propagation and damage propagation under ball-on-rod impact, edge-on-impact on plates, and shock loading are discussed. Crack propagation perpendicular to a bimaterial interface (i.e. composite materials), fragmentation in chemically strengthened glasses and crystalline glass ceramics, strength of damaged glass, and measurement of failure wave velocities under shock wave loading are presented.

Chapter 8 presents emerging research directions on dynamic behavior of ultra-hard structural ceramics as well as challenges in experimental and computational domains. With a large variability in polymorphic and stoichiometric domains, icosahedral boron-rich solids have immense potential for material design with customizable properties. They are good contenders to be examined by "Materials Data Science" (MDS), an emerging concept in materials discovery. The importance of quantum mechanical calculations and molecular dynamic simulations in deciphering deformation mechanisms at the fundamental level is emphasized, and directions for designing of ceramics with tailored properties to withstand dynamic loads is outlined. Rapid advances in additive manufacturing and 3D printing offer unprecedented economic benefits and the ability to fabricate ceramic parts in complex shapes. 3D-printed ceramics are promising materials for protection applications including body armor. Many of these concepts are in their infancy, and the necessity for a comprehensive approach in connecting fabrication, characterization, predictive modeling, and property-optimization is emphasized.

References

Awasthi, A. and Subhash, G. (2020). Deformation behavior and amorphization in icosahedral boron-rich ceramics. *Progress in Materials Science* 112: 100664.

Callister, W.D. (2013). *Materials Science and Engineering: An Introduction*. Wiley.

Ghosh, D., Subhash, G., and Bourne, G.R. (2009). Inelastic deformation under indentation and scratch loads in a ZrB_2–SiC composite. *Journal of the European Ceramic Society* 29: 3053–3061.

Ghosh, D., Subhash, G., and Orlovskaya, N. (2010). Slip-line spacing in ZrB_2-based ultrahigh-temperature ceramics. *Scripta Materialia* 62: 839–842.

Grady, D. (1998). Shock-wave compression of brittle solids. *Mechanics of Materials* 29: 181–203.

Grady, D. (2017a). *Physics of Shock and Impact*, vol. 1. IOP Publishing.

Grady, D. (2017b). *Physics of Shock and Impact*, vol. 2. IOP Publishing.

Griffith, A.A. (1921). The phenomena of rupture and flow in solids. *Philosophical Transactions of the Royal Society of London, Series A*, Containing Papers of a Mathematical or Physical Character 221: 163–198.

Haney, E. and Subhash, G. (2013). Damage mechanisms perspective on superior ballistic performance of spinel over sapphire. *Experimental Mechanics* 53: 31–46.

Heard, H. and Cline, C. (1980). Mechanical-behavior of polycrystalline BeO, Al_2O_3 and Ain at high-pressure. *Journal of Materials Science* 15: 1889–1897.

Irwin, G.R. (1957). Analysis of stresses and strains near the end of a crack traversing a plate. *Journal of Applied Mechanics* 24: 361–364.

McCauley, J., Strassburger, E., Patel, P. et al. (2013). Experimental observations on dynamic response of selected transparent armor materials. *Experimental Mechanics* 53: 3–29.

Meyers, M.A. (1994). *Dynamic Behavior of Materials*. New York: Wiley.

Ravi-Chandar, K. (2004). *Dynamic Fracture*. Elsevier.

Reddy, K., Liu, P., Hirata, A. et al. (2013). Atomic structure of amorphous shear bands in boron carbide. *Nature Communications* 4: 2483.

Rice, P. (2015). *Pottery Analysis*. University of Chicago Press.

Staehler, J.M., Predebon, W.W., Pletka, B.J., and Subhash, G. (2000). Micromechanisms of deformation in high-purity hot-pressed alumina. *Materials Science and Engineering: A* 291: 37–45.

Subhash, G. and Nemat-Nasser, S. (1993). Dynamic stress-induced transformation and texture formation in uniaxial compression of zirconia ceramics. *Journal of the American Ceramic Society* 76: 153–165.

Subhash, G., Maiti, S., Geubelle, P., and Ghosh, D. (2008). Recent advances in dynamic indentation fracture, impact damage and fragmentation of ceramics. *Journal of the American Ceramic Society* 91: 2777–2791.

Van Vooren, A., Borg, J., Sandusky, H., and Felts, J. (2013). Sand penetration: a near nose investigation of a sand penetration event. *Procedia Engineering* 58: 601–607.

Vandiver, P., Soffer, O., Klima, B., and Svoboda, J. (1989). The origins of ceramic technology at Dolni-Vestonice, Czechoslovakia. *Science* 246: 1002–1008.

Weibull, W. (1939). A statistical theory of the strength of materials. *Proceedings of the Royal Swedish Institute of Engineering Research* 151: 1–45.

Wright, J.G. (1951). Process for making puttylike elastic plastic, siloxane derivative composition containing zinc hydroxide. Google Patents.

Zhou, F. and Molinari, J. (2004). Stochastic fracture of ceramics under dynamic tensile loading. *International Journal of Solids and Structures* 41: 6573–6596.

2

High-strain-rate Experimental Techniques

2.1 Introduction

The damage evolution in brittle solids such as ceramics and glasses during a dynamic event is a complex event comprised of multiple spatially and temporally varying responses and deformation mechanisms. The evolution of inelastic failure mechanisms in intact as well as damaged ceramics not only depends on microstructural features such as material composition, crystal anisotropy, texture, and internal defects, but also on loading type, stress state, strain rate, geometry, and design parameters (Medvedovski 2010). Therefore, an in-depth understanding of the influence of all these aspects is fundamental to design impact-resistant materials and structures. However, material designers typically rank ceramics for applications in dynamic environments based on a few material properties, such as compressive strength or hardness, usually obtained under quasistatic testing conditions. Such performance indices based solely on material properties could be misleading because of the complex interaction between material properties and failure mechanisms governed by design parameters and loading conditions. Therefore, material properties alone cannot fully predict the performance of a ceramic in a given application, and the mechanisms of deformation should also be considered. For example, boron carbide is the third hardest ($\sim$30 GPa Vickers hardness) and one of the lightest structural ceramics (2.52 g/cm^3 density). Based on these two measures alone, it appears to be an ideal ceramic for lightweight protective systems. However, boron carbide undergoes a structural weakening mechanism referred to as "amorphization" that precludes its use in high-velocity impact applications (Chen et al. 2003; Dandekar 2001; Grady 1994). Focusing only on material properties such as strength, hardness, and density while neglecting the inherent inelastic deformation mechanisms, could lead to an unexpected inferior performance of the ceramic in certain scenarios. Similarly, sapphire (single-crystal aluminum oxide) has many mechanical properties that are superior to spinel (a glass-ceramic), but its ballistic performance has consistently been

Dynamic Response of Advanced Ceramics, First Edition.
Ghatu Subhash, Amnaya Awasthi, Dipankar Ghosh.

found to be inferior to that of spinel (Haney and Subhash 2013; Patterson et al. 2009; Strassburger 2009; Swab et al. 2000). Both of these issues will be discussed in detail in the ensuing chapters.

If a single test or a single material property is insufficient to accurately predict the performance of a given ceramic material in dynamic environments, how do we design impact-resistant materials? It is impractical to subject every ceramic to every imaginable dynamic loading situation. Researchers use well-characterized, laboratory loading conditions to observe inelastic deformation mechanisms that could correlate to the failure mechanisms under complex loading conditions. A popular approach is to select an appropriate high-strain-rate test that best captures the essence of the stress state and failure processes associated with an intended application. Testing at a higher loading rate is necessary because many inelastic deformation mechanisms that evolve under high-rate loading conditions cannot be simulated under quasistatic loading conditions. For example, high-rate loading on a brittle material causes multiple cracks, extensive fragmentation, and pulverization that is not typically seen in quasistatic loading. Such failure mechanisms influence the overall performance of the structure. Different experimental methods for high-strain-rate testing are categorized as (Normandia 2004):

a. Materials characterization experiments
b. Phenomenological experiments
c. Design/performance-oriented experiments.

Materials characterization experiments provide fundamental, property-based information such as failure strength, fracture toughness, indentation hardness, flexural strength, and dynamic yield strength or Hugoniot elastic limit (HEL), as well as the rate-dependent constitutive response. They provide valuable insight into structure–property relationships. To obtain these properties, experiments are conducted under well-defined loading conditions (such as uniaxial compression or tension and three- or four-point bending) so that the derived property-based information can be used for structural design, comparison with other materials, and in computational models. However, very few standards exist in the realm of high-strain-rate testing methods, even for well-established methods like the Split Hopkinson Pressure Bar (SHPB).

Phenomenological experiments on the other hand, focus on a specific phenomenon of interest, so as to gain a fundamental understanding of the underlying physics of the process. For example, to investigate damage evolution in a transparent material due to impact-induced wave propagation, experiments such as ball-on-rod and edge-on-impact tests are typically filmed with high-speed cameras to capture visual damage mechanisms (Haney and Subhash 2012; McCauley et al. 2013). Other examples, such as dwell-penetration tests or Lundberg tests (Anderson et al. 2011; Lundberg et al. 2000) or long-rod depth-of-penetration tests

(Orphal et al. 1996; Subramanian and Bless 1995) provide significant insights into the interactions between a projectile and a ceramic target.

Finally, design- or performance-oriented tests are employed to evaluate a specific design concept to determine the optimum configuration of the protective system to achieve maximum effectiveness (e.g. better penetration resistance) or to evaluate threshold requirements. An example is impact of a projectile on a ceramic protective system consisting of front and back plates held in a confined configuration. However, the performance-oriented experiments are expensive and time-consuming, and the interplay between various components of the system toward overall performance is often difficult to interpret. Isolating the influence of an individual component is nontrivial in these experiments. A good summary of the discussions on phenomenological and design-oriented tests can be found in Normandia (2004).

With this introduction, we now describe various high-strain-rate experimental methods commonly used to investigate the dynamic behavior of ceramics. We begin by briefly reviewing wave propagation principles, which are central to designing a well-characterized, high-strain-rate experimental method. More detailed descriptions of the principles of wave mechanics as applied to high-strain-rate test methods are available in several sources (Chen and Song 2011; Kolsky 1964; Meyers 1994; Nemat-Nasser 2000; Ramesh 2008).

2.2 1D-Wave Propagation Principles

Dynamic loading inherently involves stress wave propagation through a material. Depending on the magnitude of the imposed stress or impact velocity, elastic, plastic, and shock waves can be generated in the material. Elastic stress wave propagation is critical for the design of dynamic loading experiments, and a brief discussion on elastic stress waves is presented here. The details of shock wave propagation as applied to ceramics and glasses are given in Chapters 5, 6, and 7.

2.2.1 Elastic Stress Waves

2.2.1.1 Types of Stress Waves

A brief tutorial of one-dimensional (1D) elastic stress wave propagation principles is presented here. For more in-depth information, the reader is referred to monographs by Graff (1975), Achenbach (1973), Kolsky (1954, 1964), and Meyers (1994). An elementary description of wave propagation in a 1D solid is also available in Subhash and Ridgeway (2018). From an introductory "Strength of Materials" course, we know that stress can be decomposed into normal and shear stress components. Similarly, under dynamic loads, two basic types of elastic stress waves are generated: a longitudinal stress wave (also referred to as a primary wave, P-wave, or dilatational wave) and a shear stress wave (also referred to as a secondary wave, S-wave, transverse wave, or distortional wave). A stress wave can be thought of as a propagating pocket of energy of finite duration and amplitude within a medium, and each type of stress wave is characterized by the nature of particle motion induced in the material in relation to the direction of wave motion as described below:

a. *Longitudinal stress wave*: This wave is characterized by particle motion parallel to the direction of wave propagation. There are two possibilities: the neighboring particles can come close together indicating the compressive nature of the wave, or they can be pulled apart, indicating the tensile nature of the wave.
   i. If the longitudinal wave is *compressive* in nature, the particle movement is in the *same* direction as that of the wave propagation. When a compression wave arrives, each particle exerts a force on the next neighboring particle causing a "push" or "compression" on that particle. However, the particles need not be bonded together to experience this push; they only have to sit next to each other to exert or experience the compressive force. An example is falling dominoes.
   ii. If the longitudinal wave is *tensile* in nature, the particle motion is in the *opposite* direction to that of the direction of wave propagation. Therefore, when a tensile wave arrives, two adjacent particles move away from each other causing a "pull" or "tension" in between them. If the two particles are not bonded or if the magnitude of tension in the wave is greater than the bond strength, the particles will separate, creating a crack.
b. *Shear wave*: In this case, the particle motion is perpendicular to the direction of wave propagation causing shear stress in the material points. An example is "the wave" performed at sporting events. People (the particles) move vertically while the wave propagates horizontally through the stadium or arena. Now, one can visualize why so much energy is expended in shear compared to compression. It takes significantly greater amount of energy to stand up and do the wave (shear), rather than simply push (compression) the next person (particle).

To better illustrate the wave and particle motions, consider the schematics shown in Figure 2.1, where a particle is represented by a human figurine, and the set of human figurines with similar motion constitute a wave of certain duration or length. Upon the arrival of the wave, the particles move either in the same direction (compressive wave) or opposite direction (tensile wave). In the case of shear, the particle motion is depicted as vertical displacement. Note that for compression wave propagation, the human figurines (particles) in Figure 2.1a are not holding hands (not bonded), whereas for tensile wave propagation Figure 2.1b, they hold hands (particles are bonded). This implies that tension cannot transfer to the next particle unless the two particles are bonded,

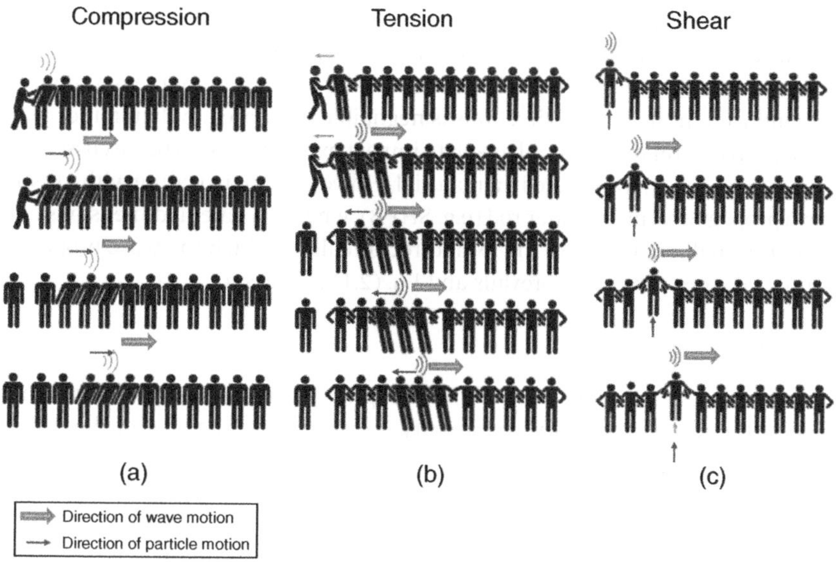

Figure 2.1 Examples of different types of stress waves with people acting as the particles: (a) compression wave, (b) tensile wave, and (c) shear wave.

whereas compression can propagate as long as the particles are next to each other, because it involves pushing (motion in the same direction as the incoming wave direction). If the magnitude of the incoming tensile wave is large enough, the hands separate (bonds break) and a crack is created. Lastly, a shear wave is depicted in Figure 2.1c. Each figure or particle must overcome the frictional resistance by the neighboring particle. Shear waves must have sufficiently high energy to cause such a motion, which requires more energy than longitudinal waves. Thus, a shear wave carries significantly greater energy than a longitudinal stress wave, leading to more extreme damage.

Other waves can also be generated during a dynamic impact including Rayleigh, plastic waves, Lamb waves, etc. Rayleigh waves are of particular interest as they relate to dynamic fracture. They propagate along the surface and carry 67% of the impact energy, whereas the shear and compression waves only carry 26% and 7%, respectively, of the total energy (Meyers 1994).

Depending on the nature of the elastic stress wave (longitudinal or shear), they travel at different velocities in a bulk solid. The longitudinal wave velocity, in a bulk solid is given by

$$C_L = \sqrt{\frac{\lambda + 2\mu}{\rho}} = \sqrt{\frac{E(1-\nu)}{\rho(1+\nu)(1-2\nu)}} \tag{2.1}$$

where λ and μ are Lamé constants, E is Young's modulus, ν is Poisson's ratio, and ρ is the density of the material. Equation (2.1) also applies to a uniaxial strain state where it is assumed that lateral dimensions of the solid are large and deformation induced in the lateral directions is negligible compared to that in the axial or wave propagation direction. Flyer-plate impact experiments, to be discussed later, impose a uniaxial strain state in a material. However, when dealing with 1D stress wave propagation in a slender rod (e.g. SHPB), where the lateral dimensions of the solid are small compared to the axial direction along which the wave propagates, a uniaxial stress condition prevails and Eq. (2.1) reduces to

$$C_L = \sqrt{\frac{E}{\rho}} \tag{2.2}$$

For ceramics with a low Poisson's ratio (typically approximately 0.2), the magnitude of the uniaxial strain wave velocity is 5% greater than that of the uniaxial stress wave. The shear wave velocity (also called the transverse wave velocity) under both uniaxial stress and strain conditions is

$$C_s = C_T = \sqrt{\frac{\mu}{\rho}} \tag{2.3}$$

where μ is the shear modulus. The longitudinal wave velocity from Eq. (2.1) and shear wave velocity from Eq. (2.3) are related by

$$C_T = C_L\sqrt{\frac{1-2\nu}{2(1-\nu)}} \tag{2.4}$$

It is important to have a feel for relative wave velocities in metals and ceramics. Assuming an elastic modulus of 200 GPa for metals (e.g. steel) and a density of 7.8 g/cm^3, we can approximate the longitudinal wave velocity to be about 5000 m/s. Similarly, for ceramics, assuming an elastic modulus of 440 GPa (for boron carbide) and density of 2.52 g/cm^3, we obtain a longitudinal wave velocity of 13,200 m/s. Thus, the longitudinal wave velocity in ceramics is almost two to three times greater than that in metals. Similarly, the shear modulus for steel is about 80 GPa and that of boron carbide is 190 GPa, and the shear wave velocities are about 3200 and 8700 m/s, respectively. It is a common practice to assume the shear wave velocity to be about 0.65 times the longitudinal wave velocity for most materials.

2.2.1.2 Interaction of Stress Waves at a Boundary

The nature of particle motion in relation to the wave propagation direction is an important consideration in understanding fracture initiation in ceramics under dynamic loads. Similarly, the behavior of the stress wave when it reaches a boundary or interface is also an important consideration. When a stress wave reaches a fixed or rigid boundary at a normal incidence, it reflects back as the same type of wave (i.e. a tensile wave reflects back as a tensile wave, and a compressive wave reflects back as a compressive wave). However, if a wave reaches a free boundary at normal incidence, the reflected wave is opposite in character (i.e. a compressive wave reflects back as a tensile wave and vice versa).

It is important to understand the interaction when a stress wave reaches an interface between two dissimilar materials. For example, if a compressive wave arrives in a direction perpendicular to an interface, the two surfaces will not separate due to the fact that the wave motion and the particle motion in the first medium are in the same direction, as seen in Figure 2.1a. The particles in the first medium push on the particles in the second medium allowing energy transfer between the two materials, even without being connected or bonded together. On the other hand, when a tensile wave arrives at the same interface, the particle motion is opposite to the wave direction. The interface may transfer the tensile wave to the next particle when the bond strength is greater than the amplitude of the stress wave, or the interface may separate causing debonding if the amplitude of the incoming wave is greater than the interfacial strength. Similarly, when a tensile wave propagates through a material consisting of heterogeneities (e.g. defects) and if the material strength at that particular location is lower than the amplitude of the incoming stress wave, damage (e.g. cracks and voids) may be nucleated. The importance of understanding the relationship between the nature of stress (compressive, tensile, or shear) and motion of the material particles within a stress wave cannot be overemphasized in the dynamic damage evolution process of brittle materials and in designing experimental methods for characterizing their deformation behavior.

Let us consider a composite block consisting of two bodies with dissimilar material properties and different dimensions. When a quasistatic uniaxial compressive load is applied to this composite block, the stress induced in each body depends solely on the area of cross-section of each block (force divided by area). Now, consider a dynamic situation where a stress wave travels at normal incidence through the composite block from one body to another. What are the magnitudes of stresses experienced by each body? The parameter that represents the induced stress under a propagating elastic stress wave is termed "impedance" (φ). Impedance is a product of density (ρ), wave velocity (C), and cross-sectional area (A) perpendicular to the direction of wave propagation:

$$\varphi = \rho CA \tag{2.5}$$

The impedance mismatch between two adjacent bodies dictates the fraction of stress wave transmitted to the adjacent material or reflected back into the same material. If the impedance of the two materials is the same (regardless of the dimensions or type of materials), an incoming compressive stress wave at normal incidence will transfer across the interface unaltered. If the impedances of the two materials are different, then only a portion of the wave is transmitted. The magnitude transmitted is dictated by the impedance mismatch. This phenomenon underlies typical experiments for evaluating the dynamic material response, which are described in the next sections and are categorized as uniaxial stress or uniaxial strain tests. For in-depth details on the influence of impedance on stress transfer across an interface, the reader is referred to Achenbach (1973), Kolsky (1963), and Graff (1975).

2.3 Uniaxial Stress Characterization Methods

As previously mentioned, the characterization tests determine either a material property (e.g. strength, fracture toughness, or indentation hardness) or the constitutive response of a material at a specific strain rate. Relevant experimental techniques that utilize uniaxial (1D) stress wave propagation in a slender rod include

a. Split Hopkinson pressure bar (SHPB) testing,
b. Dynamic indentation hardness tester (DIHT), and
c. Dynamic fracture toughness testing.

2.3.1 Split Hopkinson Pressure Bar (SHPB) Testing

SHPB is the most popular dynamic testing equipment which utilizes 1D wave propagation method to evaluate the constitutive properties of materials in tension, compression, and torsion in the strain-rate range of 10^2–10^4 s^{-1}. An excellent review of the method and its application is given in Chen and Song (2011), and numerous other articles also describe the SHPB technique for testing ductile materials (Chen and Song 2011; Gray III 2000; Kolsky 1964; Lindholm and Yeakley 1968; Meyers 1994; Nemat-Nasser 2000). In this section, we focus on its adaptation to testing brittle materials. While the traditional SHPB has been used for decades to investigate the deformation behavior of ductile materials at high-strain rates, its adaptation to ceramic testing has gained considerable interest only in the last few decades, and significant strides have been made in refining the test method to obtain accurate results. Many of the original assumptions for testing soft, ductile materials (which undergo large plastic deformations) cannot be directly adapted to testing brittle ceramics, which typically have high hardness and exhibit low failure strain (<1%). Therefore, we will first narrate the principles of the traditional SHPB test method and then discuss the assumptions and limitations for testing brittle materials.

2.3.1.1 Conventional SHPB

A conventional SHPB setup for dynamic compression testing is shown in Figure 2.2a. It consists of an incident bar, a transmission bar, a striker bar, and a gas-operated gun to launch the striker bar. The bars are aligned along an axis and supported by frictionless, self-lubricating bushings to allow axial motion during

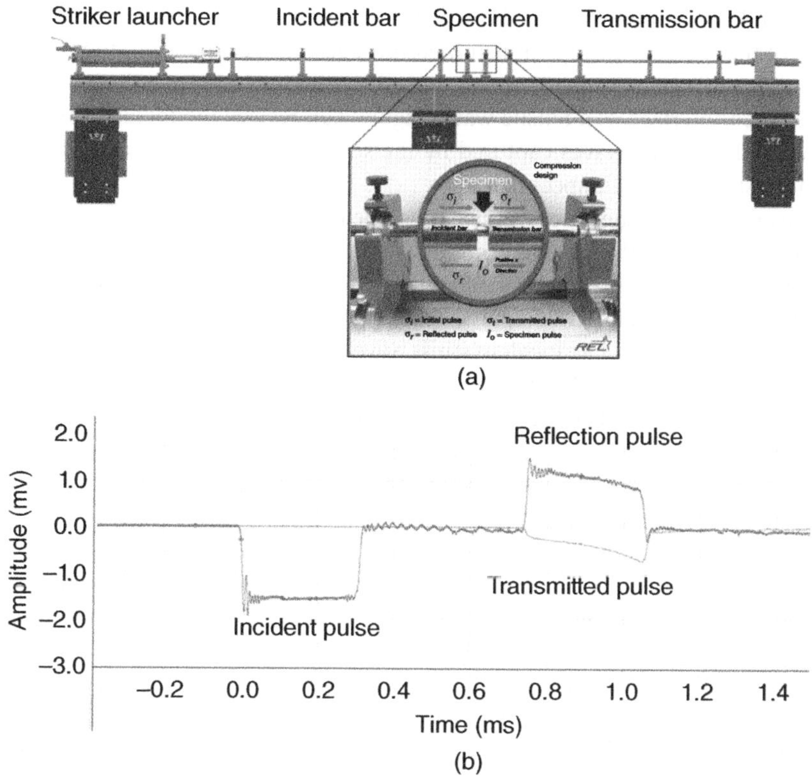

Figure 2.2 (a) A Split Hopkinson pressure bar made from maraging steel. The magnified view illustrates the specimen held between the two bars and (b) strain gage signals from incident and transmission bars during a dynamic test on a metal specimen. Source: Courtesy of REL Inc, Calumet MI, USA.

the test. The specimen is sandwiched between the incident and transmission bars and is held by friction between the surfaces. Strain gages are bonded on the surface to the middle of the incident and transmission bars to capture the elastic stress wave features (time-resolved instantaneous stress amplitude and duration) during the test. The strain gages are connected to a Wheatstone bridge, an amplifier, and a data acquisition system to record and analyze the stress wave propagation data in the bars.

A striker bar of the desired length is launched from the gas gun at a desired velocity toward the incident bar. The axial impact generates a compressive elastic stress wave (called the “incident wave”) in the incident bar and travels toward the specimen. The amplitude of the incident wave is proportional to the impact velocity of the striker, and the duration of the wave is equal to the roundtrip travel time of the longitudinal wave in the striker bar. Upon reaching the specimen, a portion of the incident wave is transmitted through the specimen into the transmission bar as a compression wave, and the rest is reflected back from the incident bar-specimen interface as a tensile wave. Figure 2.2b reveals the incident, transmission, and reflected waves captured by the strain gages on the bars during a dynamic test on a metallic specimen.

To ensure that the specimen receives a sufficiently high-magnitude stress to deform and fracture, the specimen's cross-sectional area and, consequently, its impedance, are intentionally kept smaller than those of the bars. Due to the lower impedance of the specimen compared to that of the bar, the reflected pulse is always tensile in nature. The incident stress wave transfers through the specimen, and after few initial internal reflections, the stress equilibrates (becomes uniform within the specimen) and emerges in the transmission bar. The transmitted stress amplitude is proportional to the stress experienced by the specimen during the wave transfer. Thus, the specimen is subjected to dynamic uniaxial compression for the entire duration of the incident wave.

From the 1D analysis of reflected and transmitted strains measured by the strain gages on the bars (i.e. $\varepsilon_R(t)$ and $\varepsilon_T(t)$, respectively), it has been shown that the transmitted wave strain signal is related to the time-dependent stress response ($\sigma_s(t)$), in the specimen, while the reflected wave strain gage signal is related to the time-dependent strain rate ($\dot{\varepsilon}_s(t)$) in the specimen (Chen and Song 2011; Gray III 2000; Subhash and Ravichandran 2000). The strain rate data is integrated with respect to time to obtain strain-time data ($\varepsilon_s(t)$), in the specimen. The equations relating stress ($\sigma_s(t)$), strain rate ($\dot{\varepsilon}_s(t)$), and strain ($\varepsilon_s(t)$) in the specimen to the measured strain gage signals are

$$\sigma_s(t) = \frac{A_b}{A_s} E_b \varepsilon_T(t) \tag{2.6}$$

$$\dot{\varepsilon}_s(t) = -\frac{2C_L}{l_s} \varepsilon_R(t) \tag{2.7}$$

$$\varepsilon_s(t) = -\frac{2C_L}{l_s} \int_0^t \varepsilon_R(t)\, dt \tag{2.8}$$

where A_s is the cross-sectional area, l is length, E is Young's modulus, and C_L is the longitudinal wave velocity in the bar calculated from Eq. (2.2). The subscripts b and s refer to the bar and specimen, respectively.

By combining the time-resolved signals of stress and strain data, one can construct the stress–strain response of the specimen. The following basic assumptions must be met for the data to be valid (Subhash and Ravichandran 2000):

(1) The impacting bars remain axially aligned, and their end surfaces in contact with the specimen stay flat and parallel to each other. It is also assumed that

the bars remain elastic during the entire test procedure. Recall that the strain gages bonded on the surface at the midpoint of the length of the bars measure the stress wave duration and instantaneous stress amplitude in the incident and the transmission bars. It is assumed that the measured stress on the surface of the bars is representative of the stress in interior of the bars, meaning that the instantaneous stress amplitude remains the same through any cross section of the bars. This is possible only when there is planar axial impact of the striker on the incident bar resulting in a planar stress wave. Therefore, the impacting surfaces should be perpendicular to the axis of the bars, and the impact must occur in a collinear fashion so that the incident pulse has constant stress in the cross-section of the bar.

(2) The specimen undergoes uniaxial and uniform deformation at a constant strain rate during the stress wave loading. This assumption is the same as the one we use when testing dog-bone specimens under tensile loading in quasistatic conditions, where the gage section is assumed to have uniaxial and uniform deformation.

(3) The specimen is subjected to only one incident pulse loading so that the measured deformation (change in length) of the specimen is a result of a single incident pulse. Recall that the reflected stress wave in the incident bar from the specimen is tensile in nature, and this tensile wave travels back toward the impact end (or striker end). When it reaches the striker end of the bar (which is a free surface), it reflects as a compressive wave and returns to the specimen, potentially subjecting the specimen to a second compressive load. This process can repeat several times during a test. In this situation, the measured change in length (or strain) in the specimen will not be the same as that determined from Eq. (2.8) because the measured strain on the physical, posttested specimen is a result of multiple compressive wave loadings, and the strain calculated from the reflected wave signal is from the first reflection only.

2.3.1.2 Modifications to SHPB for Testing Brittle Materials

The SHPB was originally designed to test soft ductile metals that exhibit large plastic deformations in excess of 10%. However, ceramics are hard and exhibit mostly elastic deformation prior to brittle failure. To obtain valid data, the SHPB testing procedure must be modified to meet the following criteria:

Stress Equilibrium in the Sample Prior to Failure

Ceramics are hard and brittle and undergo only a small elastic strain prior to failure (<1% strain). Therefore, the major objective of testing brittle materials with SHPB is to determine failure strength or failure strain as a function of strain rate. However, the ceramic specimen must reach stress equilibrium before this failure stress is reached, similar to having uniaxial and uniform stress states in the gage section of a tensile specimen under quasistatic loading. The specimen is being loaded by a stress wave whose duration is several times greater than the wave transit time in the specimen. When the stress wave arrives at the specimen-incident bar interface, the other end of the specimen in contact with the transmission bar is still unstressed. The stress wave travels through the specimen and reaches the other end (specimen-transmission bar interface), but the stress at the incident bar end has now reached a higher magnitude due to the incoming ramp loading by the incident stress wave. A portion of the wave that has reached the specimen-transmission bar interface reflects back due to an impedance mismatch with the transmission bar and travels to the other end of the specimen while the original incident pulse continues to arrive and load the specimen to higher amplitude. This process of wave reflections within the specimen continues until the stress builds throughout the specimen to the final amplitude. If the ceramic specimen fails during this stage of stress-build up before equilibrium is reached, the measured failure strength is invalid because the stress state was nonuniform in the specimen prior to failure. Thus, the uniformity of stress-state within the specimen must be ensured by allowing a sufficient number of stress wave reflections within the specimen prior to its failure.

A schematic of the wave propagation in the bars and specimen is shown in the distance–time (x–t) diagram in Figure 2.3a. The differences in stress amplitude between the two-ends of the specimen during this process of wave reflections are shown in Figure 2.3b (Ravichandran and Subhash 1994). In Figure 2.3, the vertical axis, $R(t)$ is the difference in stress divided by the average stress at the two ends of the specimen at each instant. Eventually, the stress in the specimen reaches a steady-state, where the stress at both ends become almost equal, and stress equilibrium is reached. It was found that this uniform stress (or stress equilibrium) state is attained throughout the specimen after the wave completes approximately four roundtrips in the specimen. The failure of the specimen should occur only after this duration for the failure strength data to be valid. Because of this stress-equilibrium requirement, the initial portion of the stress–strain response is often invalid, and the associated slope of the stress–strain response is not reflective of the Young's modulus of the specimen. Most recently, Liao and Chen (2018) have comprehensively analyzed the stress equilibration for any bar-specimen material combination, incident pulse shape and impedance mismatch, and provided robust guidelines for selection of suitable test parameters to achieve quick equilibrium in the specimen.

One may wonder why such a restriction is not imposed while testing ductile materials using SHPB. Recall that the SHPB was originally designed to extract the inelastic (plastic) response of ductile materials at high-strain rates. For metals, the failure strain can be significantly greater than the typical elastic strain limit in brittle materials, and the stress equilibrium would be attained within the

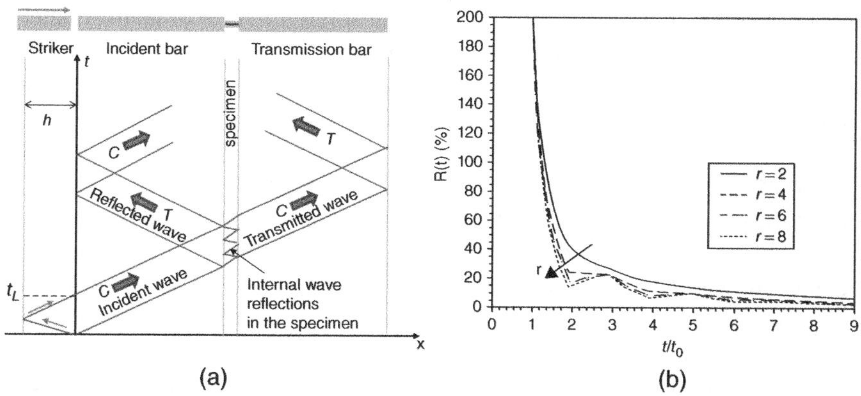

Figure 2.3 (a) Schematic of SHPB and the relevant x–t diagram revealing wave reflections in the bars and the specimen. The arrows indicate wave propagation directions with "C" for compression and "T" for tension. (b) Plot of stress difference between the two ends of the specimen as a function of normalized wave transit time in the specimen. Source: Ravichandran and Subhash (1994). © 1994, John Wiley & Sons.

first few microseconds (or first few wave reflections) of the incident pulse while the specimen is still elastically deforming. Thus, the plastic response in metals is not affected by the time constraint to reach stress equilibrium. Whereas for brittle solids, which exhibit only a small elastic strain before failure, it is imperative that failure should occur only after the specimen has reached stress equilibrium for fracture strength data to be valid.

Let us illustrate this concept with an example. Assume that a striker bar with a length, of 0.150 m (~6 in) is used to generate an incident stress wave in the SHPB. The duration of the incident stress wave is labeled in Figure 2.3a and calculated as follows:

$$t_L = \frac{2L_s}{C_L} = \frac{0.3}{5000} = 60\ \mu s \tag{2.9}$$

The wave speed in the bar is calculated with Eq. (2.2), and for steel it is approximately 5000 m/s. Assuming a typical metallic specimen of length 6 mm, it takes about 1.2 μs (=0.006 m/5000 m/s) for the stress wave to travel once through a metallic specimen. However, for a ceramic with a wave speed of 13,200 m/s, it only takes 0.45 μs to traverse the length of the specimen. As per Figure 2.3b, at least four round trips must be completed before the specimen fails. Thus, the ceramic specimen should fail only after 3.6 μs have elapsed after the incident wave starts to load. This can be verified by measuring the time at the peak value of the transmitted signal, which measures the magnitude of stress reached before the failure of the specimen. Therefore, a rapid rise of stress in the loading pulse can cause failure of the ceramic before stress equilibrium is achieved, making the strength data invalid. For metals the four-reflection duration is 9.6 μs, longer than that for a ceramic, but well below the total loading duration of approximately 60 μs. Thus, stress equilibrium in metal specimens is achieved prior to the beginning of yielding, and plastic deformation continues to occur under stress equilibrium, making the derived plastic response valid.

Specimen Subjected to Uniaxial Stress

One of the prime requirements of SHPB testing is that the specimen be in a state of uniaxial stress during deformation. Unlike metals, ceramics are hard and have significantly higher compressive strength. For example, the hardness of steel is between 5 and 8 GPa (Klecka et al. 2013; Bhattacharyya et al. 2014), whereas the hardness of ceramics such as silicon carbide or boron carbide can be as high as 30 GPa (Ghosh et al. 2012; LaSalvia et al. 2010; Pittari et al. 2015b). Accordingly, the ceramic compressive strength is several times greater than the yield strength of the SHPB steel bars; for instance, boron carbide has a compressive strength of more than 2.5 GPa (Ghosh et al. 2012; LaSalvia et al. 2010; Pittari et al. 2015b), and the yield strength of steel is approximately 0.5 GPa. Recall from Eq. (2.6) that the stress in the specimen is a function of the ratio of the cross-sectional areas of the bar and specimen. For a ceramic to reach its high failure strength, the specimen diameter must be much smaller than the bar diameter. However, testing small, hard ceramic specimens in a SHPB introduces a new problem. When a hard ceramic is compressed between two steel bars, it can indent into the steel bars and introduce stress concentrations. This indentation causes three-dimensional stress states in both the bars and sample and invalidates the uniaxial stress state requirement, rendering the extracted failure strength data inaccurate. Deformation of the bar end-surfaces also violates the assumption that the bar-ends remain flat and parallel during testing. Modification in the testing technique is thus warranted to mitigate this situation.

As discussed earlier, impedance, determined with Eq. (2.5), is an important parameter for the wave propagation behavior in materials. We will utilize this knowledge to avoid indentation of the steel bar by the hard ceramic. We could modify the geometry of the specimen to gradually transition from the specimen size to the bar diameter to reduce the stress concentration from the sudden change in diameter at the interface between the specimen and the bar. But this is an expensive proposition because defect-free machining of such a cylindrical specimen geometry from a ceramic is labor-intensive and expensive due to the brittle nature of the ceramic. An alternate solution is to insert another hard ceramic between the specimen and SHPB whose diameter is intermediate between those of the bar and specimen. However, randomly choosing a cross-sectional area could cause additional wave reflections at these new interfaces due to an impedance mismatch, complicating the interpretation of reflected and transmitted signals. Any modification of the test procedure cannot alter the wave characteristics.

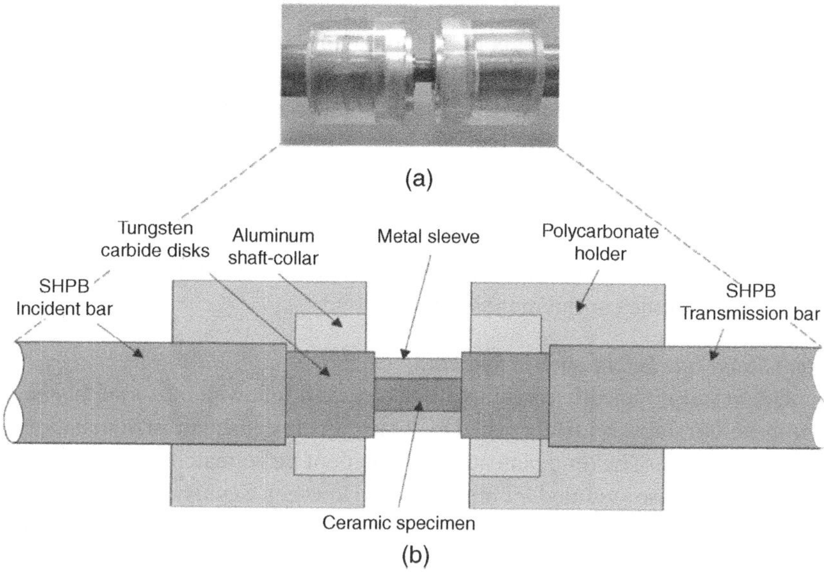

Figure 2.4 (a) Photograph of polycarbonate fixture designed to facilitate mounting of specimen assembly between the two bars in SHPB and (b) schematic of the cross section of the polycarbonate fixture and specimen assembly. Source: (a) Reproduced with permission from Shafiq and Subhash (2016), (b) Shafiq and Subhash (2016). © 2016, Elsevier.

To conform to this doctrine, we match the impedance of the insert to that of the bar material, so that the incident wave transmits through the insert without any alteration. It is a common practice to use tungsten carbide (WC) inserts due to their high hardness and strength. The insert diameter is chosen such that the impedance of the insert is the same as that of the bar:

$$(\rho CA)_{WC} = (\rho CA)_{bar} \tag{2.10}$$

Here, the wave velocity, C, is given by Eq. (2.2). The length of the insert is arbitrary and is often chosen to be about 0.5 times its diameter. The specimen is sandwiched between these WC inserts, and this entire assembly is placed between the incident and transmission bars for testing. The setup is shown in Figure 2.4. All the interfaces in the assembly are lubricated to reduce friction between the contacting surfaces.

However, simply inserting two WC platens is insufficient to protect the bars from indentation. Upon failure of the ceramic specimen during testing, the hard, sharp fragments can produce a stress concentration in the WC inserts and initiate their failure. These fragments can then indent into the bars and cause stress concentrations. To prevent failure of the inserts, they are confined in a metal collar as shown in Figure 2.4 (Subhash 2000). This entire assembly can be too heavy to be held between the SHPB bars by friction alone. Lightweight plastic jackets are then placed over the ends of the bars to house the WC insert and specimen assembly (Shafiq and Subhash 2016). The plastic jackets also assist in centering the inserts along the axis of the bar. The use of these impedance-matched WC inserts allows the ends of the steel bars to remain flat and parallel during the deformation of the specimen, one of the main requirements of SHPB testing. The specimen in Figure 2.4 is shrink fitted with a metal sleeve to impose additional confinement pressure, but similar setups with WC inserts and plastic are also employed without the confinement sleeve.

Constant Strain-rate Deformation in Specimen

The incident pulse typically rises rapidly and is then followed by a long plateau. This trapezoidal shape is ideally suited to obtain plastic deformation in metals at a constant strain rate. The long plateau of the incident pulse matches well with the ideal plastic response assumed for most metals. However, a constant strain rate for ceramics must be achieved during the small elastic deformation. If a traditional rectangular or trapezoidal incident pulse is used (as is the case for metals), the rapid rise time of this stress pulse exceeds the rate of elastic deformation, and the sample will fail before a constant strain rate is reached. The excess energy in the plateau section of the stress pulse will pulverize the specimen, eliminating the possibility of identifying the actual failure mode during axial loading (Subhash and Ravichandran 2000). The rise part of the incident stress pulse must match the slope of the rate of elastic deformation in the specimen to reach a constant strain rate until failure. Deploying a triangular-shaped incident pulse instead of a rectangular-shaped pulse accomplishes this objective. This triangular shaped stress pulse is achieved by attaching a thin metal (copper) disc called a "pulse-shaper" on the impact-end of the incident bar. Upon impact by the striker bar, the pulse-shaper deforms plastically, retarding the transfer rate of energy into the incident bar and producing a triangular-shaped (monotonically rising) incident pulse. The rise time of this pulse needs to be adjusted to suit the slope of the stress–time relationship during testing of the ceramic. Nemat-Nasser et al. (1991) were the first to implement this technique and developed an analytical procedure for copper pulse-shaper deformation. Later, Frew et al. (2002) developed more in-depth procedures for employing different pulse shapers made from various metals and polymers. Figure 2.5 illustrates an example of a triangular-shaped incident pulse generated using a copper pulse-shaper while testing a ceramic in SHPB. Most recently, Liao and Chen (2018) have conducted a comprehensive investigation of the relationship between the incident pulse rise time and stress equilibration for a range of specimen-bar impedance combinations and provided recommendations for appropriate selection of bar material and pulse rise times to achieve faster stress equilibration in a specimen.

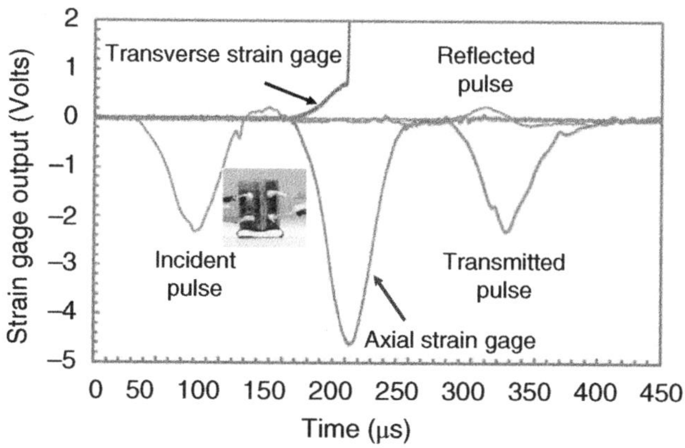

Figure 2.5 Strain gage signals measured on SHPB and on a ceramic specimen using a pulse shaper.

Single Incident-wave Loading of Specimen

While testing ceramics in SHPB to determine their failure strength, it must be ensured that the specimen failure occurs in a single incident wave loading. Recall that it is possible to have multiple compressive loadings of a specimen due to the wave reflections in the incident bar. Even if the incident compressive stress wave amplitude is less than the failure strength of the specimen, the brittle ceramics can develop damage (microcracks) at defects during the first incident pulse loading, and if the specimen is only partially damaged (without complete loss of strength) the arrival of a second compressive wave can completely fracture the already weakened ceramic. However, from Eq. (2.6), only the first transmitted wave is used to calculate the stress in the specimen, and hence, the extracted failure strength data (based on the first transmitted signal) does not correctly reflect the final status of the specimen. To avoid these issues, it should be ensured that the specimen will receive only a single, sufficiently high-amplitude incident stress wave to achieve failure. Caution should be exercised to limit the amplitude of the incident pulse. If not, the excessive energy in the pulse may obliterate the specimen thus masking the failure initiation sites or prevent observation of failure modes such as axial cracking.

To ensure that only a single compressive pulse reaches the specimen, Nemat-Nasser et al. (1991) developed a momentum-trapping (MT) technique for the SHPB. The purpose of the MT is to capture the reflected tensile wave from the specimen in the incident bar so as to prevent a second compressive wave from reloading the specimen. The impact-end of the incident bar was modified to include a "flange-sleeve-rigid mass" assembly, all together referred to as the "momentum trap," shown in Figure 2.6.

The design of the MT is based on the impedance match with the incident bar. The cross-sectional area of the sleeve is equal to that of the incident bar.

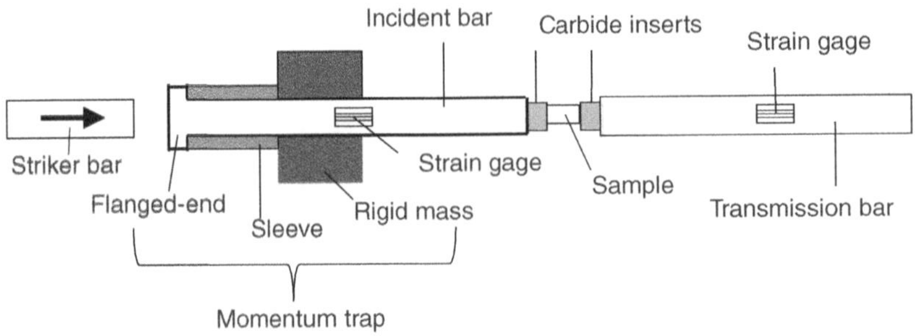

Figure 2.6 A modified Split Hopkinson pressure bar with a momentum trap on the incident bar.

Because the bar and sleeve are made of the same material, and they have the same cross-sectional areas, their impedances are now matched as calculated with Eq. (2.5). The flange diameter on the incident bar is designed to be the same as that of the outer diameter of the sleeve. The rigid mass is of significantly larger diameter than the bar (by three to four times), so the interface between the sleeve and the mass essentially acts as a rigid boundary. The length of the sleeve can be arbitrary, but it is often chosen to be equal to that of the striker bar to capture the entire incident wave in the sleeve. In-depth operational details of the MT technique are also available in publications by Nemat-Nasser (2000), Subhash (2000), and Subhash and Ravichandran (2000).

The x–t diagram for the MT incident bar is shown in Figure 2.7. With this design modification, compression stress waves are generated in both the incident bar and the sleeve upon impact of the striker. Because the impedances of the two are matched, the incident energy is split equally between them, producing equal amplitude compressive stress waves in both. The compressive wave in the incident bar travels down the bar toward the sample. The wave in the sleeve reaches the rigid mass interface and reflects as a compression wave. When this compressive wave reaches the sleeve-flange interface, it pushes the flange toward the striker and consequently generating a tensile pulse in the incident bar with the same duration as that of the original compression wave. This tensile wave (generated due to the MT) trails immediately behind the original compression wave in the incident bar. The separation between these two waves can be extended by choosing a longer sleeve length than the striker bar. The strain gage on the incident bar now records both these signals: an initial compression wave (from the original compression wave generated due to striker impact) immediately followed by a tensile wave, both of similar amplitude and duration traveling toward the specimen. Upon reaching the specimen end of the bar, the compression pulse loads the specimen, and the trailing tensile pulse retracts the bar (the particle motion in tensile waves is opposite of the wave direction) and separates the specimen from

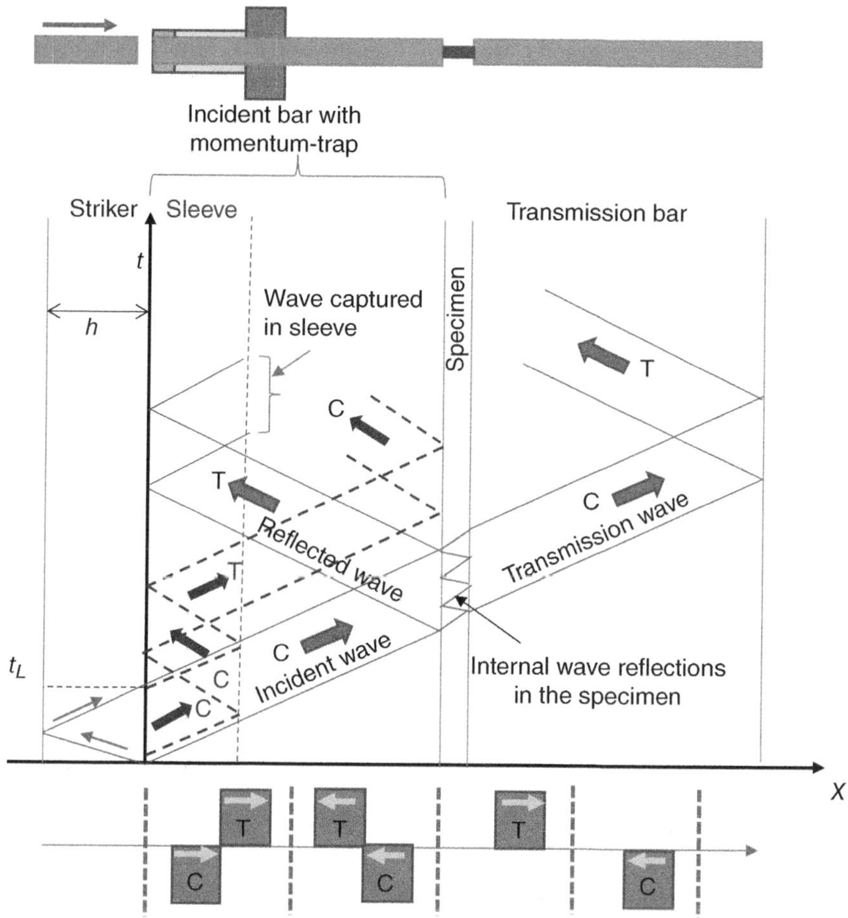

Figure 2.7 $x-t$ diagram for an MT-SHPB. The dotted lines indicate the wave reflections in the sleeve and generation of a tensile wave following the compression wave in the incident bar. The bottom schematic shows the compression (C) and tension (T) waves measured by a strain gage on the incidence bar as a function of time. The arrows indicate the direction of wave travel in the incidence bar.

the bar. Both these waves now change sign as the specimen is of lower impedance and travel back toward the MT.

In the absence of the MT, the reflected tensile wave from the specimen end of the bar will reflect back as a compression wave at the impact-end of the bar and travels toward the specimen and reloads the already weakened (partially damaged from the first compressive loading) ceramic specimen. In the modified SHPB, the impedance-matched MT now comes into play. It captures the incoming tensile pulse because the particle motion in the tensile pulse is opposite to its travel direction, and the flange–sleeve interface facilitates this transfer. The accompanying compression pulse (from the original trailing tensile wave) in the incident bar now reflects back at the MT-end as a tensile pulse and travels toward the specimen again. Since the particle motion in this wave is in the opposite direction to that of the wave, it causes further separation of the bar from the specimen. This process repeats several times until the wave is fully dissipated in the bar through friction in the supporting bushings. The result is that the specimen is subjected to a single incident wave loading. If the amplitude of the first compression pulse in the incident bar is of high enough amplitude to cause failure in the ceramic, then one can obtain an accurate failure strength of the ceramic. If the amplitude is too low, the damaged specimen can still be recovered for postmortem microstructural observations of induced damage, which can be correlated to the incident wave amplitude and duration.

Accurate Measurement of Specimen Strain

For ductile materials, where the strain to failure is large, the specimen strain is obtained from the reflected signal in accordance with Eq. (2.8). Because the failure strain in ceramics is small (<1%), the amplitude of the strain gage signal on the reflected bar is also small and could be dominated by noise in the signal. Therefore, axial and transverse strain gages are bonded directly to the ceramic specimen as shown in Figure 2.8 to capture the specimen strain *in situ*. The corresponding incident, transmitted, and reflected signals generated during a test in SHPB on a ceramic specimen are shown in Figure 2.5.

Appropriate Specimen Geometry

It is common practice in the ceramic research community to use a rectangular, parallelepiped specimen geometry with a length-to-width ratio of 1.5–2. Rectangular geometries can cause stress concentrations at the corners during loading and may lead to premature fracture of the specimen. Alternate specimen geometries such as dumbbell shaped specimens have been suggested and explored in the literature. Chen et al. (1994) analyzed stress distribution in various specimen geometries

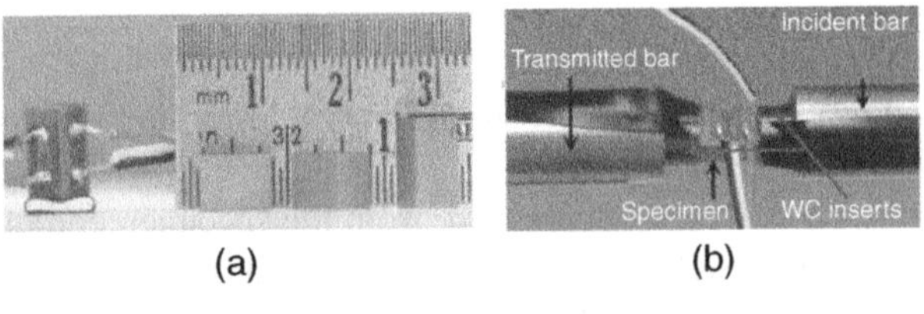

Figure 2.8 (a) Longitudinal and transverse strain gages bonded on the ceramic specimen. (b) Specimen configuration with WC inserts placed between the incident and transmission bars.

using finite element analysis and Subhash and Ravichandran (2000) summarized these specimen geometries and their effectiveness. Ideally, it is desirable to have a cylindrical specimen for compression testing and a dog-bone shaped specimen for tension testing. The latter specimen geometry reduces stress concentrations at the specimen-bar interface and potentially generates uniform and uniaxial stress state in the gage section. Early attempts to use dumbbell-shaped alumina specimens to obtain uniaxial compressive failure strength under static and dynamic conditions were well-received (Anderson et al. 1992; Blumenthal 1990; Cosculluela et al. 1992; Kanel et al. 1992). Most recently, Swab and coworkers (Swab et al. 2017; Swab and Quinn 2019; Swab et al. 2020) have resurrected this discussion, demonstrating that the use of dog-bone (dumbbell) shaped ceramic specimen geometries can indeed provide significantly higher compressive strength than the traditional ceramic specimen geometries. Numerical studies performed by Chen et al. (1994) on these dumbbell-shaped ceramic specimen geometries have indeed shown reduced stress concentration at both ends of the specimen.

However, a major problem with using these preferred geometries is the inherent brittleness of ceramics. Complex shapes are labor-intensive to produce and are prohibitively expensive as extremely low depths of cut are employed during grinding. Additional polishing operations may be required to obtain a good surface finish.

2.3.1.3 Limiting Strain Rate

With the modifications to SHPB for brittle materials testing, the dynamic failure strengths of a variety of structural ceramics (e.g. Al_2O_3, SiC, B_4C, and B_4C–SiC composites) have been determined by various researchers (Ghosh et al. 2012; Kunka et al. 2018; Pittari et al. 2015a; Subhash and Ravichandran 1998). A representative plot showing the uniaxial compressive strength of aluminum nitride versus the strain rate for quasistatic and SHPB testing is shown in Figure 2.9 (Ravichandran and Subhash, 1995). The failure strength of the ceramic does not show a noticeable strain-rate dependence below a strain rate of $10^2\ s^{-1}$ but shows marked sensitivity at strain rates between 10^3 and $10^5\ s^{-1}$. A wide variety of ceramics display similar behavior, and more detailed discussions on these issues are presented in Chapters 4 and 5.

The maximum strain rate that can be achieved in a ceramic test specimen using the SHPB is limited by the specimen length. This can be calculated by the following relationship between the strain rate and specimen length (Ravichandran and Subhash 1994):

$$\dot{\varepsilon}_l = \frac{\varepsilon_f C_L}{\alpha l_s} \tag{2.11}$$

where $\dot{\varepsilon}_l$ is the limit of the maximum strain rate, ε_f is the failure strain (~1%), C_L is the wave velocity (~13 000 m/s), α is the number of wave reflections to

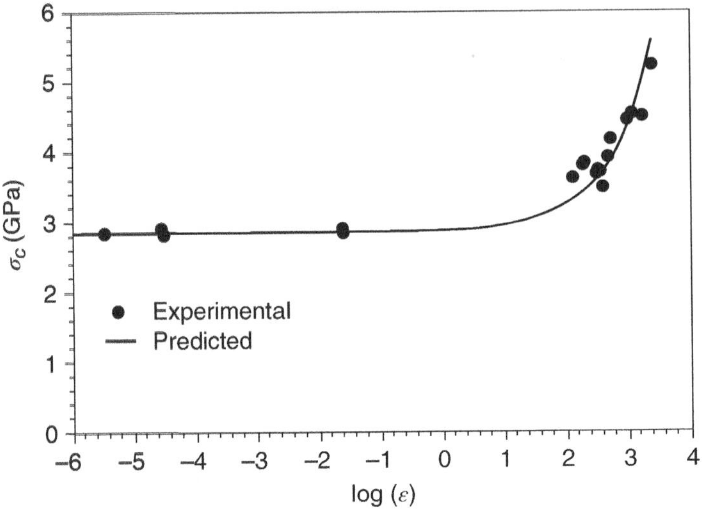

Figure 2.9 Strain rate sensitivity of failure strength of aluminum nitride. Source: Ravichandran and Subhash (1995). © 1995, Elsevier.

achieve stress equilibrium (typically four), and l_s is the length of the specimen (typically 6 mm). Assuming these values for a ceramic, the limiting strain rate is approximately 5400 s^{-1}. Many advanced structural ceramics such as SiC and B_4C have high Young's moduli (>400 GPa) and low density (3.2 and 2.52 g/cm^3, respectively) and hence have high wave velocities. By reducing specimen length, further increases in the limiting strain rate can be achieved, allowing testing of materials at greater strain rates.

2.3.2 Dynamic Indentation Hardness Tester (DIHT)

Plastic deformation beneath an indentation is one of the most widely investigated subjects in the fields of solid mechanics and materials science. Indentation by a rigid indenter is one of the simplest, nondestructive experimental methods that tests only a small sample volume with minimal specimen preparation. The data obtained from the test is also simple to analyze, requiring only the measurement of the applied load and the dimensions of the plastic impression created on the material surface. Tabor (1948) was the first to discover a relationship between the hardness H, and bulk yield strength σ_y, of metals as $H = 3\sigma_y$. The indentation hardness technique has been widely extended to a variety of materials including metals, ceramics, rocks, transparent materials, thin films, polymers, metallic glasses, biomaterials, and nanomaterials. The method is also scalable, meaning that indentations as small as nanometers to as large as millimeters in depth can be performed to determine the hardness of materials. Nanoindentation uses extremely low loads, and this method is ideally suited to extract not only bulk properties of materials but also properties of interfaces, nanophases, thin films, and nanopillars (Fischer-Cripps 2011; Oliver and Pharr 2004). Hardness is also the single-most relevant property to potentially predict the penetration resistance of a ceramic target material (Karandikar et al. 2009; Swab 2004). While the above features have made indentation an attractive characterization technique, it is still unclear what material property is represented by indentation hardness. Simply stated, hardness reflects the resistance of a material to localized plastic deformation or penetration. Based on this generic definition, it is easy to surmise that the indentation response of a material is related to a combination of physical properties (e.g. density) as well as mechanical properties such as Young's modulus, yield strength, work hardening behavior, and fracture toughness.

Despite this emphasis on hardness and its relevance to material properties, it is rarely recognized that hardness is also a strain-rate-dependent property. A majority of the literature focuses on static hardness and attempts to relate this static property to a variety of dynamic properties and impact performance. A typical static hardness test is conducted over a period of several seconds, whereas a typical dynamic event, such as a projectile impacting a target or a grinding wheel interacting with a material surface, occurs in a few tens of microseconds. Such short temporal events activate a rate-dependent material response that is not captured in a static indentation test. Thus, a dynamic indentation hardness device is required that can deliver an indentation load in a short period (similar to that in dynamic events) to reach strain rate magnitudes on par with those encountered in dynamic testing techniques such as SHPB.

The DIHT method discussed here (Koeppel and Subhash 1997; Subhash 2000; Subhash et al. 1999) relies on the same principles as the static indentation hardness testing method: the sharp indenter is brought in contact with the specimen surface, and a single indentation load is applied normal to the specimen surface over a duration of few hundred microseconds. This results in a deformation (strain) rate in the range of 10^2 to 10^3 s^{-1}. The parallels between static and dynamic indentation testing methods, except the loading duration (15 s versus 100 μs), allow a fair comparison of material hardness across the corresponding strain rate regimes.

To accomplish dynamic indentation within a duration of a few hundred microseconds, we again utilize 1D elastic stress wave propagation principles in a slender bar. A schematic of the DIHT is shown in Figure 2.10. The DIHT consists of a slender bar (similar to an incident bar) with a MT on one end and an indenter mounted at the other end. The transmission bar in SHPB is replaced by a rigid support. However, this replacement warrants an alternative way to measure the indentation load which is accomplished by mounting a high-frequency load cell on the rigid support. The specimen is placed on this load cell, and the indenter tip is moved to come in contact with the specimen so that the specimen can be

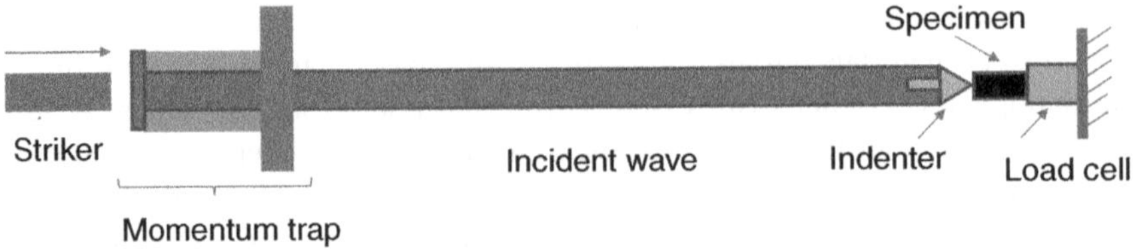

Figure 2.10 Schematic of a dynamic indentation hardness tester and corresponding measurement devices.

held by friction between the indenter tip and load cell. This is similar to the static indentation procedure where the indenter tip is made to come in contact with the specimen surface prior to application of load.

The striker bar is launched from the gas gun, and upon impact, the generated compressive pulse travels down the incident bar to the indenter tip, causing the indentation on the specimen. The indenter then retracts due to the MT capturing the reflected tensile pulse from the indenter. Thus, the specimen experiences only a single indentation load. The load transmitted through the specimen is measured by the load cell, and the corresponding load signature is captured on an oscilloscope. The indentation imprint created on the specimen is measured posttesting using an optical microscope. Similar to static hardness, the indentation hardness is calculated by dividing the maximum load, over the projected area of the indenter. For Vickers indentation, the indentation hardness is given by

$$HV = 0.01819 \frac{P}{d^2} \tag{2.12}$$

where HV is the Vickers indentation hardness (GPa), P is the load (kg), and d is the average diagonal of the indentation imprint on the ceramic (mm). Unlike in uniaxial tests such as SHPB, both the strain and strain rate within the subsurface indentation volume are not constant but vary spatially and temporally. Therefore, an average strain rate, (s^{-1}), is defined as follows:

$$\dot{\varepsilon}_{avg} = \frac{V_{indenter}}{d} \tag{2.13}$$

where $V_{indenter}$ is the velocity of the indenter and d is the length of the indentation diagonal. The velocity of the indenter can be measured using a high-speed camera. Typical strain rates of dynamic indentation are on the order of 10^2 to 10^3 s^{-1}.

Using this method, Koeppel and Subhash (1997, 1999), Subhash et al. (1999) and Subhash (2000) verified that Tabor's relation between the hardness and yield strength is also valid within the above dynamic strain rate range for metals. It was also noted that the dynamic indentation hardness was greater than the static indentation hardness, indicating strain-rate hardening. A typical result of a SiC-based ceramic revealing an increase in hardness under dynamic loads is shown in Figure 2.11.

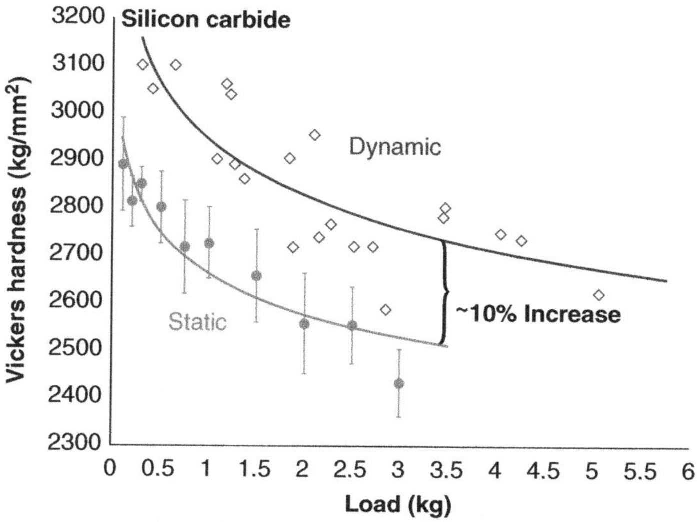

Figure 2.11 Comparison of static and dynamic indentation hardness values for SiC. Source: Klecka and Subhash (2010). © 2010, John Wiley & Sons.

However, unlike static hardness data, the dynamic hardness data has significantly greater scatter, which is characteristic of this technique. Therefore, it has become common practice to plot load (*P*) versus the diagonal square (d^2) for all the indents and extract the slope of the line. The slope is then multiplied by 0.01819 from Eq. (2.12) to obtain an average hardness from all the indentations. This is called "slope hardness," a term first coined in (Pittari et al. 2015b). Since then, the method has been adopted for a variety of ceramics including 3D-printed alumina (DeVries et al. 2018), silicon carbide, and boron carbide composites (Pittari et al. 2015b). This method works well for ceramics with inherent variability in their microstructural response during dynamic indentations.

2.3.3 Dynamic Fracture Toughness Testing

Fracture toughness is another important material property that characterizes the cracking resistance of a material. While static fracture-toughness testing is used widely to study fracture behavior under slowly applied loads, we focus here on determining the dynamic fracture toughness. A good review of dynamic fracture toughness test methods is available in Jiang and Vecchio (2009). A SHPB setup with a MT can also be used to deliver a dynamic flexural load and thus measure the rate-dependent fracture toughness of a ceramic. The detail of the setup is illustrated in Figure 2.12. Similar to static fracture toughness testing, a four-point flexural test adhering to ASTM 1421-16 (ASTM C 1421-10) is employed

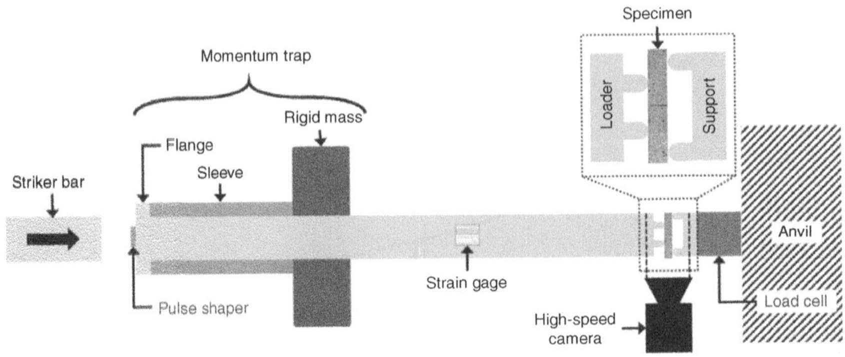

Figure 2.12 Schematic of a dynamic fracture toughness testing setup. Source: Pittari et al. (2015b). © 2015, Elsevier.

in dynamic experiments. The specimen end of the incident bar is retrofitted with the two inner supports of a four-point bend fixture, called the "loader." The longer side of the flexural fixture with two outer supports (termed "support") rests on a load cell anchored to a rigid anvil. The sample is placed between these two frames and held lightly by friction. A V-notched specimen is loaded with the sharp end of the V-notch facing the tensile side such that a stable crack is initiated from this pointed end.

In dynamic fracture toughness testing, the load is applied rapidly, in approximately a few hundred microseconds. Once again, equilibration of stress in the specimen is necessary to obtain accurate fracture toughness values. Enough time must elapse to allow the stress wave to transfer through the supports on either side of the specimen and reach a steady state (equal force at four support points) before crack growth initiates from the V-notch. This condition limits the rate at which the force can be applied because of the finite wave speed in the specimen. Higher loading rates can cause oscillations in the applied load and make stress equilibration difficult to achieve (Jiang and Vecchio 2009). One can validate the steady-state assumption by conducting a finite element analysis of the specimen with the experimental load-time profile (from the load cell) as input (Walter et al. 2013). Examples are shown in Figure 2.13. By comparing the force profiles developed at the two supports on either side of the specimen, one can determine whether the specimen reached equilibrium. One should also monitor that contact is maintained between the supports and the specimen throughout the experiment with high-speed imaging.

Another prime requirement for a valid fracture toughness test is to ensure that stable crack growth is achieved during the test. To track this crack growth, the specimen surface is coated with random speckles from paint or copy toner as seen in Figure 2.14a. Digital image correlation (DIC) is used to track the speckle pattern

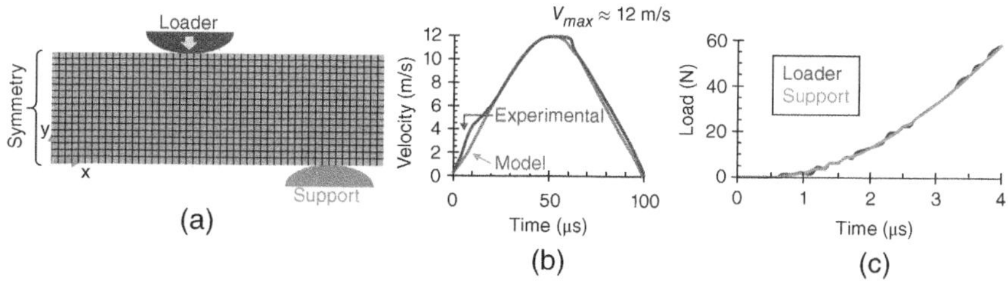

Figure 2.13 Finite element analysis of a specimen during dynamic fracture toughness testing to ensure stress equilibrium prior to failure. (a) Schematic of a 2D-model geometry, (b) comparison of experimental and model velocity profiles applied to the loader, and (c) comparison of reaction forces at loader and support. Source: Pittari et al. (2015b). © 2015, Elsevier.

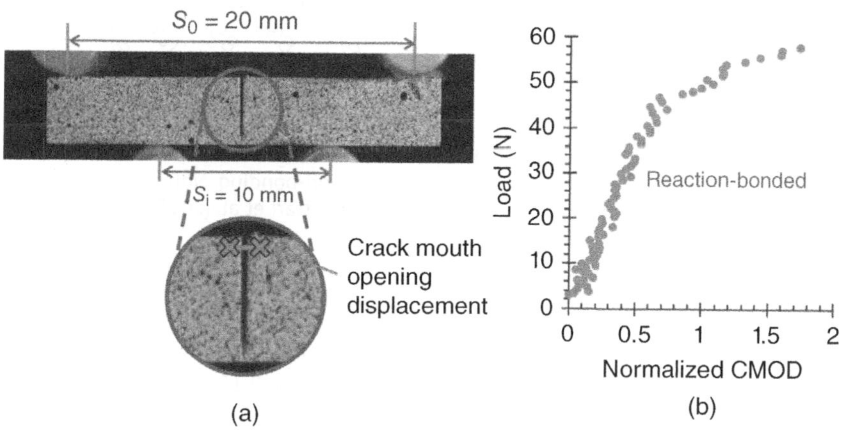

Figure 2.14 (a) V-notched SiC specimen coated for DIC analysis and (b) CMOD plot revealing stable crack growth. Source: (a) Reproduced with permission from Pittari et al. (2015b), (b) Pittari et al. (2015b). © 2015, Elsevier.

across the notch during crack growth. As the crack length increases, the distances between the speckles and their orientations change, and from these changes, the crack mouth opening displacement (CMOD) can be derived. A nonlinearity in the curve beyond the elastic response is an indication of stable crack growth as shown in Figure 2.14b (Pittari et al. 2015b).

The fracture toughness is given by (C28.01, A.S. 2009):

$$K_{I_{vb}} = \frac{Y^*_{min}\left[P_{max}\left(S_o - S_i\right)\right]}{BW^{3/2}} \tag{2.14}$$

where S_o and S_i are the outer and inner spans of the supports, and B and W are the width and thickness of the specimen, respectively. The specimen dimensions are

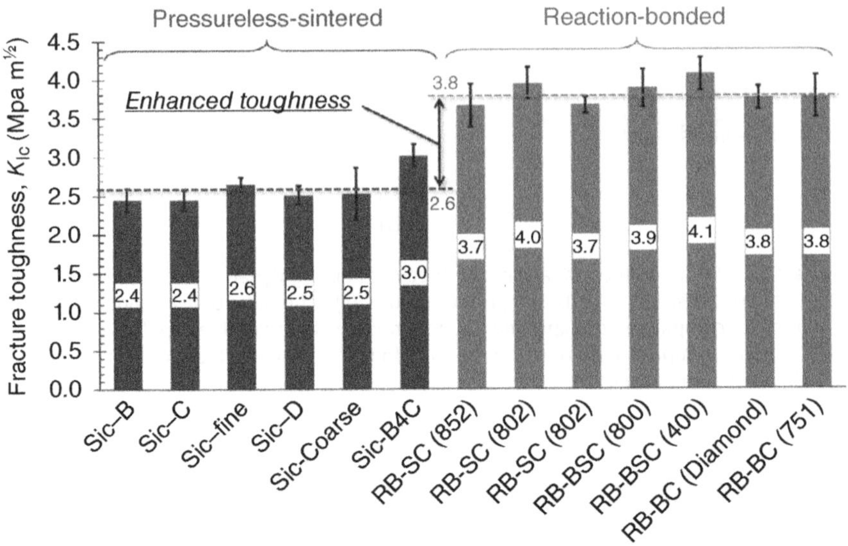

Figure 2.15 Comparison of static and dynamic fracture toughness values for several ceramics processed by pressureless-sintering and reaction-bonding. Note the increase in fracture toughness with increasing strain rates. Source: Pittari et al. (2015b). © 2015, Elsevier.

labeled in Figure 2.14a. P_{max} is the maximum load at the time of fracture initiation. Y^*_{min} is the stress intensity factor coefficient for this specific geometry, which is 4.2327 (Pittari et al., 2015b). Typical fracture toughness test results for a range of SiC and B_4C ceramic composites are shown in Figure 2.15. Similar to the compressive strength and hardness values, there is an increase in dynamic fracture toughness compared to the static values at increasing strain rates.

2.4 Uniaxial Strain Characterization Methods

In the previous section, we discussed testing methods based on a uniaxial stress condition to extract ceramic properties such as failure strength, indentation hardness, and fracture toughness. In these tests, the slender geometry of the specimen as well as the testing equipment (bars) limit the maximum stress and strain rate that can be imparted to the specimen. For ceramic specimens, the maximum strength is limited by their failure strength. Can we impose a higher stress level in the ceramic specimens to investigate their behavior at extremely high levels of stress or strain rates, similar to those encountered in a meteor impact on an object or high-velocity projectile impact? Such tests are possible by changing the geometry of the specimen (from a bar to a thin disc) and testing method. These loading configurations are based on the principles of uniaxial strain conditions. The differences between uniaxial stress and uniaxial strain experiments are:

(1) For rigid, high-stiffness materials, uniaxial stress experiments require a long cylindrical specimen. On the other hand, to achieve uniaxial strain conditions, the specimen geometry is a thin circular disk or plate with its lateral (radial) dimension significantly greater than that in the wave-propagation (thickness) direction. The large lateral dimension provides confinement to the material points in the central region of the sample, and hence, this geometry allows stress build-up to any level, several times the failure strength.
(2) In uniaxial stress testing, the slender cylindrical specimen experiences **uniaxial stress conditions** ($\sigma_x \neq 0$, $\sigma_y = \sigma_z = 0$) while the **strain is multiaxial** ($\varepsilon_x \neq 0$, $\varepsilon_y \neq 0$, and $\varepsilon_z \neq 0$). Here, the loading direction is assumed to be along the x-axis. In **uniaxial strain** experiments ($\varepsilon_x \neq 0$, $\varepsilon_y = \varepsilon_z = 0$), the **strain is one dimensional** in the specimen, but the **stress is three dimensional** ($\sigma_x \neq 0$, $\sigma_y \neq 0$, and $\sigma_z \neq 0$). These stress and strain quantities can also be considered as the principal components.
(3) In uniaxial stress loading, a specimen can only be subjected to a stress level no greater than the ultimate strength of the material. Lack of lateral confinement limits the maximum stress that can be imposed under this stress state. On the other hand, uniaxial strain loading can impose any level of stress on the specimen. The lateral confinement by the large lateral dimensions and the resulting hydrostatic compression at the center allows the specimen to withstand high levels of stress and undergo microstructural changes that are not possible under uniaxial stress loading. Thus, one can determine the dynamic yield strength called the Hugoniot elastic limit (HEL) of the material and investigate plastic deformation as well as phase changes that occur at extremely high pressures. The "spall strength" of the specimen can also be determined when the compression wave reflects as a tensile wave at the rear end of the specimen. Spall strength is also referred to as the hydrostatic tensile strength of the material. These issues will be discussed in greater detail in Chapter 5.
(4) The material points in the center of the specimen are subjected to both high hydrostatic compression, where $(\sigma_x + \sigma_y + \sigma_z)/3$, and shear stress, $(\sigma_x - \sigma_y)/2$ subscripts denote the principal stresses in the axial (x) and lateral (y and z) directions.
(5) The impedance in uniaxial stress conditions is given by Eq. (2.5), where the longitudinal wave velocity under uniaxial stress conditions are given by Eq. (2.2). For uniaxial strain tests, the area is omitted due to the large lateral dimension of the specimen, and the wave velocity is given by Eq. (2.1).
(6) In uniaxial stress experiments, the specimen length (or time required for the wave to travel its length) is significantly smaller than the incident loading

pulse width (or duration), so that the specimen eventually reaches stress equilibrium (uniform and uniaxial stress state) after multiple internal wave reflections as seen in Figure 2.3. This condition allows the constitutive response of the specimen to be obtained from the test. Ceramics subjected to compression SHPB testing tend to exhibit cracks along the loading direction and axial splitting. Under uniaxial strain experiments, the width of the stress pulse (or duration) is smaller than the specimen thickness (or time required for the wave to travel its thickness). At any instant, only a portion of the specimen thickness will be in a stressed state as opposed to the entire specimen undergoing uniform deformation as in uniaxial stress tests. As the wave propagates through the specimen in the uniaxial strain condition, it may cause temporally and spatially varying microstructural changes. Due to the short loading duration, microstructural changes such as point defect generation, nucleation of voids and cracks, and phase changes are often observed.

(7) In uniaxial stress experiments, the loading duration is in the order of few tens of microseconds giving rise to a strain rate of 10^2–10^5 s^{-1} in the specimen. In uniaxial strain experiments, a loading duration of few nanoseconds is imposed resulting in strain rate above 10^6 s^{-1}.

The uniaxial (1D) strain experiments can be broadly classified based on the stress state (normal stress or a combination of normal and shear stress) achieved in the specimen as follows:

a. Normal plate impact
b. Oblique plate impact

2.4.1 Normal Plate Impact Experiments

A normal plate impact experiment consists of a thin cylindrical specimen (a flat plate) impacted by a "flyer plate" of similar lateral dimensions. The impact occurs at normal incidence such that only a normal stress is created. Numerous works provide detailed descriptions of the experimental method (Kumar and Clifton 1977; Espinosa et al. 1992; Raiser and Clifton 1993; Raiser et al. 1990; Smith et al. 2007; Nemat-Nasser 2000) and only a brief description of the test principles and results are presented here. In normal plate impact experiments, the specimen is held at rest and the flyer plate is launched at a high velocity from a gas propelled gun toward the specimen to cause a planar impact. All critical surfaces that come in contact are lapped flat to the wavelength of sodium light (589 nm), so that there is minimal loss of wave energy from one plate to the next. The normal impact ensures that only a planar longitudinal stress wave is generated in the specimen. Upon impact, the uniaxial strain wave propagates through the thickness of the specimen and reaches the end of the specimen. The duration of the stress wave in the specimen is equal to the roundtrip travel time of the longitudinal wave in the flyer plate (see Eq. (2.9)). The particle motion at the center of the rear surface of the specimen is monitored using a laser velocity interferometer system for any reflector (VISAR). The lateral dimension of the specimen is chosen such that the release waves from the lateral boundaries do not arrive and interact with the material at the center of the rear surface of the specimen during measurement.

The material properties of the flyer are chosen such that its impedance closely matches that of the specimen (within 5%) for complete momentum transfer during impact. This configuration is called symmetric impact, and, under these conditions, the generated axial stress wave amplitude in the specimen is

$$\sigma_x = \frac{1}{2}\rho CV \tag{2.15}$$

where ρC is the impedance of the flyer plate and V is the velocity of the flyer plate at the time of impact. Due to the large lateral dimension of the specimen, confinement stress is also generated in the center of the specimen:

$$\sigma_y = \sigma_z = \frac{\nu}{1-\nu}\sigma_x \tag{2.16}$$

A material point in the center of the specimen thus experiences high shear and hydrostatic stress, given by

$$\tau_{max} = \frac{(\sigma_x - \sigma_y)}{2} \text{ and } P = \frac{(\sigma_x + 2\sigma_y)}{3} \tag{2.17}$$

where, an axisymmetric condition, $\sigma_y = \sigma_z$, is assumed.

If the impact velocity of the flyer is low, leading to a low magnitude compressive pulse, the specimen may survive the initial passage of the compression wave. This wave then reflects as a tensile wave at the rear surface. The stress wave reflections and the resulting stress state in the specimen as a function of time are illustrated graphically in a distance–time (x–t) diagram as shown in Figure 2.16a. As ceramics are weaker in tension, the passage of a tensile wave of a high enough magnitude creates damage. Upon reaching the impact surface, the tension wave reflects again as a compression wave and travels through the damaged specimen to reach the rear surface. By comparing the particle velocity measurements at the rear surface during the first and second wave arrivals, one can obtain qualitative information on the induced damage in the specimen. Postmortem microstructural analysis of the specimen can reveal the nature of damage due to these tensile waves. If the amplitude of the impact velocity is sufficiently high, phase transformation and intense damage could occur in the ceramic during the passage of the compressive wave.

In Figure 2.16a, the stress states in the specimen at various wave transit times are indicated by *C*, *T* and *0*, for regions experiencing compression, tension or zero stress, respectively. The specimen initially undergoes a compressive stress state upon impact, and when the wave reaches the rear surface, the stress state close to the rear surface becomes zero (indicated by "0"). The compressive stress is negated

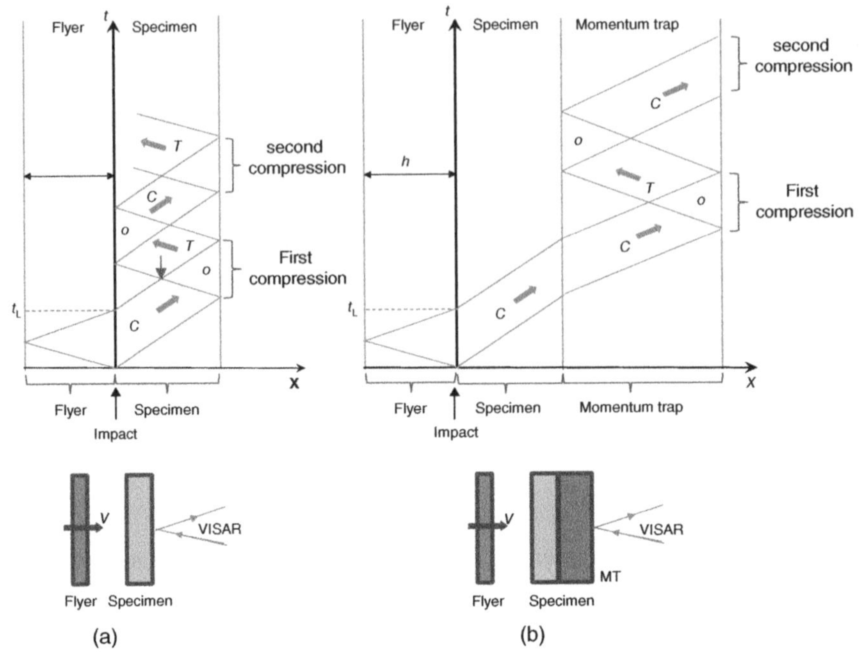

Figure 2.16 (a) $x-t$ diagram for normal plate impact experiments where a specimen can be subjected to repeated compression and tension loads due to multiple wave reflections. The broad arrows indicate the direction of wave travel and the downward arrow indicates the location of spall fracture. (b) $x-t$ diagram where a specimen is loaded with a only single compression pulse when an impedance matched momentum trap (MT) is placed behind the specimen.

by the reflected tensile stress; as the two waves pass one another in opposite directions, they momentarily cancel each other out (recall that the particle velocity on the rear surface doubles during this period). The tensile wave reappears in the interior of the bulk specimen (labeled T) after the passage of the tail of the compression wave. If the tensile stress amplitude is sufficiently high, it can cause spall fracture in the specimen. The plane of the specimen where spall fracture occurs is indicated by the red arrow in Figure 2.16a. In-depth discussions on spall fracture in brittle materials are provided in Chapters 5 and 6.

The test configuration corresponding to the x–t diagram in Figure 2.16a allows multiple wave reflections within the specimen making it difficult to correlate microstructural changes to the magnitude and duration of the incident wave. To ensure that the specimen is subjected to a single compression wave of known duration and amplitude, an impedance-matched momentum trap (IMMT) is placed behind the specimen. The compression wave is then fully transferred to the IMMT after loading the specimen. Once again, a laser interferometer is used to monitor the amplitude and duration of the compression wave on the backside of IMMT. For ceramics, lithium fluoride (LiF) is often used as the MT material because it retains its transparency (refractive index) to very high compressive shock stress levels (Carter 1973; Davis et al. 2016; Marsh 1980). It acts as a window and allows the particle velocity to be monitored either on its back surface or at the intersection of the specimen and the MT.

The relevant x–t diagram of how an IMMT ensures a single pulse during plate impact is shown in Figure 2.16b. Upon impact by the flyer, a compression wave of equal magnitude is generated both in the specimen and the flyer, each traveling in opposite directions. The compression wave generated in the flyer travels to its back-end and reflects as a tensile wave. Upon reaching the interface between the specimen and the flyer, it separates the flyer from the specimen and ends the loading phase in the specimen. Simultaneously, the front of the generated compression wave in the specimen travels toward the back-end of the specimen. If the MT is fully impedance-matched and the contact surfaces are well matched, then the entire incoming wave will transfer into the MT. When the wave front reaches the back end of the MT, it reflects as a tensile wave and reaches the specimen–MT interface and separates the two. The specimen is now loaded only by a single compression pulse. From elastic wave theory, the theoretical duration of the compression stress wave (ΔT) generated in the specimen is equal to the roundtrip travel time in the flyer:

$$\Delta T = \frac{2h}{C_L} \tag{2.18}$$

where h is the thickness of the flyer plate. The amplitude of the stress wave is given by Eq. (2.15).

Noting again that ceramics are weaker in tension than in compression, the IMMT flyer plate impact test has been further modified to generate a tensile pulse in the specimen of only a few nanoseconds. A gap of a few microns wide between the specimen and the MT is created by depositing droplets of gold of a known thickness at four corners of the specimen. This modification, although seemingly trivial, is an ingenious way to create a tensile wave of the desired duration without significantly altering the test setup. An x–t diagram in Figure 2.17 illustrates the principle of tensile stress wave generation and the relevant stress states in the specimen as a function of time. The impact by the flyer generates a compression wave that travels through the specimen to the MT–specimen interface. Upon reaching the gap, the initial compressive wave reflects as a tensile wave. The gap width is designed so that only a small portion of the wave is reflected as a tensile wave, and the remaining compression wave pushes the specimen toward the MT and closes the gap. The remaining compressive wave then transfers into the

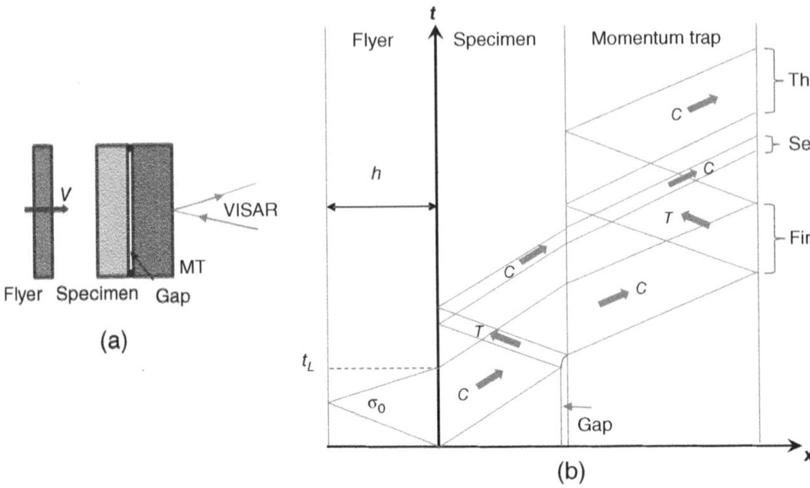

Figure 2.17 (a) Plate-impact test configuration with a momentum trap (MT) and a gap between the specimen and MT to create a tensile wave. (b) $x-t$ diagram for flyer plate impact with momentum trap. In (b), C is for compression and T is for tension. Arrows indicate the direction of wave propagation. The wave arrival sequence at the back surface of the momentum trap is indicated on the right where the rear surface particle velocity is measured by VISAR.

MT. The abbreviated, reflected tensile wave in the specimen travels toward the flyer–specimen interface but induces damage as it traverses through the ceramic. It reaches the front-end of the specimen and reflects as a compression wave that travels through the now damaged specimen and into the MT which is now fully in contact with the specimen. The compression wave reaches the back end of the MT and reflects as tension and reaches the specimen–MT interface and separates the two. Thus, the specimen is subjected to one compression (long duration) and one tension (short duration) wave and its reflected compression waves, all with same stress amplitudes.

The thicknesses of the flyer, specimen, and the MT are carefully chosen such that the original compression wave (from the impact), the reflected tension wave from the gap between the MT and specimen, and the reflected remainder of the compression wave all reach the rear of the MT in that sequence to be tracked by the interferometry measurements without ambiguity. The theoretical amplitudes of each of these signals and their arrival times can be calculated from the elasticity theory and are then compared against the measured signals on the rear surface to establish the nature and severity of the damage created in the ceramic. Raiser et al. (Raiser and Clifton 1994) conducted such experiments to investigate the damage evolved in alumina specimens by comparing the magnitude and duration of particle velocities at the rear end of the MT. These profiles, shown in Figure 2.18, can provide qualitative information of induced damage. For example, the first and

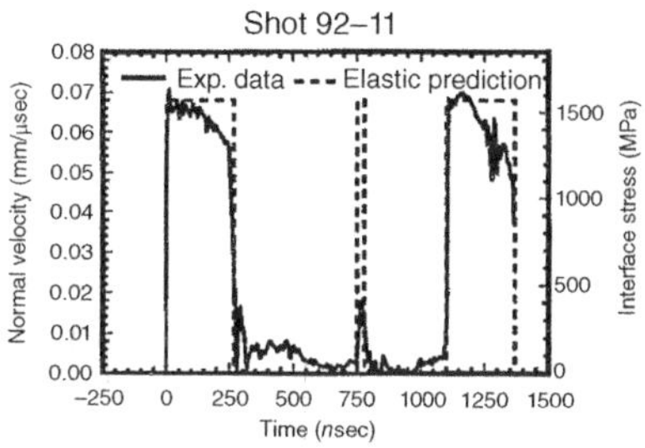

Figure 2.18 Particle velocity measurements from a typical flyer plate impact test on alumina when a small gap is introduced between the specimen and the momentum-trap. The elastic wave calculations are shown by the dashed lines, and the experimental measurements are shown in solid lines. The narrow signal at 750 ns is the tensile wave reflection from the gap and measured as a compression wave at the rear surface. Source: Raiser et al. (1994). © 1994, AIP Publishing.

third profiles are similar and illustrate elastic response and plane wave loading. The second profile indicates that damage has occurred during the tensile wave propagation because the subsequent compression wave amplitude is attenuated as it traverses the damaged specimen.

2.4.2 Oblique Plate Impact Experiments

Normal plate impact experiments induce 1D strain state in a specimen, but oblique plate impact experiments are designed to induce both normal and shear stresses. The intent is to investigate material response under combined loads by initially subjecting the specimen to normal stress (due to its higher longitudinal wave velocity) followed by a shear stress (greater wave energy). Similar to the normal plate impact experiment, a flyer plate inclined at the desired angle to travel direction is launched from a gas gun toward a specimen inclined at the same angle as the flyer. Upon impact, both a normal stress wave and a shear wave are induced in the specimen and the flyer. The fast-moving normal stress wave may induce microstructural changes in the specimen, which can influence the trailing shear wave characteristics. By once again tracking the particle motion at the rear surface of the specimen, one can infer the induced microstructural changes. However, due to its limited application to ceramics, we will not describe this specific technique any further.

2.5 Impact Testing

In the previous two sections, uniaxial stress or uniaxial strain characterization methods were discussed. These methods allow a fundamental understanding

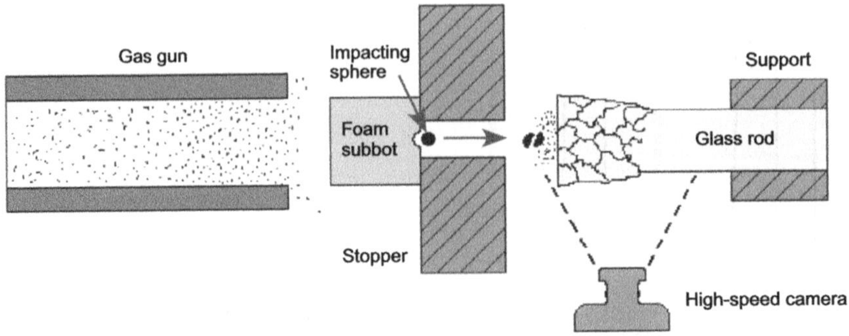

Figure 2.19 Schematic of the test setup for ball-on-rod impact experiments.

of the evolution of material deformation modes and facilitate the development of mechanistic models. In the current section, the next level of complexity of a stress state and induced damage are introduced via sphere or cylinder impact experiments on plates and rods. In these experiments, the tested material does not remain in a well-defined stress state. These experiments do not necessarily test the effectiveness of the material or design in the final application. Depending on the information to be extracted, the test is designed such that the configuration of the test and the test parameters are kept constant, but the ceramic material and microstructure are varied. These tests may be used to directly or indirectly rank materials. Examples of information sought through such tests are impact-induced damage in rods and plates, depth-of-penetration, dwell-penetration transition (to be discussed further in Chapter 3) or resistance to penetration in intact and damaged test materials.

2.5.1 Spherical Projectile Impact of Cylindrical Rods

In this ball-on-rod experiment, a long cylindrical rod is impacted with high-velocity steel or ceramic sphere with a diameter less than that of the rod. A schematic of the test setup is shown in Figure 2.19. The impacting ball is often contained in a foam sabot, which is propelled from a gas gun toward the rod. The sabot is stopped by a rigid block, allowing the ball to fly freely and impact the rod. A high-speed camera captures the impact of the ball and the induced fracture and fragmentation features in the cylindrical rod. The velocity of the ball, as well as the damage propagation velocity, can be measured from time-resolved high-speed camera images of the impact process. Details of such experimental results and evolved damage modes on transparent materials are discussed in Chapter 6.

Utilizing a long cylindrical rod as a target material allows visualization of damage evolution and propagation in one dimension. Ball impact generates a longitudinal wave followed by a shear wave traveling through the length of the rod. The faster longitudinal wave will compress the material as it arrives while the following shear wave will initiate damage within the rod. Due to its greater energy, the shear wave may cause severe fracture and fragmentation in the brittle material, similar to the one shown in Figure 1.5. Using this method, the damage induced in a range of transparent materials including spinels, strengthened glasses, and sapphire, has been investigated, and these results are present in Chapter 6. The advantage of this method is that the kinetic energy of the ball can be related to the energy dissipated in different damage and deformation modes, such as wave propagation, fracture, fragmentation, kinetic energy of the fragments, dilation of the material, and ejecta (fine debris at impact location) formation.

2.5.2 Edge-on Impact Testing

The edge-on impact (EOI) test allows visual evaluation of the spatial and temporal evolution of impact damage in a ceramic plate in two dimensions (2-D). The EOI test method was originally developed in the 1990s at the Ernst Mach Institute (EMI) in Freiburg, Germany, to visualize the dynamic fracture propagation characteristics in brittle ceramic materials (Gooch 2011). The specimen is typically in the form of a rectangular plate of relatively small thickness (compared to its other two dimensions). A short cylindrical projectile is launched toward its edge at a predetermined velocity. Damage evolves in the ceramic from the impact location and propagates into the bulk ceramic. Using a high-speed camera, the damage propagation can be imaged for various impact velocities. *In situ* visualization of damage initiation sites and growth characteristics provides insight into evolving deformation mechanics in the plate which are otherwise hidden. The EOI geometry provides an alternative means to capture the initiation and evolution of the damage in a 2D solid. In general, it has been observed that the evolved damage in a ceramic target under normal impact and EOI are similar (McCauley et al. 2013; Strassburger et al. 2008). A schematic of the test configuration is provided in Figure 2.20a, and the evolved longitudinal, shear, and damage wavefronts upon impact in a ceramic plate are depicted in Figure 2.20b (McCauley et al. 2013).

2.6 Summary And Conclusions

In this chapter, we reviewed major high-strain-rate experimental techniques to obtain dynamic properties and responses of ceramics. The choice of experimental method depends on the type of information being sought. Among these, SHPB is

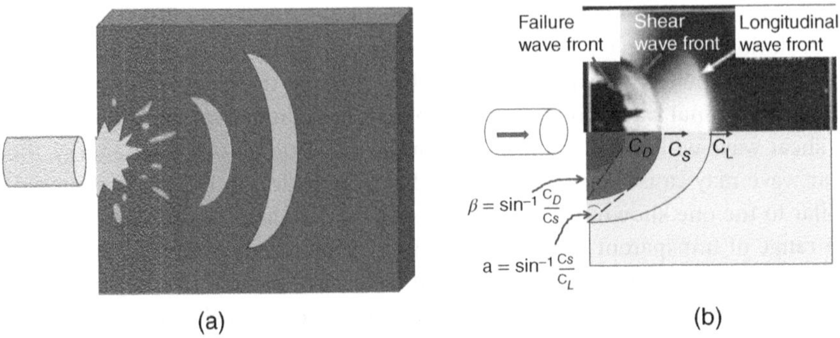

Figure 2.20 (a) Schematic of edge-on impact test. (b) Evolution of longitudinal, shear, and failure waves in a transparent material upon edge-on-impact by a cylinder. Source: McCauley et al. (2013). © 2013, Springer Nature.

the most widely used testing method, but the method can be modified to conduct dynamic indentation and dynamic fracture toughness testing. Though somewhat complex, 1D-strain testing allows materials to be subjected to significantly higher stress levels beyond the failure strength and at higher strain rates. Finally, specific characterization methods such as ball-on-rod and EOI tests provide information on specific phenomena, such as damage evolution and propagation characteristics.

The focus of this chapter was to provide a short description of dynamic characterization testing methods so that the reader has familiarity with these common test methods. In-depth discussions on ceramic properties and their failure behavior under complex loads are discussed in the ensuing chapters. In the next chapter, we turn to the specific case of ballistic impact on a ceramic target with an intent to illustrate the multiplicity of mechanisms that evolve during this complex event.

References

Achenbach, J.D. (1973). *Wave Propagation in Elastic Solids*. New York: North Holland Publishing Company.

Anderson, C., Odonoghue, P., Lankford, J., and Walker, J. (1992). Numerical simulations of SHPB experiments for the dynamic compressive strength and failure of ceramics. *International Journal of Fracture* 55: 193–208.

Anderson, C.E., Behner, T., Holmquist, T.J., and Orphal, D.L. (2011). Penetration response of silicon carbide as a function of impact velocity. *International Journal of Impact Engineering* 38: 892–899.

Bhattacharyya, A., Subhash, G., and Arakere, N. (2014). Evolution of subsurface plastic zone due to rolling contact fatigue of M-50 NiL case hardened bearing steel. *International Journal of Fatigue* 59: 102–113.

Blumenthal, W. (1990). *High-strain-rate compression and fracture of B sub 4 C-aluminum cermets.* NM: Los Alamos National Lab.

C28.01, A.S (2009). Standard test methods for determination of fracture toughness of advanced ceramics at ambient temperature. *ASTM* C1421: 624–655.

Carter, W. (1973). Hugoniot equation of state of some alkali halides. *High Temperatures-High Pressures* 5: 313–318.

Chen, W.W. and Song, B. (2011). *Split Hopkinson (Kolsky) Bar Design, Testing and Applications.* New York: Springer.

Chen, W., Subhash, G., and Ravichandran, G. (1994). Evaluation of ceramic specimen geometries used in a Split Hopkinson Pressure Bar. *Dymat Journal* 1: 193–210.

Chen, M., McCauley, J., and Hemker, K. (2003). Shock-induced localized amorphization in boron carbide. *Science* 299: 1563–1566.

Cosculluela, A., Cagnoux, J., Collombet, F. et al. (1992). *Two Types of Experiments for Studying Uniaxial Dynamic Compression of Alumina, Shock Compression of Condensed Matter–1991*, 951–954. Amsterdam: Elsevier.

Dandekar, D.P. (2001). *Shock Response of Boron Carbide*, 1–34. Aberdeen Proving Grounds, MD: US Army Research Laboratories.

Davis, J.-P., Knudson, M.D., Shulenburger, L., and Crockett, S.D. (2016). Mechanical and optical response of [100] lithium fluoride to multi-megabar dynamic pressures. *Journal of Applied Physics* 120: 165901.

DeVries, M., Subhash, G., Mcghee, A. et al. (2018). Quasi-static and dynamic response of 3D-printed alumina. *Journal of the European Ceramic Society* 38: 3305–3316.

Espinosa, H., Raiser, G., Clifton, R., and Ortiz, M. (1992). Performance of the star-shaped flyer in the study of brittle materials – 3-dimensional computer-simulations and experimental-observations. *Journal of Applied Physics* 72: 3451–3457.

Fischer-Cripps, A.C. (2011). *Nanoindentation.* New York, NY: Springer.

Frew, D., Forrestal, M., and Chen, W. (2002). Pulse shaping techniques for testing brittle materials with a Split Hopkinson Pressure Bar. *Experimental Mechanics* 42: 93–106.

Ghosh, D., Subhash, G., Zheng, J., and Halls, V. (2012). Influence of stress state and strain rate on structural amorphization in boron carbide. *Journal of Applied Physics* 111: 063523.

Gooch, W.A. Jr., (2011). Overview of the development of ceramic armor technology: past, present and the future. In: *Advances in Ceramic Armor VII* (eds. J.J. Swab, S. Widjaja and D. Singh).

GRADY, D. (1994). Shock-wave strength properties of boron-carbide and silicon-carbide. *Journal De Physique IV* 4: 385–391.

Graff, K.F. (1975). *Wave Motion in Elastic Solids*. Columbus, OH: Ohio State University Press.

Gray, G.T.R. III, (2000). Classic Split Hopkinson Pressure Bar testing. In: *ASM Handbook, Mechanical Testing and Evaluation*, vol. 8 (eds. H. Kuhn and D. Medlin), 462–476. ASM International.

Haney, E. and Subhash, G. (2012). Edge-on-impact response of a coarse-grained magnesium aluminate spinel rod. *International Journal of Impact Engineering* 40–41: 26–34.

Haney, E. and Subhash, G. (2013). Damage mechanisms perspective on superior ballistic performance of spinel over sapphire. *Experimental Mechanics* 53: 31–46.

Jiang, F. and Vecchio, K.S. (2009). Hopkinson bar loaded fracture experimental technique: a critical review of dynamic fracture toughness tests. *Applied Mechanics Reviews* 62: 060802.

Kanel, G., Rasorenov, S., and Fortov, V. (1992). *Shock Wave and High-Strain-Rate Phenomena in Materials* (eds. M.A. Meyers, L.E. Murr and K.P. Staudhammer), 775. New York, NY: Marcel Dekker.

Karandikar, P.G., Evans, G., Wong, S. et al. (2009). A review of Ceramics for Armor Applications. In: *Advances in Ceramic Armor IV*, 163–175. Wiley.

Klecka, M. and Subhash, G. (2010). Rate-dependent indentation response of structural ceramics. *Journal of the American Ceramic Society* 93: 2377–2383.

Klecka, M., Subhash, G., and Arakere, N. (2013). Microstructure-property relationships in M50-NiL and P675 case-hardened bearing steels. *Tribology Transactions* 56: 1046–1059.

Koeppel, B. and Subhash, G. (1997). An experimental technique to investigate the dynamic indentation hardness of materials. *Experimental Techniques* 21: 16–18.

Koeppel, B. and Subhash, G. (1999). Characteristics of residual plastic zone under static and dynamic Vickers indentations. *Wear* 224: 56–67.

Kolsky, H. (1954). The propagation of longitudinal elastic waves along cylindrical bars. *Philosophical Magazine* 45: 712–726.

Kolsky, H. (1963). *Stress Waves in Solids*. Courier Corporation.

Kolsky, H. (1964). Stress waves in solids. *Journal of Sound and Vibration* 1: 88.

Kumar, P. and Clifton, R. (1977). Star-shaped flyer for plate-impact recovery experiments. *Journal of Applied Physics* 48: 4850–4852.

Kunka, C., An, Q., Rudawski, N. et al. (2018). Nanotwinning and amorphization of boron suboxide. *Acta Materialia* 147: 195–202.

LaSalvia, J., Campbell, J., Swab, J., and McCauley, J. (2010). Beyond hardness: ceramics and ceramic-based composites for protection. *Jom* 62: 16–23.

Liao, H. and Chen, W.W. (2018). Specimen-bar impedance mismatch effects on equilibrium and rate constancy for Kolsky Bar experiments. *Experimental Mechanics* 58: 1439–1449.

Lindholm, U. and Yeakley, L. (1968). High strain-rate testing – tension and compression. *Experimental Mechanics* 8: 1.

Lundberg, P., Renstrom, R., and Lundberg, B. (2000). Impact of metallic projectiles on ceramic targets: transition between interface defeat and penetration. *International Journal of Impact Engineering* 24: 259–275.

Marsh, S.P. (1980). *LASL Shock Hugoniot Data*. University of California Press.

McCauley, J., Strassburger, E., Patel, P. et al. (2013). Experimental observations on dynamic response of selected transparent armor materials. *Experimental Mechanics* 53: 3–29.

Medvedovski, E. (2010). Ballistic performance of armour ceramics: influence of design and structure. Part 1. *Ceramics International* 36: 2103–2115.

Meyers, M.A. (1994). *Dynamic Behavior of Materials*. New York, NY: Wiley.

Nemat-Nasser, S. (2000). Introduction to high strain rate testing. *ASM International* 8: 427–428.

Nemat-Nasser, S., Isaacs, J., and Starrett, J. (1991). Hopkinson techniques for dynamic recovery experiments. *Proceedings of the Royal Society-Mathematical and Physical Sciences* 435: 371–391.

Normandia, M. (2004). Impact response and analysis of several silicon carbides. *International Journal of Applied Ceramic Technology* 1: 226–234.

Oliver, W. and Pharr, G. (2004). Measurement of hardness and elastic modulus by instrumented indentation: advances in understanding and refinements to methodology. *Journal of Materials Research* 19: 3–20.

Orphal, D., Franzen, R., Piekutowski, A., and Forrestal, M. (1996). Penetration of confined aluminum nitride targets by tungsten long rods at 1.5–4.5 km/s. *International Journal of Impact Engineering* 18: 355–368.

Patterson, M.C.L., DiGiovanni, A.A., Gilde, G., and Roy, D.W. (2009). Spinel armor – clearly the way to go. Annual Cocoa Beach Conference on Advanced Ceramics and Composites: A: Ceramic Engineering and Science Proceedings, 441–446.

Pittari, J., Subhash, G., Trachet, A. et al. (2015a). The rate-dependent response of pressureless-sintered and reaction-bonded silicon carbide-based ceramics. *International Journal of Applied Ceramic Technology* 12: E207–E216.

Pittari, J., Subhash, G., Zheng, J. et al. (2015b). The rate-dependent fracture toughness of silicon carbide- and boron carbide-based ceramics. *Journal of the European Ceramic Society* 35: 4411–4422.

Raiser, G. and Clifton, R. (1993). High-strain rate deformation and damage in ceramic materials. *Journal of Engineering Materials and Technology-Transactions of the Asme* 115: 292–299.

Raiser, G. and Clifton, R.J. (1994). Failure waves in uniaxial compression of an aluminosilicate glass. *American Institute of Physics Conference Series* 309: 1039–1042.

Raiser, G., Clifton, R., and Ortiz, M. (1990). A soft-recovery plate impact experiment for studying microcracking in ceramics. *Mechanics of Materials* 10: 43–58.

Raiser, G., Wise, J., Clifton, R. et al. (1994). Plate impact response of ceramics and glasses. *Journal of Applied Physics* 75: 3862–3869.

Ramesh, K.T. (2008). *High Rates and Impact Experiments, Springer Handbook of Experimental Solid Mechanics*, 929–960. Boston, MA: Springer US.

Ravichandran, G. and Subhash, G. (1994). Critical-appraisal of limiting strain rates for compression testing of ceramics in a Split Hopkinson Pressure Bar. *Journal of the American Ceramic Society* 77: 263–267.

Ravichandran, G. and Subhash, G. (1995). A micromechanical model for high-strain rate behavior of ceramics. *International Journal of Solids and Structures* 32: 2627–2646.

Shafiq, M. and Subhash, G. (2016). Dynamic deformation characteristics of zirconium diboride–silicon carbide under multi-axial confinement. *International Journal of Impact Engineering* 91: 158–169.

Smith, R., Eggert, J., Jankowski, A. et al. (2007). Stiff response of aluminum under ultrafast shockless compression to 110 GPA. *Physical Review Letters* 98: 061101.

Strassburger, E. (2009). Ballistic testing of transparent armour ceramics. *Journal of the European Ceramic Society* 29: 267–273.

Strassburger, E., Patel, P., McCauley, J.W. et al. (2008). *High-Speed Transmission Shadowgraphic and Dynamic Photoelasticity Study of Stress Wave and Impact Damage Propagation in Transparent Materials and Laminates Using the Edge-On Impact (EOI) Method*, 1–16. US Army Research Laboratories.

Subhash, G. (2000). Dynamic Indentation Testing. In: *ASM Handbook* (eds. H. Kuhn and D. Medlin), 519–529. ASM International.

Subhash, G. and Ravichandran, G. (1998). Mechanical behaviour of a hot pressed aluminum nitride under uniaxial compression. *Journal of Materials Science* 33: 1933–1939.

Subhash, G. and Ravichandran, G. (2000). Split Hopkinson Pressure Bar testing of ceramics. In: *ASM Handbook* (eds. H. Kuhn and D. Medlin), 497–504. ASM International.

Subhash, G. and Ridgeway, S. (2018). *Mechanics of Materials Laboratory Course*. Morgan and Claypool Publishers.

Subhash, G., Koeppel, B., and Chandra, A. (1999). Dynamic indentation hardness and rate sensitivity in metals. *Journal of Engineering Materials and Technology-Transactions of the ASME* 121: 257–263.

Subramanian, R. and Bless, S. (1995). Penetration of semi-infinite AD995 alumina targets by tungsten long rod penetrators from 1.5 to 3.5 km/s. *International Journal of Impact Engineering* 17: 807–816.

Swab, J. (2004). Recommendations for determining the hardness of armor ceramics. *International Journal of Applied Ceramic Technology* 1: 219–225.

Swab, J.J., Gilde, G.A., Patel, P.J. et al. (2000). Fracture analysis of transparent armor ceramics. In: *Fractography of Glasses and Ceramics IV: Proceedings of the Fourth Alfred Conference on the Fractography of Glasses and Ceramics* (eds. J.R. Varner and G.D. Quinn), 489–508. Alfred, NY: Alfred University.

Swab, J., Meredith, C., Casem, D., and Gamble, W. (2017). Static and dynamic compression strength of hot-pressed boron carbide using a dumbbell-shaped specimen. *Journal of Materials Science* 52: 10073–10084.

Swab, J.J., Chen, W., Hogan, J. et al. (2020). Dynamic compression strength of ceramics: what was learned from an interlaboratory round robin exercise. *Journal of Dynamic Behavior of Materials*. 7: 34–47.

Tabor, D. (1948). A simple theory of static and dynamic hardness. *Proceedings of the Royal Society of London Series a-Mathematical and Physical Sciences* 192: 247–274.

Walter, T., Subhash, G., Sankar, B. et al. (2013). A novel method for dynamic short-beam shear testing of 3D woven composites. *Experimental Mechanics* 53: 493–503.

3

Brief Overview of Deformation Mechanisms During Projectile Impact on a Confined Ceramic

3.1 Introduction

This chapter describes projectile impact on a confined ceramic, which serves as an example scenario that exhibits many fundamental dynamic deformation phenomena. The basic understanding of failure mechanisms that evolve during this projectile–target interaction event forms the foundation for concepts covered in subsequent chapters of this book. Due to the low tensile strength of a ceramic, the test configuration of the ceramic specimen often contains high level of lateral confinement during ballistic loading. The initial impact by a high-velocity projectile on the target generates high-amplitude compressive stress waves within the ceramic, termed "shock." These stress waves travel at velocities in excess of 12 000 m/s in the ceramic and carry the initial shock pressure away from the impact site over long distances. When the resulting stress amplitude exceeds the strength at that location, various forms of damage are initiated in the ceramic. The penetration resistance of the target is initially a function of its virgin state (properties of intact and bulk material), but with evolution of damage in the target ahead of the penetrator, the properties of the damaged ceramic become more relevant. When the kinetic energy (KE) of a projectile is sufficiently high to overcome the surface resistance of the ceramic target, damage initiates close to the tip (nose) of the impact location and the ceramic becomes highly pulverized (comminuted). Farther from the penetrator nose, the ceramic is fragmented, and fracture extends into the bulk material. As the penetration progresses, a cavity is created in the target, and the fractured and comminuted target material ahead of the projectile is compressed against the undamaged target. This compressed and comminuted material is eventually forced out of the cavity in the form of "ejecta," small, sharp-edged, hard ceramic particles, which can further erode and decelerate the incoming projectile as they flow in the opposite direction to that of the penetrator. Thus, there is a continuous loss in mass and a reduction in the velocity of the projectile, meaning a loss in KE. If the ceramic is initially

Dynamic Response of Advanced Ceramics, First Edition.
Ghatu Subhash, Amnaya Awasthi, Dipankar Ghosh.

confined, the level of confinement will change once damage initiates and will alter boundary conditions as well as the kinetics of evolving damage.

This chapter details projectile–target interaction events from mechanistic and materials perspectives. First, the ballistic phenomenon is discussed in generic terms and is then followed by a description of the sequence of events and mechanisms that activate in the ceramic target during the projectile impact. The projectile–target interactions can be separated into two phases:

1. Nonpenetration phase and
2. Penetration phase.

The nonpenetration phase refers to the time interval during which the projectile nose encounters the ceramic target (at time $t = 0$) and may fully erode at the surface or partially erode for a short duration ($t > 0$) at the surface before penetrating the ceramic. Two important penetrator-target interaction phenomena occur during this phase: "dwell" and "interface defeat." The penetration phase begins when the impactor overcomes the initial resistance of the ceramics and enters into the interior of the target. In the penetration phase, a range of inelastic deformation mechanisms can occur including cracking, plasticity, phase transformations, crack propagation, fragmentation, comminution, and ejecta formation. To understand the relevance of ceramic properties for impact resistance, the terminology and nature of damage that evolves during various phases of projectile–target interactions are introduced. The influence of the thickness of the ceramic plate, material properties, and confinement level are considered.

3.2 Damage Evolution During the Nonpenetration Phase

3.2.1 Dwell

When a high-velocity projectile approaches the target, the hard ceramic momentarily resists its forward motion. This nonpenetration phase of the target-penetrator-interaction process where the tip of a high-velocity projectile comes to a temporary stop upon reaching the ceramic surface is called "dwell." However, the back-end of the projectile is still in motion and continues to move toward the target. If the ceramic maintains its integrity, the tip of the projectile is continuously eroded through instantaneous fracture and fragmentation. These fragments flow radially outwards as the projectile grinds against the ceramic target. During this phase, the ceramic target may experience a high level of Hertzian contact stress and suffer internal damage even though the projectile has not yet penetrated the target.

The duration of the dwell phase is measured as the interval between the instant the projectile encounters the front surface of the target and the time when it begins to penetrate the ceramic. In other words, it is the period of the nonpenetration phase of the projectile-target interaction where the penetrator is trying to overcome the initial resistance offered by the ceramic target. The duration of dwell may last a few tens of microseconds depending on the penetrator material, impact velocity, ceramic properties (e.g. hardness and fracture toughness), and confinement level of the target. Using high-speed photography and X-ray shadowgraphs of small arms projectiles impacting ceramic targets, the phenomenon of dwell was first identified by Wilkins and coworkers in 1967 (Wilkins 1967). An illustration of the dwell phase through a set of high-speed photographs is presented in Figure 3.1 (Lundberg et al. 2000). Upon reaching the ceramic target, the penetrator starts to erode, and the fragments flow radially outwards on the ceramic surface, a phenomenon aptly named "mushrooming." The debris from the projectile is mostly in the form of fine particles and fragments, and its failure mechanism is mostly erosion (Anderson and Walker 2005). The dwell period contributes to the loss of KE of the projectile by: (i) deceleration of the projectile on the surface of the target and (ii) mass loss of the projectile due to erosion. Thus, dwell is a vital phase in the target–penetrator interaction.

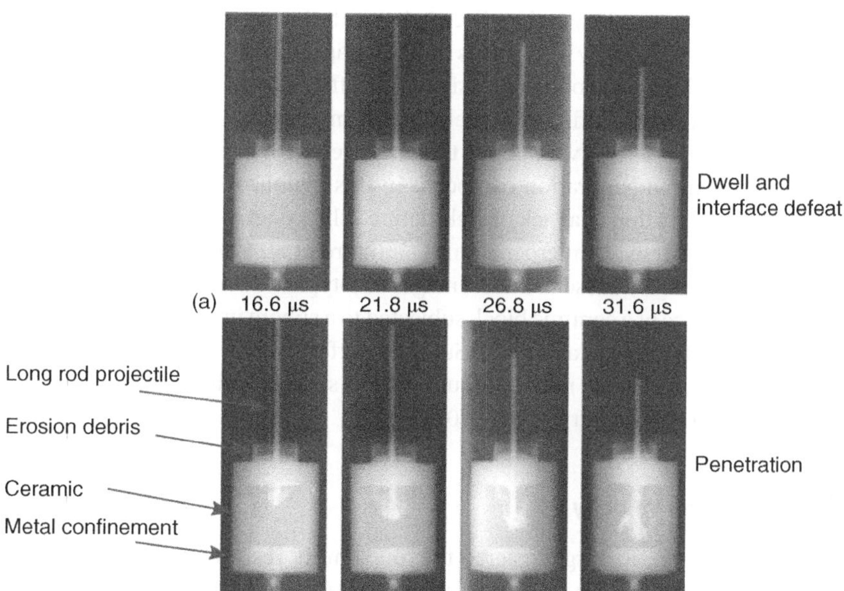

Figure 3.1 X-ray images of impact and penetration of a long steel rod projectile onto a confined ceramic target: (a) Illustration of dwell phase at the surface at 1615 m/s; and (b) penetration into the ceramic at 1805 m/s. Source: Reproduced with permission from Lundberg et al. (2000).

3.2.2 Interface Defeat

During the dwell phase, the role of the confined, hard ceramic is to fully resist the advance of a projectile. If the projectile is short or the impact velocity is below a critical level, the ceramic target can maintain its structural integrity and can completely erode the incoming projectile at the penetrator-target interface (Anderson and Walker 2005). Accordingly, this phase was termed "interface defeat" by Hauver et al. (2005). Since the projectile was completely defeated, no penetration of the target occurs, but the target may have suffered some internal damage due to Hertzian stresses. The top row of images in Figure 3.1 illustrates a projectile undergoing interface defeat.

3.2.3 Damage in Ceramic During Dwell

Although the projectile never penetrates the ceramic target during dwell and interface defeat, stress can buildup beneath the surface of the ceramic target and cause internal damage (due to Hertzian contact stresses) in an area referred to as the "Mescall zone" (Shockey et al. 1990). The Hertzian nature of contact between the rounded projectile tip and flat ceramic target generates high magnitudes of shear and hydrostatic stresses in a region directly beneath the impact location (Johnson 1985). Thus, at a sufficiently high-stress level, damage initiates in the interior of the ceramic, but the high level of compressive stress offered by the surrounding intact material and external confinement helps resist macroscopic fracture initiation. At sufficiently high stress (well above the stress required to initiate macroscopic fracture), dislocation generation can occur in this region. As ceramics have low dislocation mobility due to a lack of well-defined slip systems, the stress builds up further to cause extensive microcracking (Lankford et al. 1998) that leads to pulverized material in the Mescall zone. Images of a typical Mescall zone developed beneath the surface of a ceramic block subjected to a sphere impact are provided in Iyer (2007) and Shockey et al. (1990). Such subsurface zones of high deformation can also be seen in bearing materials subjected to similar high-intensity Hertzian contact stresses (Bhattacharyya et al. 2014).

3.2.4 Transition Velocity

The confined ceramic target exerts the maximum resistance to penetration during the dwell phase, but it pays a price of internal damage including plastic flow and microcracking in the Mescall zone. Note that the Mescall zone does not nucleate on the surface, but at a depth below the surface because the maximum shear stress occurs at a depth below the surface. Since the material on the surface is still intact, a low-velocity penetrator does not yet penetrate the ceramic target. If the KE of the projectile is sufficiently high, the diameter of the Mescall zone (or the internal damage zone) grows larger during the dwell phase and reaches the surface of the ceramic target, thus allowing penetration of the projectile. The velocity at the instant when dwell is no longer maintained and projectile penetration begins is called the "dwell–penetration transition velocity" (v_t) (Das et al. 2018). In a well-matched target–penetrator interaction, there is always a finite duration of the dwell phase, and the penetrator does not enter the target until the stress is sufficiently high and the Mescall zone boundary reaches the surface. Obviously, the v_t is less than the impact velocity due to the loss of KE during the dwell phase. The transition velocity is a function of the projectile material properties, impact velocity, target properties, and confinement pressure. For a given penetrator-target system, however, a range of velocities exist that result in penetration into the armor due to statistical variations in material properties and design parameters (Lundberg et al. 2000). Therefore, the transition velocity from interface defeat to penetration is not a singular value for an armor-penetrator combination but is instead a range of values encompassing the lowest velocity at which penetration occurs to the highest velocity at which the projectile has been defeated at the interface.

The dwell–penetration transition velocity can be correlated to the maximum surface stress that a ceramic can resist. This dwell–penetration transition stress is an important material characteristic that can be utilized to rank different ceramics for their ability to resist ballistic impact. It can also quantify the influence of confinement (prestress) on ballistic efficiency (Anderson Jr et al. 2009). Anderson and Walker (2005) and Anderson Jr et al. (2009) conducted reverse ballistic impact experiments using long-rod steel projectiles on ceramic disks backed by aluminum plates. They utilized flash X-ray radiographs to measure the position of the nose and tail of a projectile as a function of time during penetration and estimated the projectile penetration velocity (u) and the projectile consumption (erosion rate) velocity (v_c). Their results are shown in Figure 3.2. In Figure 3.2a, the measured nose and tail velocities of the projectile indicate that while the tail velocity continuously decreases from the time of impact, the nose velocity remains at almost zero, indicating erosion of the projectile during the dwell phase. In Figure 3.2b, the position of the tail changes but not that of the nose, which remains the same for nearly 20 μs. During this dwell phase, the Mescall zone gets larger and eventually reaches the surface to allow penetration. The projectile nose velocity now increases and catches up with the tail velocity as seen in Figure 3.2a. Consequently, the distance between the positions of the nose and tail remain constant for the reminder of the target–projectile interaction duration, shown in Figure 3.2b, indicating that the erosion of the projectile has ended and its penetration into ceramic has begun, during which the projectile remains rigid (no more mass loss). Accordingly, Figure 3.2c illustrates the loss of projectile

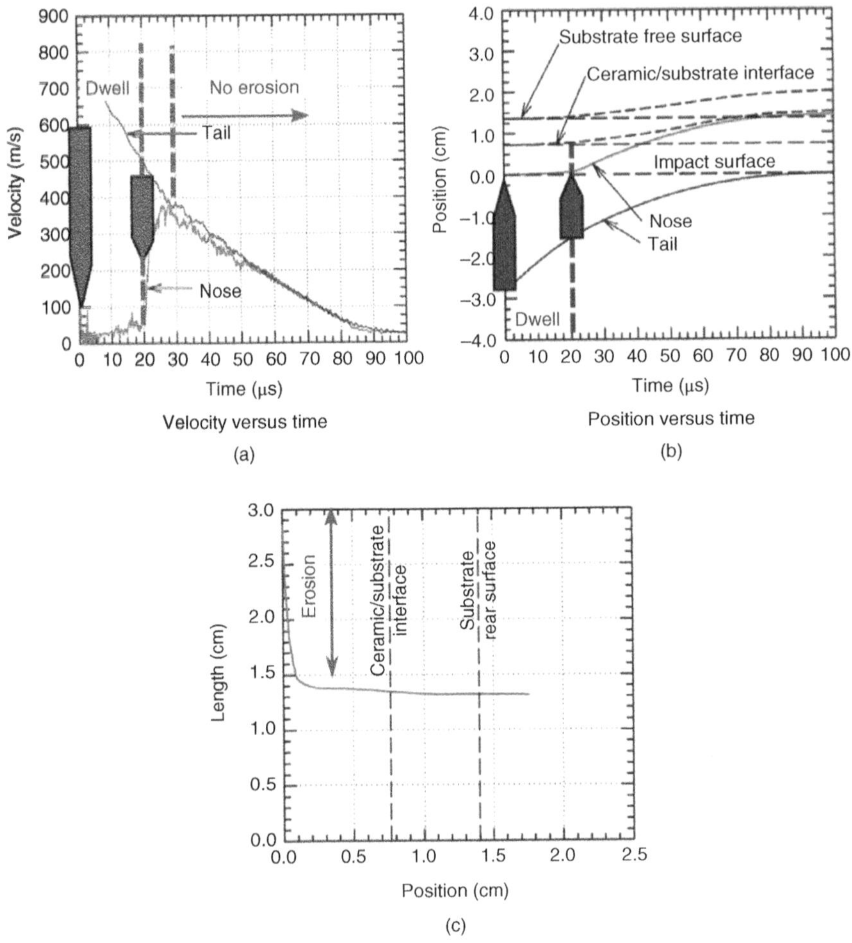

Figure 3.2 (a) Nose and tail velocities, (b) positions of a projectile, and (c) projectile length changes during impact. In (a) the dwell phase is seen to persist for about 20 μs, and then the nose and tail velocities become the same. In (b), after dwell phase, the distance between the two ends remains the same, indicating a rigid penetration of the projectile. In (c) there is a reduction in projectile length due to erosion during the dwell phase followed by its near constant length during the penetration phase. Source: Reproduced with permission from Anderson and Walker (2005). © 2005, Elsevier.

length by almost 50% due to erosion during the dwell phase, and thereafter, the length remains constant.

The dwell and interface defeat mechanisms are often observed in a well-matched projectile–target interaction. However, with an increase in KE of the projectile, the Mescall zone forms closer to the surface. This zone may not even form at higher velocities if the ceramic is initially overpowered by the KE of the projectile. In this situation, the induced stress in the ceramic due to the projectile impact far exceeds its failure strength, causing immediate fracture on the surface of the ceramic allowing penetration of the projectile.

3.3 Penetration Phase

At sufficiently high projectile velocities, the damage initiated within the ceramic during the dwell phase weakens the target, allowing penetration of the projectile. The velocity with which the projectile penetrates through the target is termed "penetration velocity." For example, in Figure 3.2a, the initial penetration velocity is around 380 m/s, whereas the impact velocity is 700 m/s (velocity of the projectile just before encountering the target surface). For short projectiles, the penetration velocity gradually decreases, as seen in Figure 3.2a. For long-rod projectiles, the velocity decreases sharply upon impact as it encounters the target, and then remains practically constant for most of the penetration process. It decreases sharply again at the end of the penetration process as the KE is gradually dissipated by many deformation and fracture mechanisms in the ceramic. Most of the KE loss in this case is due to erosion during the penetration process (Anderson and Walker 2005; Lundberg and Lundberg 2005).

The duration of the penetration phase can be much longer than the nonpenetration (dwell) phase (see Figure 3.2). Multiple damage mechanisms evolve during this phase which include creation of a comminuted zone close to the impact site, damage and fracture on the ceramic surface, conoid fracture formation from the impact site (Iyer 2007), plasticity, phase transformations, back plate deflection and failure, and cracking of the ceramic from reflected tensile waves at interfaces and edges. Modeling this phase of projectile–target interaction is more complicated than the dwell phase due to the continued loss of mass and velocity of the penetrator and the evolution of complex combination of spatially and temporally varying damage mechanisms in the ceramic target. However, Bavdekar et al. (2019) recently modified the Anderson and Walker (2005) model using a dynamic expanding cavity model coupled with an extended Mohr–Coulomb model applicable for high confinement pressures (Bavdekar and Subhash 2018; Shafiq and Subhash 2016) to evaluate the nose and tail behavior of a long-rod projectile. For example Figure 3.3a illustrates the nose and tail velocities of a long-rod (length to diameter ratio of 20) tungsten projectile impacting a confined aluminum nitride (AlN) ceramic at a velocity of approximately 3 km/s. The nose velocity of the rod quickly diminishes upon impact, and then the penetration occurs at a constant velocity of about 1.75 km/s. The corresponding split in loss of kinetic energy due

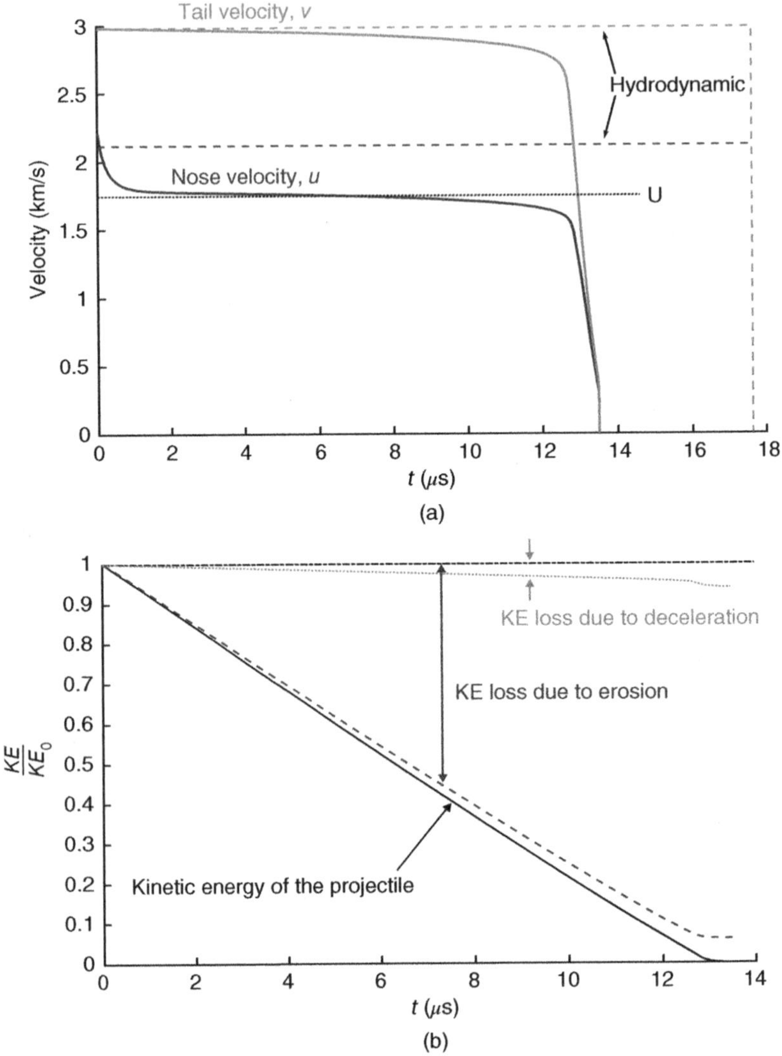

Figure 3.3 (a) Calculated nose and tail velocities of a long-rod projectile using the extended Mohr–Coulomb model during the penetration into a ceramic target at an impact velocity of 3 km/s and (b) normalized kinetic energy loss due to projectile deceleration and erosion during the penetration process. Source: Reproduced with permission from Bavdekar et al. (2019). © 2019, Elsevier.

to deceleration and erosion are shown in Figure 3.3b, where erosion is shown to be the primary mechanism that dissipates the KE of the projectile.

3.3.1 Damage within the Ceramic

The damage evolution in the bulk ceramic is governed by the local stress state, impact velocity, and the level of confinement. The projectile impact produces a high stress concentration at the impact site on the front surface of the ceramic. Postimpact analysis of a ceramic revealed that the damaged state of the ceramic is basically a combination of grain-level microcracks (both grain boundary and transgranular cracks) and large-scale macrocracks in the form of radial and/or cone cracks (Woodward et al. 1994). The stress state as well as damage evolved in the ceramic target are similar to those produced during surface indentation on ceramic by a sharp object (Hertzian contact) (Johnson 1985; Lawn and Wilshaw 1993). These damage mechanisms have been well-characterized by Iyer (2007), discussed more in Chapter 4. The high level of shear stress may cause plastic flow and an extensive network of microcracks to form in a small region beneath the impact site, leading to pulverization of the target material. The damage then extends into the bulk ceramic in the form of cone cracks (Iyer 2007), followed by fragmentation as additional cracks evolve within the cone crack envelope (Sherman and Ben-Shushan 1998). Microcracks are generally contained in a small volume in the ceramic directly beneath the impactor, and this volume continues to expand with an increase in impact velocity and stress. Grain-level microcracks also start to initiate within the regions contained by large macrocracks, leading to fragmentation and eventual perforation of the ceramic target.

The impact by a high-velocity projectile also generates a compression stress wave which quickly travels to the rear surface and reflects as a tensile wave. If the amplitude of the tensile wave is greater than the tensile strength of the ceramic at that location, damage initiates at the back surface of the ceramic. Thus, damage starts to evolve both at the front face (due to impact) and at the rear face of the ceramic plate (due to the reflected tensile wave). In addition, a thin ceramic plate bends due to the centrally applied load on the front face. This flexural load creates tension on the rear surface further aiding the damage initiation and propagation toward the front surface of the ceramic plate.

The thickness of the front ceramic plate has a strong influence on the type of cracks formed due to impact. The effect of thickness is akin to that of the confinement of a ceramic. Figure 3.4 schematically illustrates the nature of crack formation in thin (or unconfined) and thick (or confined) ceramic plates due to projectile impact. When the confinement level is low or the ceramic thickness is small (defined as a "thin-tile configuration" where the thickness, "t," is comparable to the projectile diameter, "D"), long radial cracks emerge from the back surface

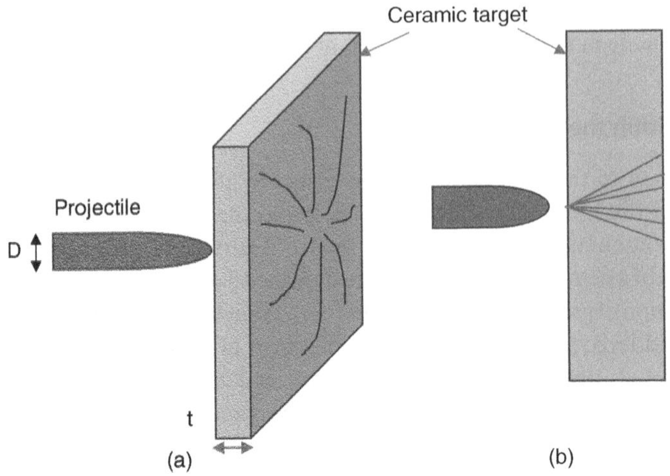

Figure 3.4 Schematic of impact induced (a) radial cracking on the rear surface of a thin ceramic plate ($t \approx D$) and (b) conoid damage development in a thick ceramic target ($t \gg D$).

of the ceramic and grow toward the impact site forming large conical damage (Anderson and Walker 2005; Sherman 2000). A schematic of this crack formation is presented in Figure 3.4a. Due to the small ceramic thickness, the cone cracks do not form an apex at the impact location. Therefore, the damage zone is in the form of a frustum of a cone with its larger base on the rear surface of the ceramic plate. This damage allows easy perforation of the projectile into the ceramic.

On the other hand, when the thickness of the ceramic plate is large ($t \gg D$) or when the confinement level is high, cone cracks initiate from the back surface of the ceramic plate and grow toward the top surface to form a fracture conoid whose vertex is at the point of impact. This is sketched in Figure 3.4b. Regions within the conoids are heavily fractured, whereas the material outside the cone is almost intact. The damaged conical envelope (containing microcracks and macrocracks) plays a critical role in projectile penetration because once this conoid is formed, the rest of the penetration occurs through this damage cone. Thus, in thick ceramic tiles, cone cracking is a major damage mechanism that governs failure during ballistic impact (Sherman 2000; Wilkins et al. 1969; Wilkins 1978; Wilkins and Laible 1980). Upon cone crack formation, the penetration depth is influenced by the progressive growth of the cone cracks as well as the fragmentation within this cracked region. Once this damage zone (from the rear surface) reaches the front surface and grows to a size comparable to the projectile diameter, penetration into the ceramic plate is initiated, provided the projectile has not been fully eroded.

In thick ceramic targets, the damage conoid formation as seen in Figure 3.4b distributes the impact energy over a larger area of the backing material. During the penetration process, the damaged material ahead of the penetrator is pushed forward and is compressed between the projectile and the intact part of the ceramic. The damaged material reduces the projectile penetration velocity because the projectile must push on a greater mass ahead of it than if there was a cylindrical damage zone of the same initial radius. With the decrease in penetration velocity, the depth of penetration (the distance from the surface to where the penetrator reached within the interior of the ceramic) is also decreased. Further fragmentation and comminution of the already damaged ceramic occurs ahead of the impactor nose, distributing the impact energy over a larger area and reducing the intensity of the stress in the material. As the projectile pushes against this ceramic debris, it forces the material to escape from the ceramic plate through a burrow formed behind the projectile nose. This ejection of debris removes material ahead of the projectile and allows further penetration into the ceramic. However, the ejected ceramic debris erodes and decelerates the remainder of the incoming projectile. All of these processes play a key role in defeating the projectile by consuming projectile KE. Thus, penetration resistance of the ceramic is a not only a strong function of initial bulk properties but is also fragmentation characteristics and properties of the damaged material. The initial bulk properties influence the dwell and interface defeat behaviors, whereas both the bulk and fragmented material properties dictate the penetration depth of the projectile into the ceramic.

3.3.2 Impact Stress and Transition Pressure

After having discussed the mechanisms of deformation and fracture in a ceramic target, we now focus on a few relevant mechanics concepts. Assuming that the projectile is a linearly elastic and perfectly plastic material and that the effects of yield strength and compressibility are small relative to inertia, the approximate maximum pressure (p_o) exerted on the ceramic surface along the axis of symmetry of the projectile during the dwell phase is (Lundberg et al. 2000):

$$p_o \approx q_p \left(1 + \frac{q_p}{2K_p}\right) + 3.27\sigma_{yp} \tag{3.1}$$

where the subscript p refers to the projectile, σ_{yp} is the yield strength, and K_p is the bulk modulus. The stagnation pressure of an ideal fluid (the projectile), q_p, is given by

$$q_p = \frac{1}{2}\rho_p v_p^2 \tag{3.2}$$

The stagnation pressure is related to the density, ρ_p, and impact velocity, v_p. The projectile is treated as a stationary jet with a certain compressibility.

From the superposition of the Boussinesq's elastic stress field solution for a point load on a semi-infinite elastic half-space and the pressure induced by a water jet impinging on a rigid surface (Reich 1926), the relationship between the maximum pressure, p_o, and the maximum shear stress, τ_o, generated within the ceramic is

$$p_o = (2.601 + 2.056\nu)\tau_o \tag{3.3}$$

where ν is the Poisson's ratio of the ceramic. From numerical calculations of the impact process, it was determined that the maximum shear stress occurs along the axis of symmetry at a depth, z_o, below the surface, given by

$$z_o = (0.622 + 0.546\nu)a \tag{3.4}$$

where a is the projectile diameter (Lundberg et al. 2000). Assuming the Poisson's ratio of a ceramic to be 0.2, the depth of maximum shear stress is about 0.7 times the projectile diameter. In general, harder ceramics have lower Poisson's ratios, and the maximum shear occurs at even shallower depths. Equating the maximum shear stress (τ_o) to the shear yield strength (τ_y) of the ceramic target, an approximate lower bound pressure (p_o^{lower}) for the transition from interface defeat to normal penetration is obtained as follows:

$$p_o^{lower} = (2.601 + 2.056\nu)\tau_y \tag{3.5}$$

Here the shear yield strength of the ceramic is treated as the strength corresponding to the initiation of microplasticity and comminution of the ceramic. Equation (3.5) provides the lower bound of the penetration pressure (and velocity) required to overcome interface defeat and penetrate the ceramic.

The upper bound for the pressure required to transition from interface defeat to normal penetration is estimated assuming complete plastic deformation, such as in indentation, and is expressed as (Hencky 1923; Hill 1998; Johnson 1985; Shield 1955):

$$p_o^{upper} = 5.70\tau_y \tag{3.6}$$

Therefore, the approximate transition pressure interval is expressed as follows:

$$p_o^{lower} \ll p_o \ll p_o^{upper} \text{ or } (1.30 + 1.03\,\nu)\sigma_y \ll p_o \ll 2.85\sigma_y \tag{3.7}$$

Here, σ_y is the yield strength of the ceramic in uniaxial compression and is equal to $2\tau_y$.

However, Birkhoff et al. (1948) realized that the impact pressure due to an incoming high-velocity projectile is significantly greater than the strength of the material and treated the interaction as fluid behavior by applying Bernoulli's equation. Because the projectile (fluid jet) is free to flow laterally in an unconstrained manner, Alekseevskii and Tate (Alekseevskii 1966; Birkhoff et al. 1948; Tate 1967), determined the applied pressure, p, on the target as follows:

$$p = 1/2\,\rho_P(u - v)^2 + Y_P = 1/2\,\rho_t u^2 + R_t \tag{3.8}$$

where ρ is density, u is penetration velocity, v is impact velocity, Yp is the yield strength, and R_t represents target resistance. The subscripts p and t stand for projectile and target, respectively. The projectile is assumed to deform at a constant flow stress during the dwell period, and the effects of strain hardening and strain-rate hardening are neglected.

The target resistance R_t in Eq. (3.8) can be used to compare various ceramics for their effectiveness against a projectile threat. However, R_t is not a material property and is time- and projectile velocity-dependent because of the evolving damage in the target ahead of the projectile. In its simplest form, it is the axial pressure at the interface during penetration. Hence, efforts to determine the term $(R_t - Y_P)$ have been carried out through expanding cavity models by Sathapathy and coworkers (Bless et al. 1999; Satapathy 2001; Satapathy and Bless 1996), and later through experimental work based on SiC impacted by spheres (Normandia 2004) and long-rod penetrators impacting confined ceramics (Orphal and Franzen 1997; Orphal et al. 1997; Orphal et al. 1996; Subramanian and Bless 1995).

3.4 Influence of Properties on Performance

One would now want to identify which properties of a ceramic relate to dwell and interface defeat of a projectile. Toward this end, LaSalvia et al. (2001) extended the model developed by Lundberg (Lundberg et al. 2000) to include the influence of microstructural and mechanical properties on dwell and penetration transition velocities. Anderson (2002), on the other hand, studied the influence of material properties and geometry by conducting numerical simulations of the penetration process. The interaction between a protective system and an incoming projectile is a complex event because of the numerous interrelated influencing factors. The event initiates deformation mechanisms in both target and projectile, and these are not only material-dependent (properties and microstructure) but also are dependent on the velocity of impact and level of confinement. Additional complications arise from geometry and design criteria such as the thickness of the plate, backing material, confinement, and number of interfaces, to name a few. Depending on the velocity of a projectile, one may also need to consider thermal and inertial effects. Thus, the evaluation of potential performance of a confined ceramic target system is riddled with complexities, but we will briefly cover relevant mechanistic issues related to performance. It is useful to introduce relevant vocabulary at this juncture.

Areal density (AD): The product of plate thickness and density, and has units of g/cm^2. It is a simple quantitative measure for comparing materials.

Residual velocity (RV): Exit velocity of the projectile after complete penetration of the target material.

Ballistic limit: Velocity at which an incoming projectile penetrates the target but stops within the target before reaching the back surface, i.e. no full perforation of the target and hence zero residual velocity. Projectiles traveling at velocities above the ballistic limit can penetrate fully through the target, meaning that the target is ineffective against the projectile.

V50: Projectile velocity at which there is a 50% probability that a projectile will fully penetrate the target. Obviously, V50 is a statistical measure determined from many tests.

The use of lightweight, high-strength materials such as ceramics and composites have resulted in a continuous decrease in the AD of armor by 70% between World War I and the Gulf war. Anderson (2002) noted that while ductile metals require AD values anywhere between 7 and 17 g/cm^2, a composite armor consisting of a ceramic front plate and a metal backing plate (e.g. B_4C/Al) requires an AD of less than 4 g/cm^2 to achieve the same ballistic outcome. Why is the ballistic performance of composites on par with that of metals with a much lower AD? To answer this question, Anderson (2002) conducted a systematic numerical study and investigated the influence of material properties of the ceramic and backing plate as well as their thicknesses on the ballistic performance. The results are summarized here.

3.4.1 Effect of Ceramic Plate Thickness

The penetration efficiency of a projectile is governed by the mode of damage in the ceramic, which is a function of plate thickness, as evident in Figure 3.4. In general, the extent of damage increases as the thickness of the ceramic layer decreases. The rate of conoid damage growth in thinner plates is high, and the target is defeated early (Sherman 2000). Sherman and Ben-Shushan (1998) conducted impact tests under similar conditions on confined alumina tiles with thicknesses of 5 and 15 mm. Both ceramic tiles were supported by a steel backing plate. Upon impact, the 5-mm-thick ceramic tile completely disintegrated into fragments, whereas the 15-mm-thick tile was cracked but not fragmented. The thick alumina tile exhibited primarily radial cracking and limited cone cracking, as seen in Figure 3.5. When the ceramic plate is "thin," the initial impact subjects the armor to local bending and induces large tensile hoop stresses at the back surface. Thus, a cracked zone develops in the form of radial cracks from the back surface of the ceramic. These radial cracks propagate toward the edges and appear on the front surface only after they have extended to the edge of the plate.

A thin ceramic plate is preferred for lightweight armor, but its benefits should be weighed against the consequences in terms of penetration efficiency. Thinner ceramics result in shorter dwell periods, higher nose velocity of the bullet, and

Figure 3.5 Front surface damage in a 15-mm-thick alumina tile confined and supported on the back by a semi-infinite steel plate. Radial cracking and cone cracking is observed. Source: Reproduced with permission from Sherman and Ben-Shushan (1998).

lower erosion of projectile. A thicker ceramic offers a greater dwell period and reduces the KE of the projectile by erosion and deceleration, decreasing the penetration velocity of the projectile much more rapidly than a thin ceramic. Moreover, for a thick ceramic, not only do the nose and tail velocities become the same at a much later time (when the projectile erosion has subsided) as compared to a thin ceramic, but also the penetration velocity inside the target continues to decrease. On the other hand, the penetration velocity remains almost steady in the case of a thin ceramic, indicating an inability to resist the projectile. Therefore, increasing the thickness of the ceramic can greatly enhance the ballistic performance of a protective system. However, the downside is that it increases the overall weight of the system.

3.4.2 Effect of Confinement

Figure 3.6 reveals typical impact-induced radial cracking with limited cone cracking in an 8-mm-thick alumina tile on the front and back surfaces (Sherman and Ben-Shushan 1998). Here the ceramic configuration was different than that in Figure 3.5. The ceramic tile was constrained from four sides, but no confinement pressure was applied. External constraint without additional confining pressure clearly alters the ceramic response. The back surface shows several radial cracks, some of which were arrested before progressing to the top surface. This damage pattern illustrates that radial cracks originate from the back face and then propagate to the front surface. Due to formation of these radial cracks, the ceramic tile becomes more compliant, causing the impact stresses to become even more localized and larger in magnitude. This leads to an accelerated evolution of local damage, rapid penetration, and eventual target failure. Thus, back-face radial cracking is a major failure mechanism in a thin ceramic plate and governs the transition from dwell to penetration (LaSalvia et al. 2010; Sherman and Ben-Shushan 1998).

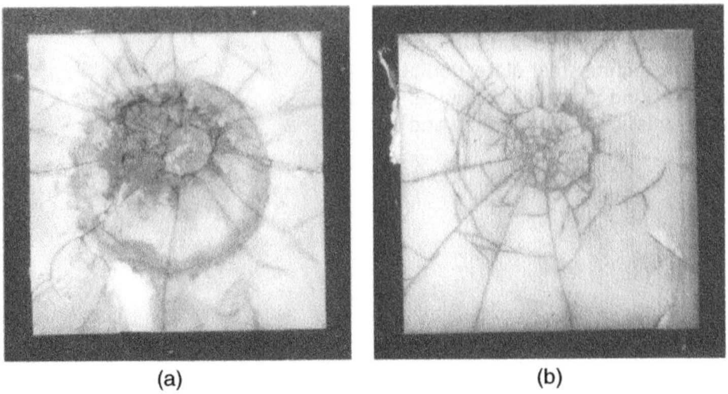

Figure 3.6 Impact-induced damage in a ceramic tile on the (a) front surface and (b) back surface. Source: Reproduced with permission from Sherman and Ben-Shushan (1998).

When the failure is directed by a tensile stress field (e.g. radial cracking), penetration proceeds much faster as compared to cone cracking. However, the tensile cracks are strongly influenced by the external confinement (Sherman and Ben-Shushan 1998). Radial cracks are tensile cracks generated by biaxial bending of ceramic plate during impact (Sherman and Ben-Shushan 1998). These types of cracks can be suppressed by confining the ceramic to restrict bending of the armor system. Figure 3.7 shows a small number of partially developed radial cracks from impact on the back surface of a 5-mm-thick confined alumina tile (Sherman and Ben-Shushan 1998). This is the same thickness of tile that completely disintegrated under a similar ballistic impact in the absence of

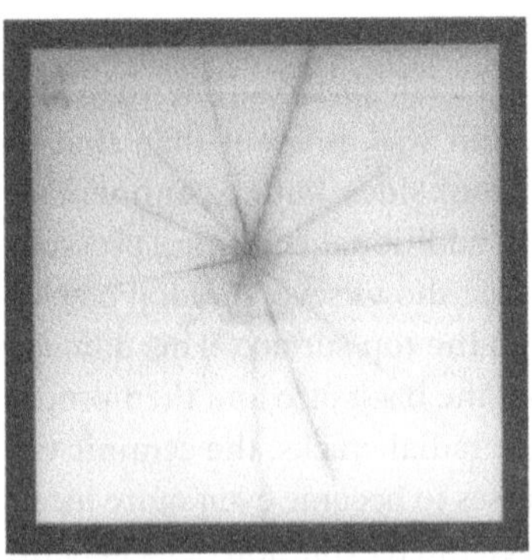

Figure 3.7 Radial cracks on the back surface of a 5-mm-thick alumina tile with steel backing and 200 MPa confinement. Source: Reproduced with permission from Sherman and Ben-Shushan (1998).

confinement. This observation clearly illustrates that the extent of damage can be reduced dramatically by confinement.

3.4.3 Effect of Ceramic Material Properties

Numerous material properties are relevant to the enhanced performance of a ceramic plate against a projectile impact. While high-tensile strength is required to prevent cracking, high hardness (or yield strength) promotes dwell, and high fracture toughness promotes interface defeat. Although it may be impossible for a ceramic to possess all the desired properties, it is helpful for a designer to realize the relative contributions of individual properties on the ballistic performance. Efficient assessment of these influences using experiments alone is prohibitively expensive and sometimes practically impossible. Sophisticated constitutive models along with efficient numerical approaches are needed to fully characterize the influence of properties on ballistic performance of ceramic plates.

3.4.3.1 Tensile Strength

Anderson (2002) performed finite element (FE) analyses of a ceramic-metal-backing-plate protective system and varied the material properties of the front ceramic plate. The model had been validated earlier for a known ceramic-metal armor system. The tensile strength of the ceramic was varied from 0.3 MPa to 1.5 GPa, and the thickness of the ceramic front plate required to defeat a projectile was noted. A fivefold increase in tensile strength of the ceramic only produced a 16% increase in ballistic performance. A similar result was found in Bavdekar et al. (2017); a 100% increase in tensile strength increased the target resistance of a ceramic by less than 2%. These results indicate that a dramatic increase in the tensile strength of a ceramic alone does not significantly increase the ballistic efficiency of a protective system. The low benefit of increased tensile strength is due to the initial shock generated by the impact process, which overpowers the ceramic by a huge margin (Anderson 2002). The effect is further amplified in a thinner ceramic due to lower bending stiffness.

3.4.3.2 Hardness

Among all the properties of a ceramic, hardness has been identified as the most relevant property to provide resistance against penetration (LaSalvia et al. 2010). Hardness is important in promoting dwell, resulting in greater projectile deceleration and reduction in KE (Anderson 2002; LaSalvia 2005a,2005b). The harder the material, the longer the dwell phase, and the greater is the deceleration and loss in KE of the projectile.

Hardness is also related to the depth-of-penetration (DOP) and ballistic efficiency of the armor system. Woodward et al. (1994) compared the residual depth of penetration and ballistic efficiency in Al_2O_3, ZrO_2, and TiB_2 ceramics and glass materials impacted with sharp and blunt projectiles. Interestingly, both DOP and ballistic efficiency were related to the hardness of the ceramic. Ceramics with greater hardness values resulted in shallower depths of penetration and, therefore, greater ballistic efficiencies. For hard ceramics, the penetration depth is small because of their ability to blunt the projectile nose and create an inefficient penetrator. From the depth of penetration and KE of the bullet, a mean resisting pressure of the ceramic plate was calculated and this pressure was found to be greater for harder materials. For alumina and glass, the calculated mean resisting pressures under a pointed projectile impact are comparable to the hardness values. This is expected because the hardness measured by a pointed tool is the resistance of a material against penetration. However, this is only true while the ceramic is undamaged. Once fracture initiates, the mean resisting pressure will be lower compared to the hardness because of the fractured material's reduced ability to support shear stress.

3.4.3.3 Fracture Toughness

Fracture and fragmentation are major energy dissipative mechanisms in a ceramic target. A higher fracture toughness (keeping all other material properties the same) will allow more energy to be absorbed in the target and thus further decrease the KE of the penetrating projectile. Delaying the fracture initiation process (through high fracture toughness) extends the dwell phase and reduces the transition velocity. Once penetration initiates, the hardness of a material (which is related to shear strength) will have less influence on the remainder of the target performance, and instead fracture toughness will be of greater importance. Therefore, fracture toughness is a crucial material property to consider for target characteristics and the overall ballistic efficiency.

LaSalvia (2002, 2005a,b) highlighted the influence of fracture toughness on the transition velocity by premising that the lower bound for ballistic penetration is associated with an unconstrained growth of microcracks when the pressure exceeds a critical value in the comminuted region beneath the surface. Utilizing Horii and Nemat-Nasser's (Horii and Nemat-Nasser 1986) ductility parameter, LaSalvia (2002) rationalized the transition from dwell to penetration phase during an impact event. The ductility parameter, Δ, is defined as the ratio of normal stress, σ_c, at the initiation of crack growth to the yield strength, τ_y, of the material:

$$\Delta = \frac{\sigma_c}{\tau_y} = \frac{K_{IC}}{\tau_y \sqrt{\pi c}} \tag{3.9}$$

where K_{IC} is the fracture toughness and c is the initial crack length. Utilizing the wing-crack growth model (Horii and Nemat-Nasser 1986), the ductility parameter was further expressed in terms of impact pressure, σ_1, and confinement pressure,

σ_2, as:

$$\Delta = \frac{\left(\frac{\sigma_1}{\tau_y}\right)\left[\left(1-\frac{\sigma_2}{\sigma_1}\right)-\mu\left(1+\frac{\sigma_2}{\sigma_1}\right)-\sqrt{2}\pi\left(\frac{\sigma_2}{\sigma_1}\right)\left(\sqrt{\frac{l_t}{c}}\right)^2\right]}{\sqrt{2}\pi\sqrt{\frac{l_t}{c}}} \tag{3.10}$$

where

$$\frac{\sigma_1}{\tau_y} = \frac{2\left\{1-\frac{2}{\pi}\sin^{-1}\left[1/\left(1+\frac{l_t}{c}\right)\right]\right\}}{\left(1-\frac{\sigma_2}{\sigma_1}\right)-(2\mu/\pi)\left(1+\frac{\sigma_2}{\sigma_1}\right)\sin^{-1}\left[1/\left(1+\frac{l_t}{c}\right)\right]} \tag{3.11}$$

and l_t is the length of tensile wing crack and μ is the friction coefficient.

Recall from Eq. (3.4) that during the impact process, the maximum shear stress beneath a point of application of load occurs along the axis of symmetry at a depth, z_o, and it is expected that damage first initiates from this point and grows larger with the continued application of force. The damage zone within a confined ceramic during the dwell phase is fully surrounded (confined) by the intact material, so this damage zone must grow and reach the surface in order for the projectile in the dwell state to initiate penetration into the ceramic (transition from dwell to penetration). The critical pressure (P_{crit}) for the dwell–penetration (D-P) transition is related to the yield strength of the material and is taken as $2.85\sigma_Y$. Relating the mean pressure, p_m, to material properties, LaSalvia (2002) derived the following relationship:

$$\frac{p_m}{2.85\sigma_Y} = \frac{0.93\pi\Delta^*}{(1-2v)-\mu(3+2v)-(1+2v)\pi\sqrt{\left(\frac{\lambda_1}{c}\right)_{crit}}} \tag{3.12}$$

And the penetration velocity, V_P, is related to mean pressure as

$$V_P = \sqrt{\frac{2K_p}{\rho_p}}\left[-1+\sqrt{1+2\left(\frac{P_m-3.27\sigma_{Y_p}}{K_p}\right)}\right] \tag{3.13}$$

where K_p is the bulk modulus of the penetrator.

McCauley et al. (2012) plotted the penetration velocity found from Eq. (3.13) as a function of yield strength for various ductility parameters and noted that greater ductility parameters give greater transitional velocity values for a given yield strength. Therefore, by improving the ductility parameter, one can also increase the transition velocity of a ceramic. From Eq. (3.9), increasing the ductility parameter requires that the flaw size, c, or σ_y should be decreased or that K_{IC} must be increased. Let us examine these three options further. From a processing perspective, the flaw size can be controlled and decreased. Decreasing yield strength (and consequently deceasing hardness) is not a preferred strategy as it has other implications for armor performance (e.g. decreasing dwell). It is possible, however, to improve the fracture toughness of a ceramic. Using the above analysis, LaSalvia (2002) derived a relationship between fracture toughness, mean pressure, and yield strength as follows:

$$\frac{(\Delta K_{IC})_{max}}{K_{IC}^{A}} = \frac{2.85\sigma_{Y}^{A}}{P_{m}^{A}} - 1 \tag{3.14}$$

This equation is useful for understanding the effect of fracture toughness and is best illustrated with an example. Boron carbide has a yield strength of 15.8 GPa (assuming a hardness of 35 GPa and a conversion factor of 2.8 between yield strength and hardness for ceramics) and a fracture toughness of 2.8 MPa-m$^{1/2}$ (DeVries et al. 2016). If the target must withstand a projectile velocity of 1450 m/s, corresponding to a mean pressure of 23 GPa (Eq. (3.12), the change in fracture toughness (ΔK_{IC}) required is 2.7 MPa-m$^{1/2}$, giving an overall K_{IC} of $2.8 + 2.7 = 5.5$ MPa-m$^{1/2}$. This value is quite high and may not be easily achieved in boron carbide without compromising other properties. Regardless, this exercise allows one to recognize the relative importance of fracture toughness in improving the ballistic efficiency of a ceramic protective system.

3.5 Summary and Conclusions

In this chapter, we presented various phases that evolve during projectile impact of a ceramic target. The relationship between the "nonpenetration" phase (dwell and interface defeat) and material properties (hardness, fracture toughness, tensile strength, and yield strength) is highlighted. The evolved fracture modes in ceramic plates due to projectile impact are introduced. We conclude this chapter by emphasizing that ballistic efficiency of a ceramic protective system is a complex combination of many mechanical properties and geometrical features, each acting either individually or collectively, concurrently or sequentially during armor–penetrator interaction phases. Even when dramatic changes are made to a specific geometrical feature (e.g. thickness) or improvements to a mechanical property (e.g. tensile strength), only marginal improvements in ballistic performance are attained. Modeling is crucial to understanding the complex interplay between ceramic plate geometry and material properties to meet the desired performance requirements. In the reminder of this book, we will dwell into in-depth analysis of the mechanisms of deformation from experimental, analytical, and computational perspectives.

References

Alekseevskii, V. (1966). Penetration of a rod into a target at high velocity. *Combustion, Explosion and Shock Waves* 2: 63–66.

Anderson, C.E. (2002). Developing an ultra-lightweight armor concept. In: *Ceramic Armor Materials by Design: Ceramic Transactions* (eds. J. McCauley et al.), 485–498. The American Ceramic Society.

Anderson Jr, C., Behner, T., Holmquist, T., Orphal, D., and Wickert, M. (2009). Dwell, interface defeat, and penetration of long rods impacting silicon carbide. SwRI Report 18.12544/008, prepared for RDECOM-TARDEC. AMSRD-TAR-R, WarrenMI.

Anderson, C. and Walker, J. (2005). An analytical model for dwell and interface defeat. *International Journal of Impact Engineering* 31: 1119–1132.

Bavdekar, S. and Subhash, G. (2018). Comparison of pressure-sensitive strength models for ceramics under ultrahigh confinement. *International Journal of Impact Engineering* 118: 60–66.

Bavdekar, S., Parsard, G., Subhash, G., and Satapathy, S. (2017). An improved dynamic expanding cavity model for high-pressure and high-strain rate response of ceramics. *International Journal of Solids and Structures* 125: 77–88.

Bavdekar, S., Subhash, G., and Satapathy, S. (2019). A unified model for dwell and penetration during long rod impact on thick ceramic targets. *International Journal of Impact Engineering* 131: 304–316.

Bhattacharyya, A., Subhash, G., and Arakere, N. (2014). Evolution of subsurface plastic zone due to rolling contact fatigue of M-50 NiL case hardened bearing steel. *International Journal of Fatigue* 59: 102–113.

Birkhoff, G., MacDougall, D.P., Pugh, E.M., and Taylor, S.G. (1948). Explosives with lined cavities. *Journal of Applied Physics* 19: 563–582.

Bless, S., Satapathy, S., and Normandia, M. (1999). Transverse loads on a yawed projectile. *International Journal of Impact Engineering* 23: 77–86.

Das, S., Ronan, W., Wadley, H., and Deshpande, V.S. (2018). Penetration of confined ceramics targets. *Extreme Mechanics Letters* 18: 45–57.

DeVries, M., Pittari, J., Subhash, G. et al. (2016). Rate-dependent mechanical behavior and amorphization of ultrafine-grained boron carbide. *Journal of the American Ceramic Society* 99: 3398–3405.

Hauver, G.E., Rapacki, E.J. Jr., Netherwood, P.H., and Benck, R.F. (2005). *Interface Defeat of Long-Rod Projectiles by Ceramic Armor*. Aberdeen Proving Ground, MD: Army Research Lab, Weapons And Materials Research Directorate.

Hencky, H. (1923). Über Einige Statisch Bestimmte Fälle Des Gleichgewichts In Plastischen Körpern. *ZAMM – Journal of Applied Mathematics and Mechanics/Zeitschrift für Angewandte Mathematik und Mechanik* 3: 241–251.

Hill, R. (1998). *The Mathematical Theory of Plasticity*. Oxford University Press.

Horii, H. and Nemat-Nasser, S. (1986). Brittle failure in compression – splitting, faulting and brittle-ductile transition. *Philosophical Transactions of the Royal Society a – Mathematical Physical and Engineering Sciences* 319: 337–374.

Iyer, K. (2007). Relationships between multiaxial stress states and internal fracture patterns in sphere-impacted silicon carbide. *International Journal of Fracture* 146: 1–18.

Johnson, K.L. (1985). *Contact Mechanics*. Cambridge, New York: Cambridge University Press.

Lankford, J., Predebon, W., Staehler, J. et al. (1998). The role of plasticity as a limiting factor in the compressive failure of high strength ceramics. *Mechanics of Materials* 29: 205–218.

LaSalvia, J. (2002). Recent progress on the influence of microstructure and mechanical properties on ballistic performance. *Ceramic Transactions* 134: 557–570.

LaSalvia, J. (2005a). A Predictive Model for the Dwell/Penetration Transition Phenomenon. *Proceedings of the 22nd International Symposium on Ballistics*, Vancouver, Canada, **2**, 717–25.

LaSalvia, J. (2005b). Effect of ceramic thickness on the dwell/penetration transition phenomenon. Ballistics 2005: 22nd International Symposium on Ballistics.

LaSalvia, J.C., Horwath, E.J., Rapacki, E.J. et al. (2001). *Fundamental Issues and Applications of Shock-Wave and High-Strain-Rate Phenomena* (eds. K.P. Staudhammer, L.E. Murr and M.A. Meyers), 437–446. Oxford, UK: Elsevier Science, Explomet 2000, June 18–22, 2000, Albuquerque, New Mexico.

LaSalvia, J., Campbell, J., Swab, J., and McCauley, J. (2010). Beyond hardness: ceramics and ceramic-based composites for protection. *The Journal of The Minerals, Metals & Materials Society* 62: 16–23.

Lawn, B. and Wilshaw, T.R. (1993). *Fracture of Brittle Solids*. Cambridge University Press.

Lundberg, P. and Lundberg, B. (2005). Transition between interface defeat and penetration for tungsten projectiles and four silicon carbide materials. *International Journal of Impact Engineering* 31: 781–792.

Lundberg, P., Renstrom, R., and Lundberg, B. (2000). Impact of metallic projectiles on ceramic targets: transition between interface defeat and penetration. *International Journal of Impact Engineering* 24: 259–275.

McCauley, J.W., Crowson, A., Gooch, W.A. Jr., et al. (2012). *Ceramic Armor Materials by Design*. Wiley.

Normandia, M. (2004). Impact response and analysis of several silicon carbides. *International Journal of Applied Ceramic Technology* 1: 226–234.

Orphal, D. and Franzen, R. (1997). Penetration of confined silicon carbide targets by tungsten long rods at impact velocities from 1.5 to 4.6 km/s. *International Journal of Impact Engineering* 19: 1–13.

Orphal, D., Franzen, R., Piekutowski, A., and Forrestal, M. (1996). Penetration of confined aluminum nitride targets by tungsten long rods at 1.5–4.5 km/s. *International Journal of Impact Engineering* 18: 355–368.

Orphal, D., Franzen, R., Charters, A. et al. (1997). Penetration of confined boron carbide targets by tungsten long rods at impact velocities from 1.5 to 5.0 km/s. *International Journal of Impact Engineering* 19: 15–29.

Reich, F. (1926). *Umlenkung eines freien Flüssigkeitstrahles an einer senkrecht zur Strömungsrichtung stehenden ebenen platte*. VDI-Verlag, GmBH.

Satapathy, S. (2001). Dynamic spherical cavity expansion in brittle ceramics. *International Journal of Solids and Structures* 38: 5833–5845.

Satapathy, S. and Bless, S. (1996). Calculation of penetration resistance of brittle materials using spherical cavity expansion analysis. *Mechanics of Materials* 23: 323–330.

Shafiq, M. and Subhash, G. (2016). An extended Mohr-Coulomb model for fracture strength of intact brittle materials under ultrahigh pressures. *Journal of the American Ceramic Society* 99: 627–630.

Sherman, D. (2000). Impact failure mechanisms in alumina tiles on finite thickness support and the effect of confinement. *International Journal of Impact Engineering* 24: 313–328.

Sherman, D. and Ben-Shushan, T. (1998). Quasi-static impact damage in confined ceramic tiles. *International Journal of Impact Engineering* 21: 245–265.

Shield, R.T. (1955). On the plastic flow of metals under conditions of axial symmetry. *Proceedings of the Royal Society of London. Series A, Mathematical and Physical Sciences* 233: 267–287.

Shockey, D.A., Marchand, A.H., Skaggs, S.R. et al. (1990). Failure phenomenology of confined ceramic targets and impacting rods. *International Journal of Impact Engineering* 9: 263–275.

Subramanian, R. and Bless, S. (1995). Penetration of semi-infinite AD995 alumina targets by tungsten long rod penetrators from 1.5 to 3.5 km/s. *International Journal of Impact Engineering* 17: 807–816.

Tate, A. (1967). A theory for the deceleration of long rods after impact. *Journal of the Mechanics and Physics of Solids* 15: 387–399.

Wilkins, M. (1967). *Second Progress Report of Light Armor Program*. Livermore, USA: California University Lawrence Livermore Lab.

Wilkins, M.L. (1978). Mechanics of penetration and perforation. *International Journal of Engineering Science* 16: 793–807.

Wilkins, M.L. and Laible, R.C. (1980). Chapter 10 – Computer Simulation of Penetration Phenomena. *Methods and Phenomena* 5: 225–252.

Wilkins, M., Cline, C., and Honodel, C. (1969). *Fourth Progress Report of Light Armor Program*. Livermore: California University Lawrence Radiation Lab.

Woodward, R., Gooch, W., Odonnell, R. et al. (1994). A study of fragmentation in the ballistic impact of ceramics. *International Journal of Impact Engineering* 15: 605–618.

4

Static and Dynamic Responses of Ceramics

4.1 Introduction

Brittle materials are characterized by their elastic response prior to fracture. At the microstructure level, the lack of ductility in ceramics has been mainly attributed to two intrinsic factors:

1. Sensitivity to microstructural defects, and
2. High Peierls stress barrier (Nabarro 1997) that limits dislocation mobility (or slip) in the crystal lattice.

On the other hand, the failure behavior and fracture strength of ceramics are also a function of two extrinsic factors:

1. Applied strain rate, and
2. External confinement pressure.

Thus, the strength of a ceramic is not a singular value but depends on many factors including microstructure (grain size, size and shape of defects, porosity level, etc.), loading type (e.g. compression, tension, bending, and a combination thereof), temperature, and strain rate. In this chapter, we detail how microstructural inhomogeneity impacts the failure strength as a function of strain rate and pressure, followed by a description of various models designed capture such responses.

4.1.1 Sensitivity to Microstructural Defects

The mechanical behavior of ceramics is intimately related to inherent microstructural defects. These microstructural inhomogeneities may be either process-induced or deformation-induced. Some examples of process-induced defects are impurities, pores, grain boundaries, triple points, phase discontinuities, inclusions, twins, and cracks from thermal expansion mismatch of phases

Dynamic Response of Advanced Ceramics, First Edition.
Ghatu Subhash, Amnaya Awasthi, Dipankar Ghosh.

or grain orientations. Deformation-induced defects could be dislocations, twins, stacking faults, slip bands, void generation, cracks, and phase transformations. For example, ceramics such as aluminum oxide exhibit dislocations and twins during deformation (Staehler et al. 2000). In boron carbide, deformation-induced amorphous zones are formed under high pressure (Awasthi and Subhash 2019; Parsard et al. 2018; Subhash et al. 2016). Similarly, slip-line-like features have been observed in aluminum nitride (AlN) (Heard and Cline 1980), which also exhibits phase transformation (Kipp and Grady 1994). Similar dislocation-like features were observed in zirconium diboride under contact loading (Ghosh et al. 2010) as well. Even at low macroscopic stress levels, high-stress concentration occurs at these defects to initiate crack nucleation leading to crack growth and premature failure of the material (Lankford et al. 1998).

4.1.2 High Peierls Stress

Ceramics lack well-defined atomic slip planes and hence the force needed to move a dislocation through the atomic lattice (called Peierls–Nabarro force) is high. This frictional resistance to dislocation mobility during deformation causes stress build-up at defects or potential nucleation sites. At sufficiently high-stress level, crack nucleation is expected to occur eventually leading to crack growth and macroscopic fracture of the specimen. If crack nucleation and growth from the preexisting defects can be suppressed (by applying confinement stress perpendicular to the crack-growth direction), a high level of stress can be sustained in ceramics to potentially initiate plastic deformation modes such as dislocations, twins, and phase transformation. Lankford et al. (1998) subjected high-purity aluminum oxide to high confining pressure experiments (indentation and shock loading) and observed few dislocations to initiate near defects. Interestingly, they also noted microcracks in these deformed regions and postulated that initiation of plasticity leads to the fracture of alumina ceramics. Lack of fully activated dislocation motion limits the yield strength (σ_y) of ceramics to well below the limits set by hardness (H) for ductile materials, per Tabor's relationship $\sigma_y = H/3$. Thus, one can contrast the behavior of ceramics from that of metals by rationalizing that, in metals, yield stress is reached first followed by plasticity and fracture, whereas in brittle materials, fracture stress is reached before yielding may occur. The presence of defects facilitates this fracture process and hence, ceramics deform primarily by the initiation of cracks. So how can we delay the onset of characteristic early macroscopic brittle fracture and induce plasticity or ductile deformation mechanisms in a ceramic without modifying its microstructure or composition? The simple answer is to apply confinement pressure to suppress tensile cracking.

4.1.3 Compression and Confinement Testing of Ceramics

Suppression of premature tensile cracking allows ceramics to withstand large loads. For example, uniaxial compressive loading yields three to five times the failure strength values compared to the tensile strength of a ceramic. Further improvements in compressive strength can be achieved by applying lateral confinement on the specimen. In the following sections, we will present the effect of lateral confinement on strength and strain rate response of ceramics. It should be realized that confinement only delays the onset of cracking, and cracks will eventually evolve under local tension at the crack tip to cause the crack growth and macroscopic failure of the ceramic. Compared to tension testing, compression and confinement testing offers many interesting avenues for research on the ceramic behavior due to the following reasons:

1. The fracture strength of ceramics is sensitive to microstructural heterogeneities. Tensile testing using the conventional dog-bone shaped specimen under quasistatic (QS) loading often activates only a single critical flaw from which a crack grows perpendicular to the applied load to cause fracture of the specimen into two pieces. With increasing strain rate, more cracks may nucleate, as predicted by computations (Molinari et al. 2007; Zhou and Molinari 2004b; Zhou et al. 2004, 2005, 2006a,b,c). The stress concentration associated with a single critical flaw will initiate early fracture during the QS tensile loading, and macroscopically, the specimen exhibits only an elastic response before fracture. Accordingly, the failure strength observed is small compared to that in compressive loading. Such a failure mode often prevents activation of other inelastic deformation mechanisms such as phase transformation or generation of dislocations. Hence, the energy absorbed during the QS tensile loading is limited only to elastic deformation followed by creation of two new surface from a single crack.
2. The compressive strength of a ceramic is three to five times its tensile strength, and hence, ceramics are more effective to resist compressive loads, such as those encountered in ballistic loading. Therefore, understanding the deformation behavior under compressive loading is of particular interest.
3. A ceramic specimen often exhibits more complex deformation mechanisms under compressive loading, especially under dynamic loads and at high pressure, whereas only a single crack growth occurs in tensile loading. For example, compressive loading of a confined ceramic initiates multiple cracks and can produce fragmentation, comminution, plasticity, phase transformation, and amorphization, all of which absorb more energy than creation of two surfaces from a single crack in tension. Moreover, in tension, once the cracks are formed, the fragments separate and no further loading can be applied, whereas in confined compression, fragments of the fractured specimen remain confined and continue to sustain load.

4. The compressive strength of a ceramic increases not only with increase in applied strain rate but also applied lateral compression (confinement). This is an attractive feature that can be effectively exploited in certain applications such as armor.
5. Finally, fracture strength may be the only inelastic property that can be measured in a uniaxial tension test of a ceramic, whereas compression (confinement) tests on brittle materials activate new mechanisms and offer new avenues to effectively use ceramics under extreme environments. While the elastic response is rate-independent for structural ceramics, many of the abovementioned mechanisms are highly strain rate-dependent. Thus, compression testing of ceramics offers a wider playing field for researchers to dwell into the mechanisms of deformation.

Realizing the above advantages of compressive testing, a detailed discussion of compression-induced mechanisms and modeling efforts to describe these features are provided. In the following, we will also present the deformation behavior of intact and failed ceramics.

4.2 Strain-rate Dependence of Compressive Failure Strength of Intact Ceramics

4.2.1 Fracture at Quasistatic Strain Rates

Crack growth from preexisting defects under compression has been addressed using a sliding microcrack model, originally developed by Brace and Bombolakis (1963) and later analyzed in-depth by Nemat-Nasser and Horii (1982) and Ashby and Hallam (1986). Upon application of load, suitably oriented microcracks grow tensile wing cracks at a critical stress intensity factor (K_{IC}). These cracks grow in a stable manner under uniaxial loading in a direction parallel to the compression axis (Brace and Bombolakis 1963; Horii and Nemat-Nasser 1986). In sintered ceramics, cracks nucleate from pores due to local tension and then grow parallel to the compression axis (Sammis and Ashby 1986). Depending on the aspect ratio of pores and preexisting cracks, stable or unstable crack growth occurs. If the aspect ratio of these pores is small, they have the ability to store large elastic energy before they release this energy at higher stress intensity factor (K_I greater than K_{IC}) in a catastrophic manner leading to unstable crack growth and fracture. The lack of stable crack growth in ceramics, especially sintered ceramics, results in a rate-insensitive response of fracture strength below a critical strain rate (Ravichandran and Subhash 1995).

Subhash and Nemat-Nasser (1993) captured the initiation of multiple axial cracks in zirconia ceramics under dynamic uniaxial loading by tailoring the incident stress wave in a momentum-trapped split Hopkinson pressure bar (SHPB). They subjected a rectangular cuboid specimen to dynamic uniaxial loading successively in two orthogonal directions and observed the evolved crack patterns. Due to the high rate of loading and use of momentum trap in SHPB, they were able to control the amount of input energy (through pulse shaping) into the specimen and limit crack growth. After the first loading, a few cracks were observed to grow parallel to the loading axis. When the specimen was reloaded in an orthogonal direction, new cracks were seen growing parallel to this direction as well (i.e. perpendicular to the first loading), again confirming the axial crack growth parallel to the loading axis under uniaxial loading. The axial cracks during the first loading render the specimen transversely isotropic as far as elastic constants are concerned, and the specimen is weaker in tension in a direction perpendicular to the crack orientation. Despite this slight anisotropy, the second loading also causes axial cracking, perpendicular to the initial cracks. The fracture pattern after uniaxial loading in the two directions is shown in Figure 4.1 (Subhash and Nemat-Nasser 1993). Also, note that the dynamic loading initiates multiple cracks spaced at short distances.

Of relevance to the current study is the room temperature compression strength as a function of microstructure and strain rate. The microstructure of a ceramic depends on the processing method and can include vast variations in the nature of defects including size, shape, and distribution of grains, porosity, impurities, secondary phases, and other heterogeneities. We consider these microstructural features as intrinsic variables because a given ceramic comes with certain type of defects and their distributions. The strain rate, on the other hand, is an extrinsic variable as it can be controlled during a test. Motivated by the vast variations in the observed microstructures and the associated compressive strength values

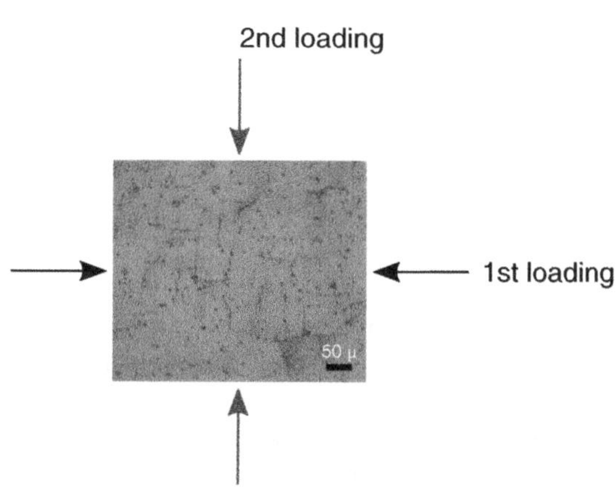

Figure 4.1 Axial crack growth during two sequential, uniaxial compression loadings in two perpendicular directions on zirconia ceramics. Note that even with a slight anisotropy developed during the first loading, cracks still grow parallel to the loading axis during the second loading. Source: Reproduced with permission from Subhash and Nemat-Nasser (1993). © 2013, Elsevier.

of different ceramics and brittle materials, we often normalize the relationship between these two parameters. Kimberley et al. (2013) and Ramesh et al. (2015) collated data for a variety of brittle materials including some geological materials and plotted the above two parameters as shown in Figure 4.2. The plot of normalized strength and strain rate reveals that at low strain rates, in the range below 10^{-1} s^{-1}, the uniaxial compressive fracture strength of ceramics is weakly strain-rate sensitive; the strength remains almost constant over several orders of magnitude of strain rate. At these low strain rates, the applied compressive loading activates only few favorably oriented weakest defects with highest stress concentration to reach their limiting stress value and initiate crack growth. The lack of sensitivity of ceramic fracture strength to strain rate is attributed to unstable crack growth from preexisting defects at stress-intensity factors well above the critical stress-intensity factor (Ravichandran and Subhash 1995). Due to the slow rise in magnitude of applied load, a large amount of elastic energy is readily available for crack growth, and upon reaching a critical stress level, these cracks grow in an unstable manner parallel to the direction of applied macroscopic uniaxial compressive loading leading to axial splitting of the specimen. As a result, in QS compressive loading, the ceramic specimens fail into a few columnar fragments.

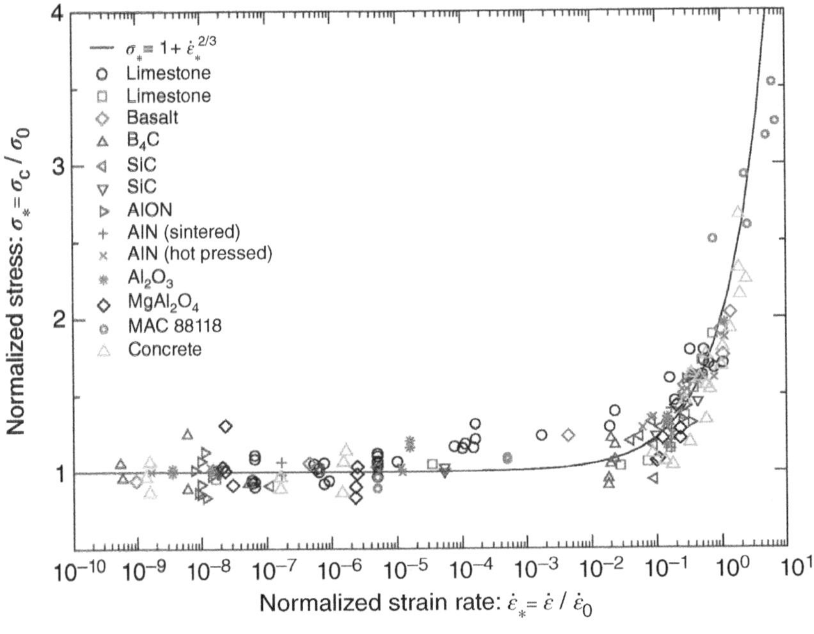

Figure 4.2 Normalized compressive stress and strain rate for various ceramics and geological materials. Source: Kimberley et al. (2013). © 2013, Elsevier.

4.2.2 Fracture at Intermediate Strain Rates

4.2.2.1 Inertia Effects

In the intermediate strain rate range between 10^{-1} s^{-1} and 10^4 s^{-1}, the compressive fracture strength increases at a faster rate. The strength increase can be as high as two to four times the QS fracture strength. In this strain rate regime, the relationship between fracture strength and strain rate can be written as

$$\overline{\sigma} = k\left(\bar{\dot{\varepsilon}}\right)^n \tag{4.1}$$

where $\bar{\dot{\varepsilon}}$ is the normalized strain rate and $\overline{\sigma}$ is the normalized compressive strength. The normalization is performed at a specific strain rate and strength value (e.g. strength σ_0 at a strain rate of 1 s^{-1}). The parameters k and n depend on the defect characteristics (size and their distribution) and material properties (Young's modulus and wave speed) (Grady and Lipkin 1980). For most brittle materials, the value of n has been determined to be close to one-third, regardless of how the ceramic is processed (Grady 1998; Grady and Lipkin 1980; Ravichandran and Subhash 1995). Lankford (1981) found this exponent to be around 0.27 for α-SiC, but this data was limited to few tests. The reason for this consistent exponential value for all the ceramics is the inertia-dominated dynamic crack growth from preexisting defects. Upon dynamic uniaxial stress wave loading, the applied stress rises faster than the rate of increase in stress intensity at the crack tip, and hence, at the onset of macroscopic fracture a higher stress value is recorded. Grady and Lipkin (1980) argued that the strength-increase exponent should be around one-third for all ceramics based on the inertial effects of crack growth at intermediate strain rates up to 10^5 s^{-1}.

Ramesh et al. (2015) recently modified the above relationship as follows:

$$\frac{\sigma_c}{\sigma_0} = 1 + \left(\frac{\dot{\varepsilon}}{\dot{\varepsilon}_0}\right)^n \tag{4.2}$$

where the normalized quantities are defined in terms of material properties (fracture toughness and wave speed) and microstructure (flaw size and flaw density). However, this normalization not only modified the relationship from Eq. (4.1) but also yielded an n value of two-third for Eq. (4.2). Such variations stem from the improved features included in the formulation. In this case, the defect distribution in some functional form (e.g. Gaussian) as well as the interactions of the defects were considered, whereas in Eq. (4.1), an isolated flaw was considered, and its average effect was assumed. This improvement in the model allows investigation of how the distributions of defect sizes impact the fracture strength and will be discussed later in this chapter (Kimberley et al. 2013). Statistical analysis to understand mechanisms leading to the rate effects on fracture strength of ceramics were investigated by Zhou and Molinari (2004b, 2005) and these concepts will be presented at the end of this chapter. Micromechanics-based wing-crack growth models (Huang and Subhash 2003; Nemat-Nasser and Deng 1994; Paliwal and Ramesh 2008; Ravichandran and Subhash 1995) were also developed to capture the above rate sensitivity of ceramics in the intermediate range of strain rates ($10^2\,s^{-1}$–$10^4\,s^{-1}$). All these models will be discussed later in this chapter.

From a microstructure perspective, the crack growth behavior in this strain rate regime is also dramatically different from that of the QS loading. Under dynamic stress wave loading, the rapid rise in stress magnitude simultaneously activates multiple preexisting defects to reach a critical stress intensity before any of them have time to react and initiate crack growth. Thus, the inertial effects play a role in delaying the crack growth associated stress release. With a continued increase in load, a large number of microcracks grow and eventually coalesce to cause fracture of the specimen into multiple small fragments. In summary, under low strain rate (QS) uniaxial compressive loading only a few defects activate, whereas under dynamic (high strain rate) stress wave loading, a large number of defects are activated to cause axial crack growth in a stable manner. Due to their proximity, the cracks quickly coalesce to result in numerous small fragments.

4.2.3 Fracture at High Strain Rates

At strain rates beyond $10^5\,s^{-1}$, the compressive fracture strength of many ceramics once again becomes insensitive to strain rate (Grady 1998). As shown in Figure 4.3, saturation in the failure strength was also supported by numerous other works (Millett and Bourne 2000; Paliwal and Ramesh 2008; Rosenberg et al. 1987, 1991). Vogler et al. (2004) also noted that the Hugoniot elastic limit (HEL) of boron carbide remained almost the same when the material was shock loaded at velocities ranging from 0.98 to 4.88 km/s. In Figure 4.3, the failure strength beyond 10^5 s^{-1} refers to the HEL values obtained using flyer plate impact shock experiments (Grady 1998), where the ceramic specimen is under uniaxial strain condition as opposed to uniaxial stress condition in the previous two lower strain rate regimes. In these flyer plate impact experiments, a material point in the specimen experiences both axial stress and lateral confinement stress of $\sigma_y = \sigma_z = \frac{\nu}{1-\nu}\sigma_x$, which is comparable in magnitude to the applied axial stress. Therefore, the fracture process transitions from an inertia-dominated crack mechanism for the intermediate strain rate regime to a crack-shutdown mechanism. The absence of favorable conditions for cracking allows initiation of new inelastic mechanisms such as point defect generation and phase transformation. Furthermore, even if one assumes frictional sliding of crack surfaces and associated crack growth, under high levels of confinement stress the frictional resistance grows rapidly, making it impossible for the crack surfaces to slide and develop sufficiently high levels of tensile stresses at the crack tip. Thus, the mechanism of failure deviates from the rate-dependent

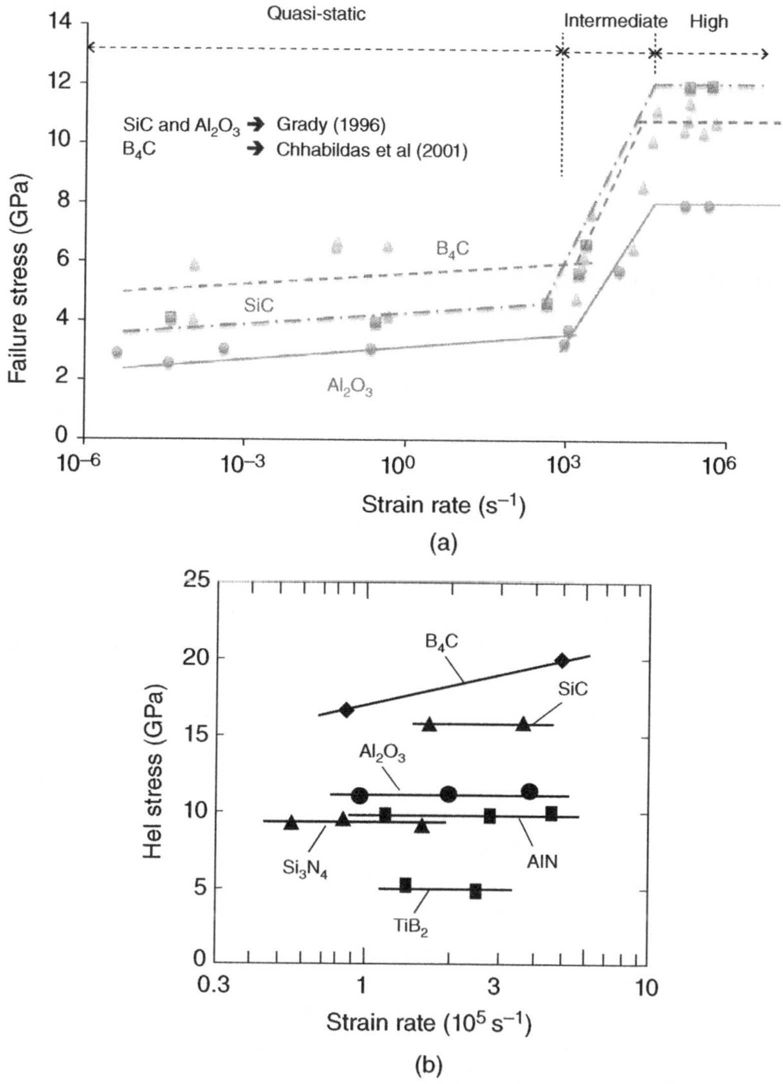

Figure 4.3 (a) Strain rate dependence of compressive strength of ceramics in the strain rate regime below 10^5 s^{-1} and (b) HEL values in the shock regime above 10^6 s^{-1}. Source: (a) (Grady 1998), Vogler et al. (2004), (b) Grady 1998. © 1998, Elsevier.

kinetics of crack growth to inelasticity-based (or plasticity-based) processes, which are less-rate sensitive. At these high stress levels, dislocations and point defects evolve throughout the specimen causing high stress concentrations and leading massive microcracking and disintegration of the entire ceramic specimen. The failure at such high strain rates in brittle materials has also been explored by instrumented shock experiments in glasses and transparent materials. Loss of tensile and shear strengths in the material behind the shock wave, as well as measured occurrence of dilatancy behind the shock front implies failure wave generation (Brar and Bless 1992; Brar et al. 1991; Kanel et al. 1998). These issues are dealt in Chapter 7 on transparent materials.

4.3 Pressure Dependence of Compressive Failure Strength of Intact Ceramics

In addition to the strain rate, fracture strength of a ceramic is also pressure- (or confinement-) dependent. Heard and Cline (1980) conducted confinement experiments on AlN and noted significant ductility in their responses. Large inelastic strains resulted from phase transformation as well as changes in the crack growth mechanisms under a high level of confinement. Analyses of crack growth from preexisting defects under uniaxial and biaxial compression were carried out by Ashby and Hallam (1986) and Horii and Nemat-Nasser (1986). Models for biaxial compressive loading under dynamic loads were developed by numerous researchers (Nemat-Nasser and Deng 1994; Paliwal et al. 2006; Ravichandran and Subhash 1995; Huang and Subhash 2003) who were also able to capture the strain-rate-sensitive behavior of ceramics in the range between $10^1\ s^{-1}$ and $10^4\ s^{-1}$. Similar approach is often employed to effectively use ceramics under extremely high compressive (pressure) loads such as those in flyer plate impact experiments and projectile impact on confined ceramic tiles.

Chen and Ravichandran (1996a,b) conducted confinement tests on ceramics under dynamic conditions by using a "shrink-fit assembly" to confine a cylindrical ceramic specimen with a metal sleeve of suitable wall thickness. The metal sleeve, whose initial inner diameter is slightly smaller than the diameter of the ceramic specimen, was heated to a predetermined temperature (to expand the inner radius), and then the ceramic specimen was forcibly inserted into the sleeve using a mechanical press. Upon cooling, the metal sleeve shrinks and exerts lateral compressive stress on the ceramic. Due to the resistance offered by the ceramic, a portion of the metal sleeve in contact with the ceramic undergoes plastic deformation. The radius of this plastic zone and the magnitude of confinement stress induced in the ceramic are a function of the properties of the metal sleeve and its thickness and the amount of shrink fit. Figure 4.4a illustrates a schematic of the dimensions of the ceramic and metal sleeve, and Figure 4.4b shows ceramic specimens in different metal sleeves.

The radial depth of plastic zone in the metal sleeve and induced residual pressure in the ceramic are derived from the thick-walled pressure vessel equations

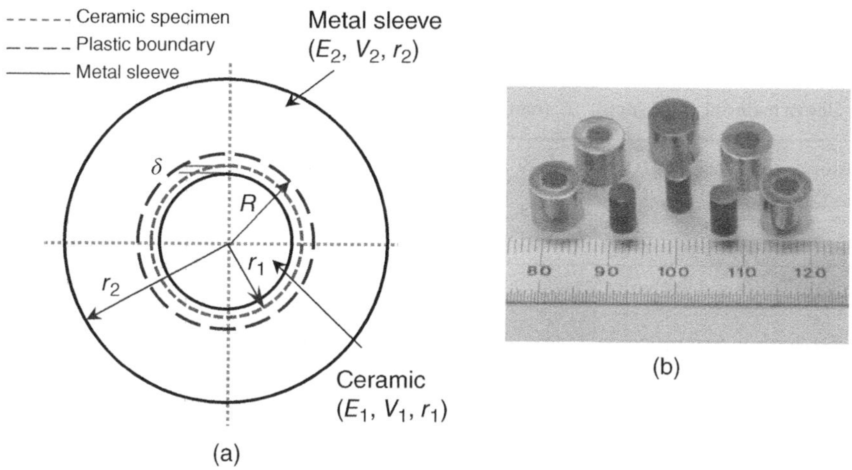

Figure 4.4 (a) Schematic of the dimensions of the ceramic specimen and sleeve, and (b) composite specimens with metal sleeve shrunk-fit over the ceramic. Source: (a) Shafiq and Subhash (2014). © 2014, Springer Nature, (b) Reproduced with permission from Shafiq and Subhash (2014).

and are given by Chen and Ravichandran (1996b) and Shafiq and Subhash (2014):

$$\frac{\delta}{2r_1} = (1+\nu_2)(1-2\nu_2)\left(\frac{\sigma_{Y_2}}{E_2}\right)\left[\ln\left(\frac{r_1}{R}\right)+\frac{1}{2}\left(\frac{R}{r_2}\right)^2-\frac{1}{2}\right]+(1-\nu_2^2)\left(\frac{\sigma_{Y_2}}{E_2}\right)\left(\frac{R}{r_1}\right)^2+(1-\nu_1)\left(\frac{\sigma_{Y_2}}{E_1}\right)\left[\ln\left(\frac{R}{r_1}\right)+\frac{1}{2}\left(1-\left(\frac{R}{r_2}\right)^2\right)\right] \quad (4.3)$$

where δ is the desired shrink-fit, σ_y is the yield stress of the sleeve material, R is the radius of the plastic boundary, r refers to the radius, and subscripts 1 and 2 refer to the ceramic and metal, respectively. Upon calculation of the radial depth of plastic zone, R, iteratively, the confinement pressure, P, can be determined from

$$P = \sigma_{Y_2}\left[\ln\left(\frac{R}{r_1}\right)+\frac{1}{2}\left(1-\left(\frac{R}{r_2}\right)^2\right)\right] \quad (4.4)$$

Shafiq and Subhash (2014) used various sleeve materials and different shrink-fit dimensions on zirconium diboride–silicon carbide (ZrB_2–SiC) ceramic specimens to induce various levels of confinement stress (pressure) listed in Table 4.1.

Utilizing the above composite specimens with different levels of lateral confinement, the QS (10^{-3} s^{-1}) and dynamic (10^3 s^{-1}) compression responses of a ZrB_2–SiC ceramic were determined (Shafiq and Subhash 2016b). These results are shown in Figure 4.5. All these composite specimens exhibited a considerable amount of ductility in their-stress–strain response. Greater levels of confinement

Table 4.1 Sleeve materials and their dimensions used to create different levels of confinement stress on zirconium diboride–silicon carbide specimens.

Sleeve material	δ (mm)	r_1 (mm)	r_2 (mm)	R (mm)	E (GPa)	ν	Pressure (MPa)
Aluminum-2024	0.02	1.99	3.92	2.64	73	0.33	180
Brass-385	0.011	1.99	3.99	2.98	98	0.32	195
Stainless steel-303	0.014	1.91	3.99	2.52	196	0.25	284

Source: Shafiq and Subhash (2014).

led to higher compressive strengths. The strength increased further at higher strain rates (see Figure 4.5b). Similar experiments were earlier conducted on AlN by Chen and Ravichandran (1996b). A summary of the compressive failure strength as a function of strain rate and confinement level for the above two ceramic materials is shown in Figure 4.6. Both QS and dynamic failure strengths increase linearly with confinement pressure in the range of pressures investigated, consistent with the Mohr–Coulomb (MC) model given by

$$\sigma = \sigma_0 + \alpha P \tag{4.5}$$

where P is the pressure, α is the slope of the data between failure strength and pressure, and σ_0 is y-intercept.

In the data presented in Figure 4.6, the dynamic compression strength is greater than the static compressive strength at all the confinement pressures investigated in this study. For AlN, the uniaxial strength without confinement is greater than that of ZrB_2–SiC, but with confinement pressure, both the static and dynamic compressive strengths of ZrB_2–SiC increase past those of AlN, thus exhibiting a higher level of strain-rate- and pressure-sensitivity of failure strength. The increase in compressive strength under confinement is attributed to delayed initiation of fracture due to the applied lateral confining pressure, which resists tensile crack growth. With lateral confinement superposed, the failure mode now transitions to shear faulting as shown in Figure 4.7 (Shafiq and Subhash 2016b). This failure mode transition in biaxial compression loading under QS conditions was first studied by Horii and Nemat-Nasser (1985, 1986) both analytically and experimentally, and later by Ashby and Hallam (1986), Horii and Nemat-Nasser (1986), and Chen and Ravichandran (1996b).

To obtain higher confinement stress levels above those provided by the metal sleeves under dynamic loads, the specimen geometry must be modified to a thin plate and subjected to high velocity impact by a similar plate. These flyer-plate impact experiments were discussed in Chapter 2. In these experiments, the confinement stress is proportional to the imposed axial stress and hence confinement pressure well beyond even HEL of the material can be applied.

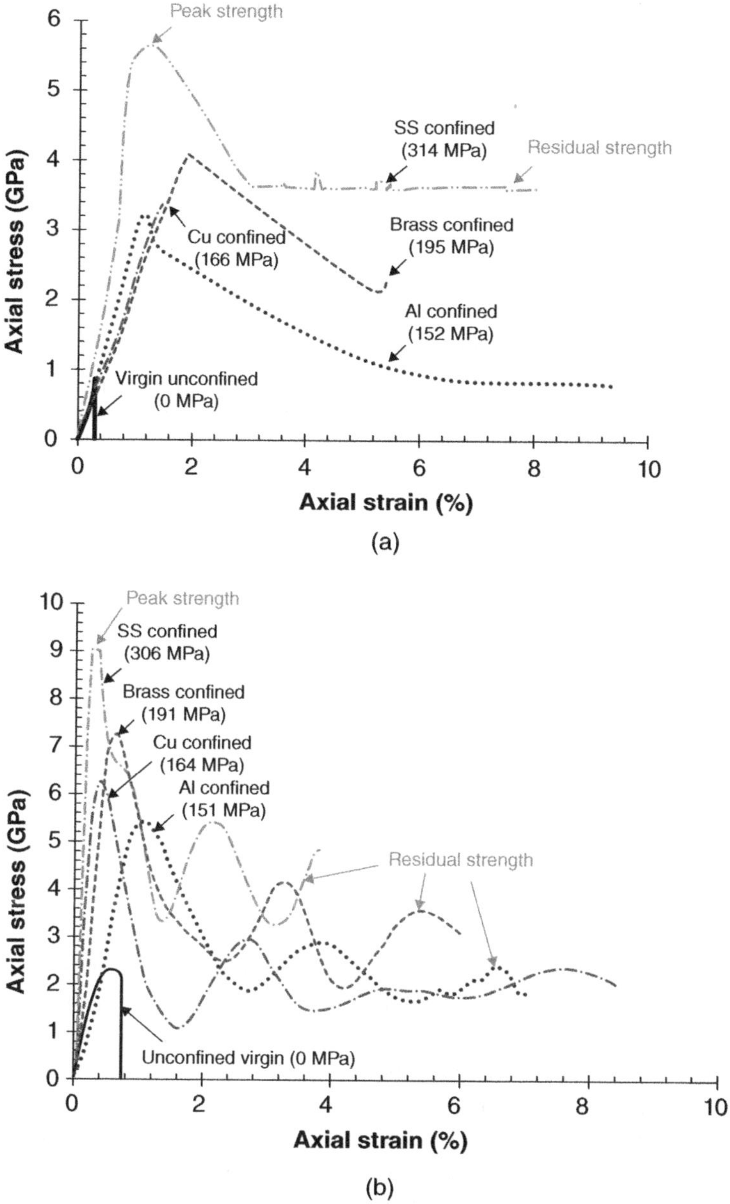

Figure 4.5 Stress–strain response of confined ceramics revealing considerable ductility during axial loading under (a) QS loading and (b) dynamic loading in SHPB. In both cases, the failed ceramic exhibits significant residual strength. Source: Shafiq and Subhash (2016b). © 2016, Elsevier.

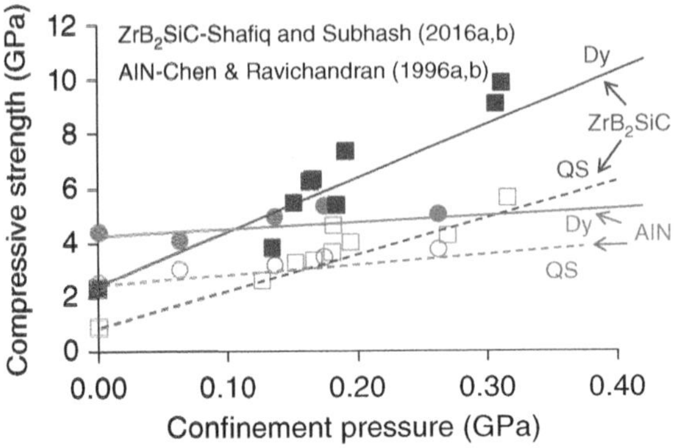

Figure 4.6 Pressure dependence of compressive fracture strength of ceramics under QS and Dynamic (Dy) axial loading.

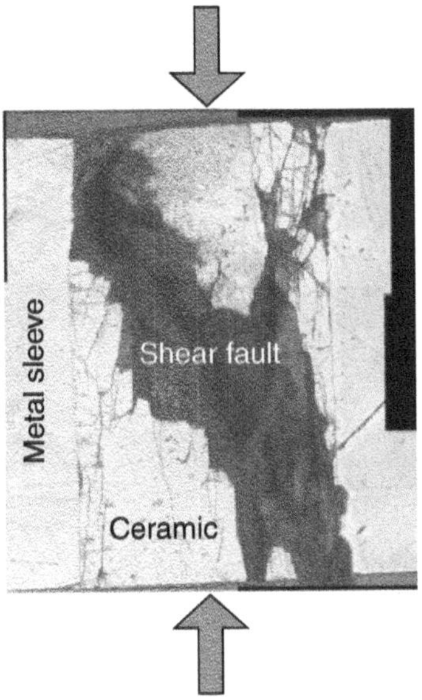

Figure 4.7 Shear faulting in confined specimens under dynamic axial compression. Source: Reproduced with permission from Shafiq and Subhash (2016b). © 2016, Elsevier.

While the ceramic compressive failure strength is a linear function of strain rate at low confining pressures (Figure 4.6), Shafiq and Subhash (2016a) noted that at high confining pressures similar to those encountered in shock loading, the ceramic failure strength is highly nonlinear with respect to the strain rate. To make the right comparison between the uniaxial stress-based failure strength and uniaxial strain-based (but multiaxial stress) HEL values, the shear strength is plotted on the *y*-axis, and the *x*-axis is represented by hydrostatic pressure. The corresponding equations for calculating shear strength (τ) and hydrostatic pressure (P) are given in terms of principal stresses (σ_1 and $\sigma_2 = \sigma_3$), respectively, by

$$\tau = \frac{\sigma_1 - \sigma_2}{2} \text{ and } P = \frac{\sigma_1 + 2\sigma_2}{3} \tag{4.6}$$

where $\sigma_2 = \sigma_3 = (\nu/(1-\nu))\,\sigma_1$ (4.6a)

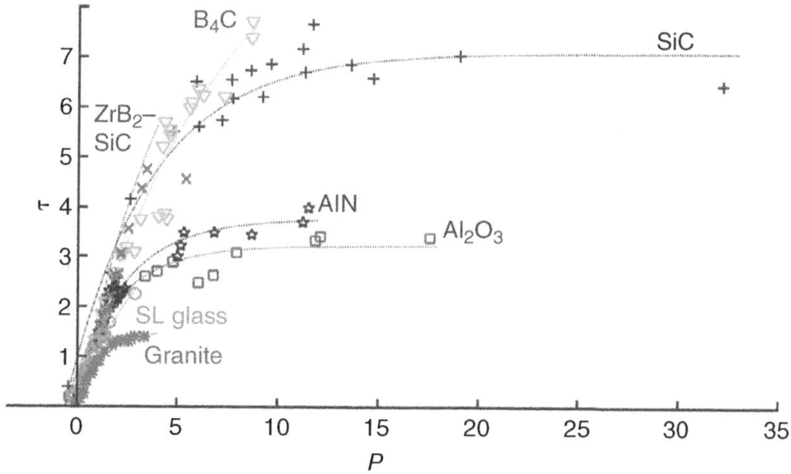

Figure 4.8 Shear strength of various ceramics as a function of hydrostatic pressure. All ceramics reveal a linear response at lower confinement pressure and a transition to a constant shear strength at higher confinement pressure. Source: Bavdekar and Subhash (2018). © 2018, Elsevier.

The data for several ceramics is shown in Figure 4.8 (Bavdekar and Subhash 2018). At a low confinement pressure, the strength is a linear function of pressure similar to Figure 4.6. At intermediate pressure ranges, the failure strength becomes nonlinear, and at higher pressures, the strength is relatively independent of pressure. This complex behavior can once again be rationalized based on the response of heterogeneity present in a ceramic microstructure. At low confining pressures, the ceramic response is dominated by the defects present in the initial microstructure. By applying a small level of lateral confinement, the crack initiation and growth processes from the preexisting defects are delayed, and an increase in failure strength is achieved. With an increase in the confinement level, the tensile crack growth from defects is further suppressed, and the failure mechanism transitions to plasticity-based deformation processes. This transition process is gradual, and the relationship between shear strength and pressure becomes nonlinear. At much higher confinement levels, a dislocation-based plasticity mechanism (or even phase transformation) initiates; however, it is limited by the lack of slip planes in ceramics, resulting in high local stress concentrations, initiation of numerous microcracks, and eventual pulverization of the material under the shock loading. Beyond a critical pressure, this mechanism dominates the failure process, and the failure strength saturates. This critical pressure limit is associated with the HEL.

By normalizing the observed experimental data for various ceramics listed in Figure 4.8 by the corresponding HEL values (i.e. shear strength by τ_{HEL} and

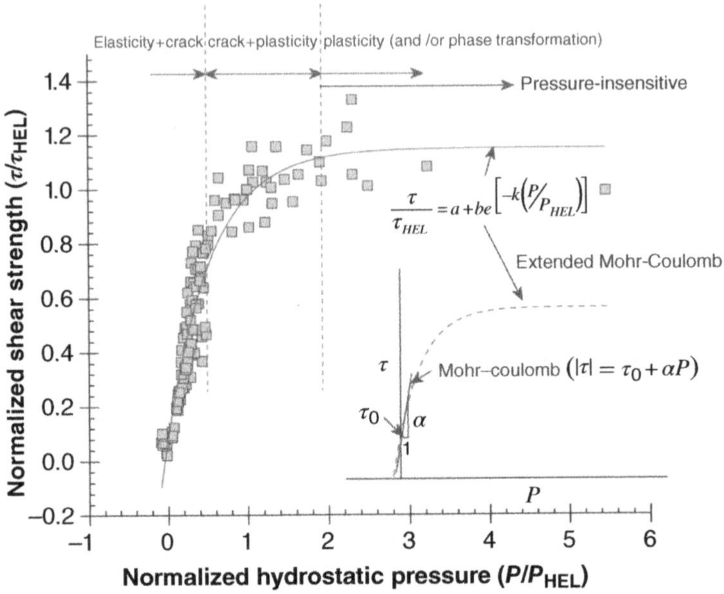

Figure 4.9 Normalized failure strength data of several ceramics revealing three distinct trends for different pressure ranges. Source: Shafiq and Subhash (2016a). © 2016, John Wiley & Sons.

hydrostatic pressure by P_{HEL}) Shafiq and Subhash (2016a) observed the data to collapse along a single curve shown in Figure 4.9. A universal relationship, named extended Mohr–Coulomb model (EMC), that captures the entire pressure (P) dependent response of shear strength (τ) of ceramics was proposed as

$$\frac{\tau}{\tau_{HEL}} = \left(a + be^{-k\left(\frac{P}{P_{HEL}}\right)}\right) \tag{4.7}$$

where a, b, and k are universal constants whose values are determined from a curve fit to the data; they are 1.15, −1.06, and 1.78, respectively. The values for the HEL, P_{HEL} and τ_{HEL} for these ceramics are given in Table 4.2.

The relationship between these three stresses is given by

$$\text{HEL} = P_{HEL} + \frac{2}{3}\sigma_{HEL} \tag{4.8}$$

where

$$P_{HEL} = \frac{K}{K + (4/3)G}\text{HEL and } \sigma_{HEL} = \frac{2G}{K + (4/3)G}\text{HEL} \tag{4.8a}$$

Here, the HEL is obtained from the shock experiments, and the hydrostats of the material equivalent stress, σ_{HEL} and τ_{HEL}, are calculated as per the procedure explained in Holmquist et al. (2001) and Westerling et al. (2001). The hydrostat

Table 4.2 Relevant HEL properties of selected ceramics.

Property (units)	B_4C	SiC	AlN	ZrB_2–SiC	Glass	Granite
HEL (GPa)	19	14.6	9.0	11.63	5.95	4.5
σ_{HEL} (GPa)	15.44	13.0	6.0	9.1	4.54	2.66
τ_{HEL} (GPa)	7.72	6.5	3.0	4.55	2.27	1.33
P_{HEL} (GPa)	8.71	5.9	5.0	5.4	2.92	2.73

Source: Shafiq and Subhash (2016a). © 2016, John Wiley & Sons.

is the nonlinear pressure–volume relationship (also called the equation of state) obtained from plate-impact experiments and will be described in detail in Chapter 5. Interestingly, there is no strain-rate dependence of strength in the model proposed in Eq. (4.7), even though the data for all the ceramics were obtained from a range of experiments encompassing both the QS and extreme dynamic strain rate regimes. In addition, at $P = 0$, one can derive the shear strength of a ceramic under pure shear ($\tau = \tau_0$) as

$$\tau_0 = \tau_{HEL}\,(a + b) = 0.09 * \tau_{HEL} \tag{4.9}$$

Thus, the shear strength of any ceramic under pure shear condition can be determined if the *HEL* is known. Note that at low-pressure the shear strength data from Figure 4.8 is linear for all the structural ceramics considered here, similar to the behavior noted earlier in Figure 4.6. Extrapolating this experimental data to zero pressure, we can obtain the elusive τ_0 (shear strength under pure shear) value for these ceramics. On the other hand, the exponential model in Eq. (4.9) can also provide the pure shear strength value if τ_{HEL} is known since the constants a and b are deemed universal for at least this class of ceramics (Bavdekar and Subhash 2018; Shafiq and Subhash 2016a). The values of τ_0 were calculated from the above two methods for various structural ceramics and are shown in Table 4.3. Despite the fact that the ceramic data was obtained from a range of materials, different experimental methods adopted by different laboratories, and also noting that considerable scatter exists in experimentally determined ceramic strength values from various laboratories and test methods, the τ_0 values from the above two methods are remarkably close.

As was noted earlier, the low-pressure shear strength data (Figure 4.6) aligns well with the MC model as given in Eq. (4.5), which suggests a linear relationship between the shear strength and pressure where the slope α is the coefficient of internal friction. One can determine this friction coefficient for ceramics from the low-pressure experimental data presented in Figure 4.8 by fitting the MC model. Also, from the Taylor series expansion of the exponential function for the EMC

Table 4.3 Comparison of experimental and model derived values for pure shear strength and coefficient of friction.

Description		B_4C	SiC	AlN	ZrB_2–SiC	Glass	Granite
Shear strength (τ_0) Gpa	Equation (4.9)	0.69	0.60	0.18	0.40	0.20	0.12
	Data extrapolation	0.59	0.86	0.28	0.32	0.29	0.14
Coefficient of friction (α)	Equation (4.10)	1.67	2.08	1.13	1.58	1.47	0.92
	Data extrapolation	1.07	1.21	1.26	1.07	0.82	0.67

Source: Shafiq and Subhash (2016a). © 2016, John Wiley & Sons.

model in Eq. (4.7), we can obtain the coefficient of the pressure term (or slope) as

$$\alpha = \frac{-bk\tau_{\mathrm{HEL}}}{P_{\mathrm{HEL}}} \tag{4.10}$$

Since all the terms on the right-hand side of Eq. (4.10) are known for the above ceramics, one can now compare these coefficient values to the measured slope (α) in Figure 4.8. For clarity, this early pressure-shear data at low pressures is plotted separately in Figure 4.10 (Shafiq and Subhash 2016a). Comparison of the slope and the calculated α values are provided in Table 4.3, which shows a reasonable agreement between the two. Note that a larger discrepancy is observed for the friction coefficient between the two methods than for the shear strength values because the values used for b and k in Eq. (4.10) are derived based on the entire data set (Eq. (4.7)) as opposed to the data based on the linear response only. Nevertheless, the failure strength data can be captured effectively using the EMC model for a variety of ceramics at a range of confining pressures. Recently, Bavdekar et al. (2017, 2019) and Bavdekar and Subhash (2018) have used this EMC formulation in a dynamic expanding cavity model (ECM) (Satapathy 2001) and showed that it is more effective than the linear MC formulation in capturing the target resistance, dwell, and penetration depth of ceramics during a projectile impact.

4.4 Strength of a Damaged Ceramic

Thus far, we have presented strain-rate and confinement dependence of the failure strength of intact ceramics. As previously described in Chapter 3, a projectile is in contact with an intact ceramic target only at time $t = 0$, and beyond that failure initiates in the ceramic ahead of the projectile. Thus, for any $t > 0$, the projectile will be in contact with a damaged ceramic as the penetration process progresses, but this damaged ceramic is confined by the surrounding intact bulk material. Hence, it is more realistic to consider the properties and the behavior of a confined, damaged ceramic in modeling the interaction between a projectile and

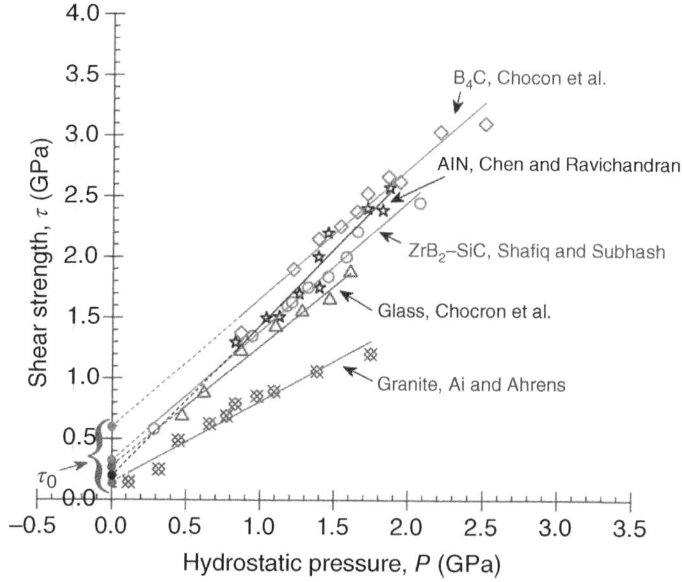

Figure 4.10 Low-pressure shear strength values for various structural ceramics revealing a typical linear relationship conforming to traditional MC criterion. Source: Shafiq and Subhash (2016a). © 2016, John Wiley & Sons.

a ceramic target. The next question is: what is the strength of a damaged ceramic? Obviously, the answer depends on the magnitude of confinement stress imposed and the severity of damage. A fragmented material has no tensile strength but can resist loads under compression if the material is confined (similar to the behavior of a sandbag during a bullet impact). In addition, if the ceramic target is externally confined, additional resistance to projectile penetration is offered by the damaged ceramic. However, one would expect the strength of a damaged ceramic to be a function of the state of damage, i.e. whether the ceramic contains a single crack, multiple cracks, or is fully comminuted material. In the following, we discuss literature available on strength of damaged materials under QS and dynamic conditions. Since it is difficult to pre-induce damage in high-strength structural ceramics, considerable work has been conducted in brittle materials such as glass to pre-damage the material and study its constitutive response. The fundamental conclusions from these studies are assumed to be applicable to damaged ceramics as well.

4.4.1 Static Strength of Damaged Glass Under Confinement

During a projectile penetration, considerable damage occurs in the target material ahead of the projectile. Chocron et al. (2007) realized the importance of properties of a damaged ceramic in their impact simulation. They conducted confined compression experiments on damaged borosilicate glass (instead of a typical ceramic), but the results are expected to be applicable to other brittle materials as well. The intact samples of borosilicate glass were confined in maraging steel sleeves. To pre-damage the glass, the composite specimen was exposed for 20 min in a furnace at 500 °C and plunged into ice water. The process was repeated twice. This method created extensive damage in the glass specimen, which was confirmed by postmortem analysis of a few specimens, but the damage was not quantified. Strain gages were bonded on the surface of the steel confinement sleeve to monitor axial and hoop strains during cyclic loading. Repeated cyclic load–unload compression experiments were conducted on some samples, and others were loaded monotonically.

A typical stress–strain response of a composite glass-steel specimen during the cyclic loading test is shown in Figure 4.11a. These curves indicate that the elastic response is followed by an inelastic response where damage growth occurs. The maximum load increased in each cycle potentially causing additional damage in the glass specimen. An important finding from these tests was that the elastic response of the damaged glass did not change dramatically even if the glass was pulverized (*in situ* comminuted). Sudden jumps in the axial and hoop strains are attributed to formation of macroscopic shear faults visible in Figure 4.11c along which the comminuted material slips. Every time the hoop strain jumps (i.e. slippage occurs), the stress level drops. Posttest sectioning of the composite specimen revealed slippage planes inclined to the loading axis, which indicate dominance of shear. The magnitude of each strain-jump increases in successive slip steps, and eventually, failure occurs with slippage. A plot of slippage points revealed that at high pressures, the stress stabilizes to a value between 1.8–2.3 GPa (Chocron et al. 2007).

From the stress–strain response, Chocron et al. (2007) determined slopes of the curves and established elastic properties in terms of Lame's constants and sleeve properties. It was observed that the elastic constants remained unchanged from intact to pre-damaged (thermally shocked) glass; even when the glass was severely damaged in later cycles, the average modulus and Poisson's ratio remained almost constant when compared with the modulus of the intact material determined from QS tests. Thus, a highly comminuted glass in confinement has a modulus similar to that of intact glass. This unexpected result can be rationalized by the fact that the comminuted material does not have any space to move and, consequently, voids are not allowed to open and hence there is no decrease in the elastic modulus. Extending these results to penetration experiments, they concluded that during ballistic penetration, inertia and confinement of the surrounding intact material keep the failed target material in place for some time. Therefore, even if the material right in front of the projectile is failed, its elastic modulus is almost unchanged.

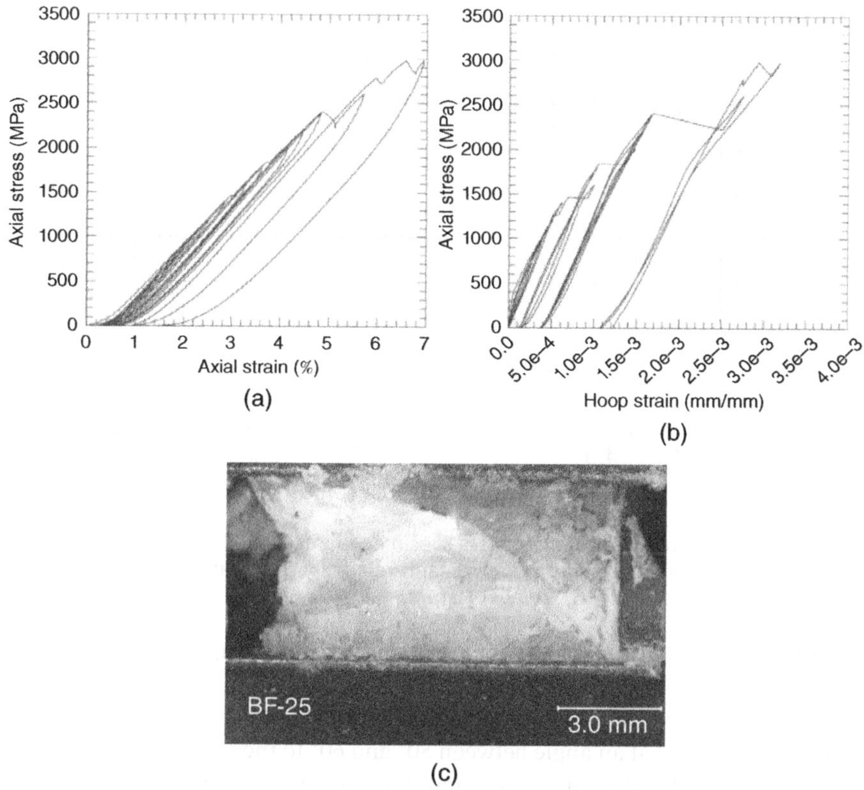

Figure 4.11 (a) Axial strain and (b) hoop-strain responses for a pre-damaged glass specimen in a steel sleeve due to cyclic compressive loading and (c) the resulting shear faulting in the specimen. Source: (a) Chocron et al. (2007). © 2007, John Wiley & Sons, (b) and (c) Reproduced with permission from Chocron et al. (2007).

In a later study, Chocron et al. (2013b) conducted additional tests on pre-damaged specimens of starphire® and borofloat glasses in a hydraulic cylinder at fluid pressures up to 500 MPa. The specimens were pre-damaged by exposing intact specimens to two thermal cycles at 500 °C and quenching in ice water. The specimens developed damage in the form of noncontiguous cracks but had maintained structural integrity and residual strength. Upon application of a QS load, both the intact and damaged specimens exhibited an initial linear response until a maximum load was reached. The responses of the confined damaged and intact materials are shown in Figure 4.12. For damaged specimens, the linear-response was followed by a load drop to a lower stress level, where the response remained either at a constant value or of saw-tooth form centered around a fixed value. During this stage, formation of a shear plane and further

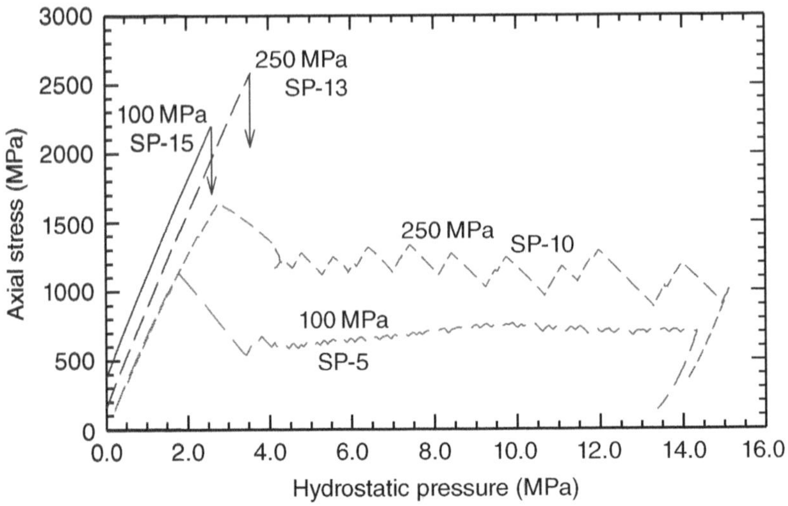

Figure 4.12 Response of a confined glass specimen in intact and damaged conditions. After damage, the stress level drops and remains relatively steady at that level. Source: Chocron et al. (2013a). © 2013, Springer Nature.

accumulation of damage occurs, but the specimen still supports a significant amount of load (residual strength) for large strains. The saw-tooth shape is indicative of relative sliding of the failure surfaces, and the shear faults were noted to be inclined at an angle between 50° and 60° to the loading axis.

Employing both Drucker–Prager and MC models, Chocron et al. (2013a) rationalized the responses of intact and damaged glasses under confinement. The intact specimens exhibited a linear dependence of failure strength with pressure (similar to Figure 4.6) when the failure data were plotted in an equivalent stress versus hydrostatic pressure space. However, the pre-damaged specimen data followed a different slope as shown in Figure 4.13. After failure, the specimens were still able to carry load (inverted triangles and open triangles). Chocron et al. (2013a) later compared these results to flyer plate impact experimental data. It was noted that there exists a maximum load-carrying capability (strength cap) for the damaged glass. Also, the failure surface for the damaged glass was found to be independent of strain rate. Interestingly, there is no significant decrease in the value of the strength cap as a function of damage, at least for the pressures range investigated here. Finally, their modeling of damaged ceramic strength using the Drucker–Prager and MC models has shown that the slope of the linear portion of data seem to be relatively independent of the damage level, whereas the cohesive strength (i.e. strength at $P = 0$) decreased with increasing damage.

From these experiments, the following behavior of damaged ceramics can be summarized:

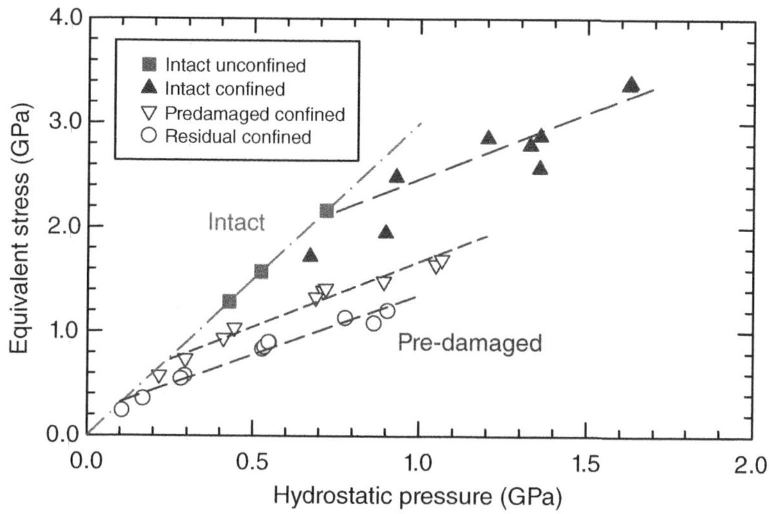

Figure 4.13 Response of intact and pre-damaged specimens under unconfined and confined conditions. Source: Chocron et al. (2013a). © 2013, Springer Nature.

1. In compressive loading with confinement, the stiffness of the damaged material is the same as the intact material.
2. Regardless of severity of damage, the strength of a damaged ceramic under confinement saturates at a level lower than that of the intact ceramic.
3. The strength of both the confined intact and confined damaged ceramic increases with the level of confinement stress but is independent of strain rate.

While the above results are limited to QS loading range, in the next section, we will describe similar experimental efforts under dynamic loads in SHPB.

4.4.2 Dynamic Strength of a Damaged Ceramic Under Confinement

The previous section focused on the strength of a damaged ceramic under QS loading. In this section, we will focus on determining the dynamic strength of a damaged ceramic. A typical interaction between a high-velocity projectile and a target lasts for only a few hundred microseconds, during which extensive fracture occurs in the ceramic target at high strain rates. Chen and Luo (2004), Chen et al. (2007), and Luo et al. (2006) developed an experimental method based on a modified SHPB to create damage in a bulk ceramic and then evaluate the response, *in situ*, of that damaged ceramic at high strain rates.

In this test method, shown in Figure 4.14, an intact ceramic specimen is inserted into a thin metal sleeve and subjected to uniaxial compressive loading (Chen and Luo 2004). The metal sleeve provides a low level of confinement and holds the

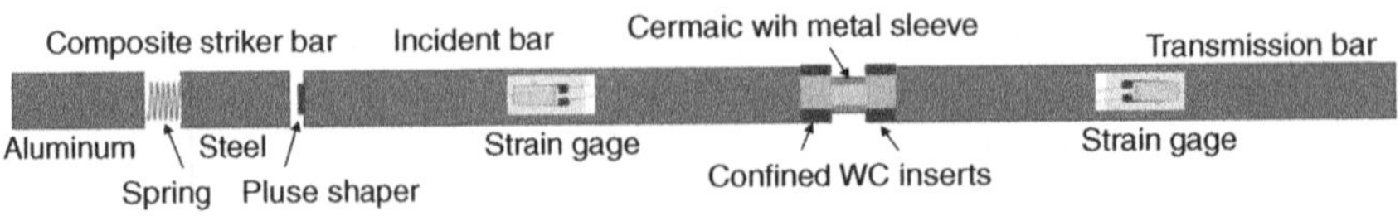

Figure 4.14 Schematic of the double-impact SHPB setup to create damage in a confined ceramic and then evaluate *in situ* the response of the damaged ceramic. Source: Modified from Chen and Luo (2004).

ceramic fragments in place after damage is induced. The striker bar is modified into a composite bar consisting of two metal rods (steel followed by aluminum) connected by a spring. Both bars are propelled by a gas gun toward the incident bar. Upon impact of the composite striker bar on the incident bar, two incident pulses are generated successively: first by the impact of the steel striker bar onto the incident bar and the second by the impact of the aluminum striker bar on to the steel striker already in contact with the steel incident bar. After fully unloading the momentum into the incident bar, the steel striker bar remains in contact with the incident bar, and then the aluminum striker bar impacts on the steel striker bar causing the transmission of another compressive stress wave into the incident bar. The metal compression-spring placed in between two striker bars physically separates the two loading pulses with a specified separation time, allowing the damaged specimen to settle before the second compressive stress wave arrives. The first stress wave generated by the impact of the steel striker creates damage the intact ceramic specimen and renders it in a comminuted state, and the second stress wave generated by the impact of the aluminum striker bar loads this damaged ceramic to a subsequent high-strain-rate compressive loading. Analyzing the transmitted and reflected signals from the second incident wave can provide the constitutive response of the damaged ceramic. However, the contribution from the sleeve must be subtracted to calculate the constitutive response of the specimen.

Using the above method, the influence of strain rate, damage level, and confinement pressure on the behavior of comminuted ceramics has been investigated in SiC–N and Alumina (Al_2O_3). In Figure 4.15a, the raw signals of the first and second incident, transmitted, and reflected signals are shown. The two incident signals (solid line) are easily identified with two different amplitudes. Recall from Chapter 2, Eq. (2.6), that the transmitted signal is a measure of the stress in the specimen. The first transmitted signal (dashed line) has a high amplitude indicating the high compressive stress level in the intact specimen, and this signal drops suddenly upon specimen failure. However, the failed specimen exhibits residual strength because it is confined in the metallic sleeve, and this is recognizable as a small stress bump next to the peak amplitude in the transmitted signal. This second transmitted wave is at a much lower amplitude than the first wave and at an almost constant voltage level, corresponding to the stress level

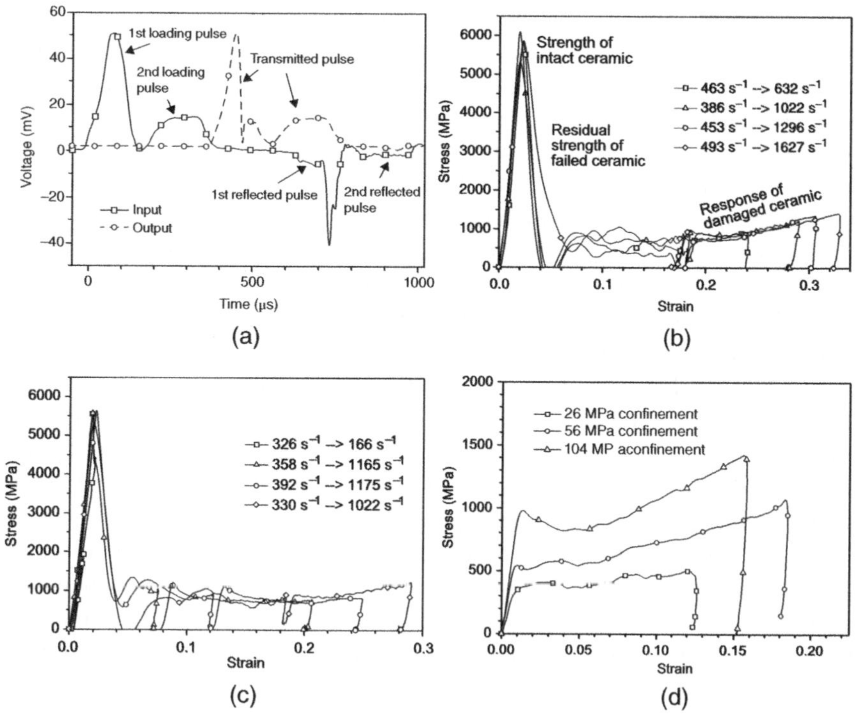

Figure 4.15 (a) Incident and transmission strain gage signals upon impact by a composite striker bar. (b) Stress–strain responses of four intact SiC–N ceramic specimens and their damaged state at a fixed confinement pressure. (c) Comparison of stress–strain responses during the second loading of one undamaged (after the first loading) and three damaged specimens. (d) Stress–strain response of damaged specimen at three confinement levels. Source: (a) Chen and Luo (2004). © 2004, Springer Nature, (b)–(d) Luo et al. (2006). © 2006, John Wiley & Sons.

reached or sustained by the damaged ceramic. Recall from Eqs. (2.7) and (2.8) in Chapter 2 that the reflected signal in the SHPB is a measure of strain rate (and hence strain upon integration) in the specimen. The failure of the intact specimen is better identified in the reflected signal where the accumulation of elastic strain in the intact ceramic is followed by a steep drop (indicating the failure) in the first reflected signal. This peak signal is followed by the response of the damaged specimen indicated as the second reflected pulse.

Figure 4.15b shows the stress–strain response of the two states (intact and damaged) in several ceramic specimens with an initial lateral confinement of 104 MPa. The intact ceramics exhibit variations in compressive strength from 5 to 6 GPa, and the residual strength of the specimens damaged during the first loading is slightly below 1 GPa. The second loading causes additional deformation in the damaged ceramic where some strain-hardening-like behavior is observed, which may be attributed to further compaction and rearrangement of fragments as well as the resistance offered by the damaged ceramic in the confinement of the metallic sleeve. Interestingly, the strain rate achieved in the damaged material was always higher than the intact ceramic, implying significant strain accumulation as seen in the stress–strain curves in Fig. 4.15b. In the strain rate range employed in this study, there appears to be no strain rate effect on the stress–strain response of the damaged ceramic. The strength of the damaged ceramic saturates, similar to the conclusion regarding QS loading of the glasses described in the previous section.

It is also interesting to compare the stress–strain responses during the second loading of an undamaged specimen with that of a damaged specimen after the first loading. Figure 4.15c illustrates this comparison of three specimens damaged during the first loading and one undamaged specimen after the second loading. The undamaged specimen is indicated with strain rates of 326 and 166 s^{-1} in this figure. Unlike the damaged specimens shown in Figure 4.15b, the second loading in the undamaged specimen in Figure 4.15c produces a significantly lower strength and lower strain rate than its first loading. This behavior implies that a significantly lower strain is accumulated at a lower strain rate in the undamaged specimen during the second loading than in the damaged specimens which exhibited a higher strain and higher strain rate. Finally, Figure 4.15d illustrates the effect of initial confinement on the response of damaged ceramic. With an increase in confinement, the stress increases in the damaged ceramic, and the hardening behavior changes from a linear response to non-linear response.

It is interesting to note that both QS compression of damaged glasses and dynamic compression of the damaged ceramics give similar responses. While Chocron et al. (2007) concluded that the stiffness of a damaged specimen under QS loading remains the same as that of the intact specimen and that strength of a damaged specimen saturates at a stress level below that of the intact specimen, additional conclusions can be drawn from the dynamic experiments on the response of a damaged ceramic as follows:

- When confined, a damaged ceramic exhibits considerable load-bearing capacity and can resist applied loads.
- Compared to intact ceramics, the damaged ceramic exhibits higher ductility and hence can undergo greater deformation.
- The greater the level of confinement of a damaged ceramic, the greater is its load-bearing capacity.
- Beyond a critical damage level, the load-bearing capacity of a damaged ceramic is not influenced by further damage.
- In the strain rate regime tested, the damaged ceramic response is independent of strain rate.

While the insights gained from the above test methods on the response of damaged ceramic have been of significant value, a major drawback of the above studies is that neither the state of damage nor the critical damage level are described empirically or statistically. It is unclear if the damaged ceramic has a few cracks (or a few large fragments), a large number of cracks (or many small fragments), or is fully pulverized. Regardless, it is interesting to note that despite the differences in experimental techniques (static testing or dynamic testing) and materials (ceramics and glasses), the strength of a damaged ceramic has a saturation value independent of strain rate and damage level.

4.5 Modeling the Failure Behavior of Ceramics

Having described the strain rate and pressure-dependent failure strength of intact and damaged ceramics, we will now embark on modeling approaches to rationalize the observed behavior and gain further insight. These modeling efforts can be classified into three categories: phenomenological, micromechanical, and statistical.

a. Phenomenological models describe the phenomenon without regard to the underlying physics that contributes to that behavior. Numerous phenomenological models have been proposed in the literature to describe the failure behavior of ceramics, and we will summarize these models and illustrate their effectiveness.
b. Micromechanical models capture the fundamental mechanics that govern the observed behavior. In the case of ceramics, the behaviors of interest include crack nucleation, microcracking, dynamic crack growth, fracture and fragmentation, strain rate and pressure dependence of strength, and phase transformation.
c. Statistical models intend to capture the inherent variability in strength that is a characteristic of any ceramic. The statistical variability in fracture and fragmentation evolves due to the variation in size, shape, and distribution of defects and their propensities for crack growth under the applied loads. Stochastic models use a probabilistic approach requiring a known preexisting defect size and defect distribution, criteria for nucleation and growth of cracks, their damage growth rate, and interaction kinetics.

This chapter will address both analytical and computational approaches that require constitutive responses of both the intact and damaged ceramics. Models for phase transformation, crack initiation, damage growth, and fragmentation are also discussed.

4.5.1 Phenomenological Constitutive Models

4.5.1.1 Mohr–Coulomb (MC) Model

The MC model (Eq. (4.5)) was briefly discussed previously in relation to the linear response (Figure 4.6) of the failure strength of ceramics with pressure. Here, we will emphasize certain aspects of this model so that the reader can get a comprehensive overview of the literature on various models. The MC model was the earliest models for describing the failure behavior of ceramics and can be traced back to Coulomb (1776) and Mohr (1900). It was recognized that the failure behavior in brittle materials is a function of applied pressure and the strength of a ceramic can be expressed in terms of shear strength (τ) as a linear function of pressure (P) given by

$$\tau = \tau_0 + \alpha P \tag{4.11}$$

where τ_0 is the cohesive strength of the material (i.e. shear strength at $P = 0$, see Figure 4.10) and α is a constant representing the sliding friction between crack surfaces (see Eq. (4.10)). Note that even under uniaxial stress loading (σ_x) a material point in a ceramic specimen has both shear ($\tau = \sigma_x/2$) and hydrostatic pressure ($P = \sigma_x/3$) components, and so experimental determination of τ_0 is nontrivial.

Equation (4.11) was extended to include equivalent flow stress, σ, by utilizing the Drucker–Prager model (Drucker and Prager 1952) given by

$$\sigma = \sigma_0 + \beta P \tag{4.12}$$

where

$$\sigma = \sqrt{\frac{1}{2}\left[\left(\sigma_1 - \sigma_2\right)^2 + \left(\sigma_2 - \sigma_3\right)^2 + \left(\sigma_3 - \sigma_1\right)^2\right]} \tag{4.13}$$

and β is a material constant related to the internal friction, similar to α in Eq. (4.11). However, as was seen in Figure 4.9, the strength of a ceramic saturates beyond a critical pressure level (about the HEL), and hence modifications have been proposed to the failure models by introducing a cut-off pressure as

$$\tau = \begin{cases} \tau_0 + \alpha P, & P < \overline{P} \\ \overline{\tau}, & P \geq \overline{P} \end{cases} \tag{4.14}$$

where $\overline{\tau}$ is the maximal shear strength and $\overline{P} = \left(\overline{\tau} - \tau_0\right)/\alpha$, and

$$\sigma = \begin{cases} \sigma_0 + \beta P, & P < \overline{P} \\ \overline{\sigma}, & P \geq \overline{P} \end{cases} \tag{4.15}$$

where $\overline{\sigma}$ is the maximal flow stress and $\overline{P} = \left(\overline{\sigma} - \sigma_0\right)/\beta$. Equations (4.14) and (4.15) can be graphically represented as shown in Figure 4.16. For most experimental conditions (such as plate impact or shock experiments) where axisymmetric specimens are used ($\sigma_2 = \sigma_3$), both these models are equivalent because

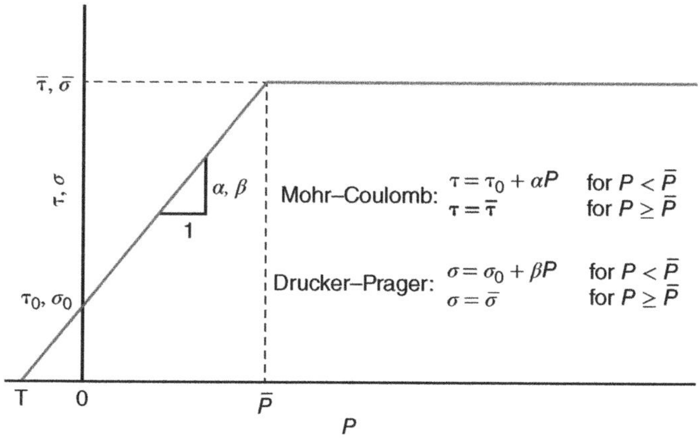

Figure 4.16 Mohr–Coulomb (MC) and Drucker–Prager (DP) models for intact ceramics. *T* is the hydrostatic tensile strength of the intact ceramic. Source: Bavdekar and Subhash (2018). © 2018, Elsevier.

Eq. (4.15) reduces to

$$\sigma = \sigma_1 - \sigma_3 = 2\tau \tag{4.16}$$

In recent years, these two models have been well explored, and their equivalency has also been established in granular soils (Jiang and Xie 2011; Rani et al. 2014).

At low confining pressures, the linear dependence of failure strength (MC criterion) has been demonstrated in numerous ceramics as presented earlier (e.g. Figures 4.6 and 4.10). The saturation of failure strength with pressure of intact ceramics (Figure 4.8) and damaged ceramics (Figure 4.12) is also well established through experiments. While the above two models (Figure 4.16) successfully capture both the linear dependence and saturation of ceramic failure strength with pressure, they fail to model the transition between these two regimes as shown in Figure 4.17, where both the shear strength and pressure have been normalized with the corresponding HEL values. The normalization allows the data to fall within a narrow band and suggests that there is a common failure mode for all these materials. To capture the observed transition from linear dependence to saturation of strength, new models have been proposed which are discussed in the following.

4.5.1.2 Johnson and Holmquist Models

Among the available phenomenological models, the Johnson–Holmquist (JH) models are the most comprehensive in terms of capturing the holistic material behavior. They include failure behavior of both intact and damaged ceramics, as well as their strain-rate and pressure dependencies. The models also capture

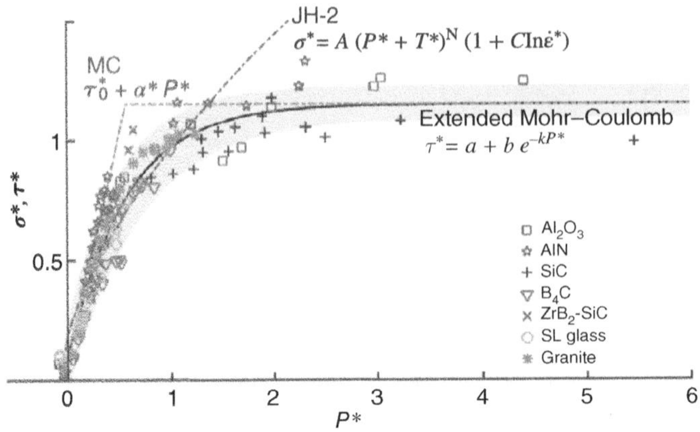

Figure 4.17 Comparison of various models with the experimental data on pressure dependence of strength. The MC model fails to capture the data in the transition regime from linear dependence to saturation. The JH-2 model does not capture the saturation. Only the EMC model captures the entire spectrum of experimental data. Source: Bavdekar and Subhash (2018). © 2018, Elsevier.

the evolving ceramic behavior during the damage process. They are named JH-1, JH-2, and Johnson–Holmquist–Beissel (JHB) models. The JH1 model (Johnson and Holmquist 1992) describes the ceramic behavior in the form of piecewise linear functions, whereas the JH2 model describes the behavior in terms of continuous functions. The JHB model incorporates phase transformation in ceramics (Johnson et al. 2003). We will describe each of these models separately.

JH-1 Model

The main features of the JH1 model (Holmquist and Johnson 2002) are the following:

1. Pressure-dependent (i.e. level of confinement) and strain-rate-dependent strength of intact and damaged ceramics, and
2. Effect of fracture in terms of the resulting bulking behavior (dilation or increase in volume).

The intact ceramic strength is described in a piecewise, linear fashion as an analytical function. The intent is to include the transition regime in addition to low pressure linear dependence and high-pressure saturation of strength. The strength is assumed to be a function of confinement pressure, strain rate, and damage level. The strength increases with confinement only up to a certain pressure level and then remains constant as shown in Figure 4.17 (Bavdekar and Subhash 2018; Shafiq and Subhash 2016a).

The increase in strength with strain rate (Figure 4.1) is expressed by the function:

$$\sigma = \sigma_o (1 + C \ln \dot{\varepsilon}^*) \tag{4.17}$$

where the normalized strain rate is similar to that defined in Eq. (4.1) and is given by $\dot{\varepsilon}^* = \frac{\dot{\varepsilon}}{\dot{\varepsilon}_o}$

The failure of the ceramic is modeled by introducing a damage-growth parameter, D, where $D = 0$ represents an intact ceramic (no damage) and $D = 1$ represents a completely damaged ceramic. No intermediate stages of damage are considered, implying that the material transitions from an intact to a damage state instantaneously, i.e. instantaneous softening. The stress is now related to pressure, strain rate, and level of damage (Holmquist and Johnson 2002), and this relationship is represented in the form of three linear segments as shown in Figure 4.18a. Here T is the maximum tensile hydrostatic stress where the intact ceramic fails. Obviously, the corresponding tensile strength for a damaged ceramic is zero, i.e. $T = 0$ for $D = 1$. The damaged-ceramic strength is reduced dramatically to another nonzero value.

Because both the strength and damage level in a ceramic are functions of the confinement pressure, the corresponding strains are also a function of pressure. It is assumed that the plastic strain associated with the failure process increases linearly up to a certain pressure level and then remains constant as seen in Figure 4.18b, similar to the strength saturation at high pressures. The failure strain is given by

$$\varepsilon_p^f = \phi (P + T) \tag{4.18}$$

where ϕ is the slope of the line. Now the damage, D, at any strain level is defined as the ratio of incremental equivalent plastic strain to the maximum equivalent plastic strain at failure:

$$D = \sum \frac{\Delta \varepsilon_p}{\varepsilon_p^f} \tag{4.19}$$

The next step is to connect pressure and failure strain due to damage. When fracture initiates, the creation of new surfaces results in a volume expansion of the specimen, and this bulking behavior is expressed in terms of a pressure–volumetric strain relationship as a function of the damage level, as shown in Figure 4.18c. For an intact ceramic, this pressure–volume relationship, also referred to as equation of state (EOS), can be determined from shock experiments (to be discussed in Chapter 5). A similar relationship is assumed for the damaged ceramic, except that the curve passes through the origin for the lack of volumetric strain at zero pressure. The hydrostatic pressure before failure is given by the following EOS:

$$P = K_1 \mu + K_2 \mu^2 + K_3 \mu^3 \tag{4.20}$$

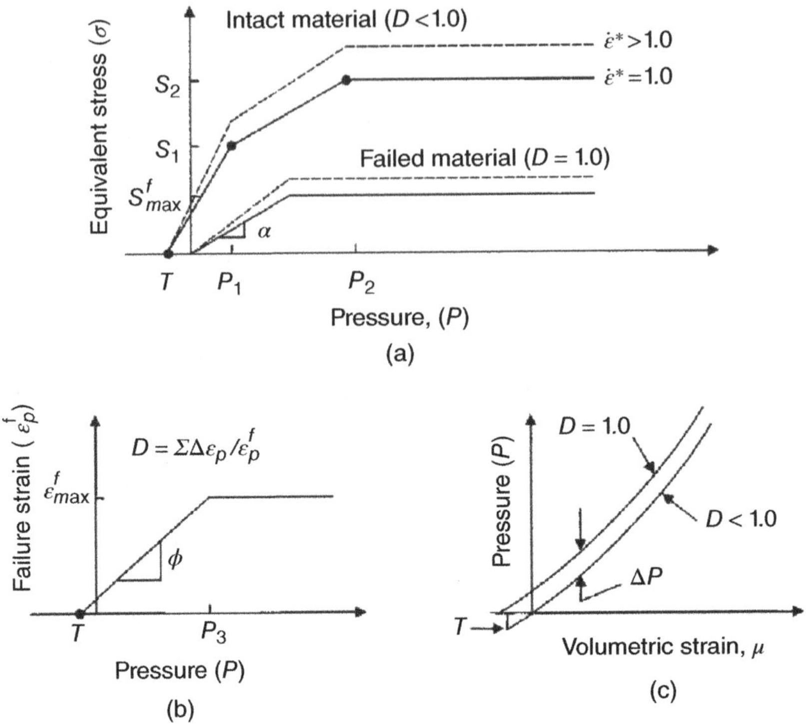

Figure 4.18 Illustration of JH-1 model (a) with pressure, strain rate, and damage dependence of ceramic strength, (b) linear strain accumulation and then saturation with increase in pressure and (c) the relationship between volumetric strain and pressure. Source: Holmquist and Johnson (2002). © 2002, AIP Publishing.

where μ is the volumetric strain, K_1 is the bulk modulus, and K_2 and K_3 are additional constants that fit the above nonlinear curve. Upon failure ($D = 1$), new fracture surfaces are created, and both the pressure and volumetric strain increase. This increase in volume, ΔV, referred to as "bulking," results in increase in pressure, ΔP, and the new pressure is given by

$$P = K_1\mu + K_2\mu^2 + K_3\mu^3 + \Delta P \tag{4.21}$$

Calculation of the increase in pressure is nontrivial, and, therefore, ΔP is determined from energy considerations. Damage causes a loss in the elastic (internal) strain energy of the stressed bulk solid, and hence the level of damage can be assessed by calculating the difference between the elastic strain energies of intact (U_i) and damaged (U_f) materials:

$$\Delta U = U_i - U_f \tag{4.22}$$

where the total elastic strain energy (deviatoric strain) is given by

$$U = \sigma^2/6G \tag{4.23}$$

Here σ is the von-Mises equivalent stress and G is the shear modulus. However, only a small fraction (β) of this elastic energy is converted to potential hydrostatic energy, given by

$$\Delta P\mu_f + \Delta P^2/2K_1 = \beta\Delta U \tag{4.24}$$

where μ_f is the volumetric strain at failure ($V_0/V-1$). The two terms on the left-hand side are the approximate potential energy for compression ($\mu > 0$) and potential energy for tension ($\mu < 0$), respectively. The rest of the energy (lost energy) can be converted to potential hydrostatic energy by adding ΔP. This energy conversion is expressed as

$$\Delta P = -K_1\mu_f + \sqrt{(K_1\mu_f)^2 + 2\beta K_1\Delta U} \tag{4.25}$$

Utilizing this model, the authors (Holmquist et al. 2001; Johnson and Holmquist 1999) simulated damage induced in a confined SiC ceramic due to a projectile impact in addition to dwell and interface-defeat phenomena. However, the model does not allow the fragmented or a comminuted ceramic target to resist penetration of the projectile, which is unrealistic, because the ceramic is assumed to have an instantaneous strength loss and no bulking until failure. Thus, the JH1 model lacks the damage growth feature. This assumption was modified by incorporating the gradual loss of strength upon failure (progressive strength degradation) in the JH-2 model.

JH-2 Model

In subsequent work by Johnson and Holmquist (1994) and Johnson et al. (2003), the JH-1 model was improved and is referred to as the JH-2 model. Instead of describing the ceramic strength in terms of linear segments, a dimensionless analytical description was provided, and the strength parameters were normalized by the dynamic yield strength or the HEL to allow for a more direct comparison between different ceramics. Similar to JH-1, the JH-2 model includes a representation of the strength of intact and fractured materials, a pressure–volume relationship that allows for bulking, and a damage model where the material transitions from an intact to a fractured state. However, unlike the JH-1 model where only intact ($D = 0$) and fully damaged ($D = 1$) states of materials are considered, the JH-2 model includes gradual softening of the strength with damage, i.e. intermediate damage levels are represented by fractional values of D ($0 < D < 1$). The bulking pressure gradually increases as damage accumulates. Therefore, parameters such as energy loss and bulking can be incrementally calculated.

The equivalent stress (σ) normalized by the equivalent stress at HEL (σ_{HEL}) is a smoothly varying function of the normalized intact strength σ_i^*, fracture strength σ_f^* (both of which are functions of strain rate), and the level of damage, D. The normalized equivalent stress for the strength is expressed as

$$\sigma^* = \sigma_i^* - D\left(\sigma_i^* - \sigma_f^*\right) \tag{4.26}$$

where

$$\sigma^* = \sigma/\sigma_{HEL} \tag{4.27}$$

and the equivalent stress is given by

$$\sigma = \sqrt{1/2\left[\left(\sigma_x - \sigma_y\right)^2 + \left(\sigma_x - \sigma_z\right)^2 + \left(\sigma_y - \sigma_z\right)^2 + 6\left(\tau_{xy}^2 + \tau_{xz}^2 + \tau_{yz}^2\right)\right]} \tag{4.28}$$

The normalized intact equivalent stress is given by

$$\sigma_i^* = A(P^* + T^*)^N\,(1 + C\ln\dot{\varepsilon}^*) \tag{4.29}$$

Here, P^* and T^* are normalized with corresponding HEL values (i.e. $P^* = P/P_{HEL}$, where P is the actual pressure and P_{HEL} is the pressure at the HEL). Similarly, the normalized maximum tensile hydrostatic pressure is $T^* = T/P_{HEL}$, where T is the maximum tensile hydrostatic pressure. In a similar manner, the normalized fracture stress of a damaged ceramic is described as follows:

$$\sigma_f^* = B(P^*)^M\,(1 + C\ln\dot{\varepsilon}^*) \tag{4.30}$$

where $\sigma_f^* \leq \sigma_{f\,max}^*$, and A, B, C, M, N, T^*, and $\sigma_{f\,max}^*$ are the material parameters. Note that T^* is not included in Eq. (4.30) as a fractured ceramic will have no tensile strength. The equivalent strain rate in the above equations is defined as

$$\dot{\varepsilon} = \sqrt{2/9\left[\left(\dot{\varepsilon}_x - \dot{\varepsilon}_y\right)^2 + \left(\dot{\varepsilon}_x - \dot{\varepsilon}_z\right)^2 + \left(\dot{\varepsilon}_y - \dot{\varepsilon}_z\right)^2 + 3/2\left(\dot{\gamma}_{xy}^2 + \dot{\gamma}_{xz}^2 + \dot{\gamma}_{yz}^2\right)\right]} \tag{4.31}$$

The normalized equivalent stress values as a function of pressure were obtained from various experimental data in the literature (Johnson et al. 2002). Utilizing this model, the constitutive response of float glass (Holmquist et al. 1995), boron carbide (Johnson and Holmquist 1999) and AlN (Holmquist et al. 2001) were successfully captured. The comparison of the data and the model are shown in Figure 4.19 for AlN and boron carbide.

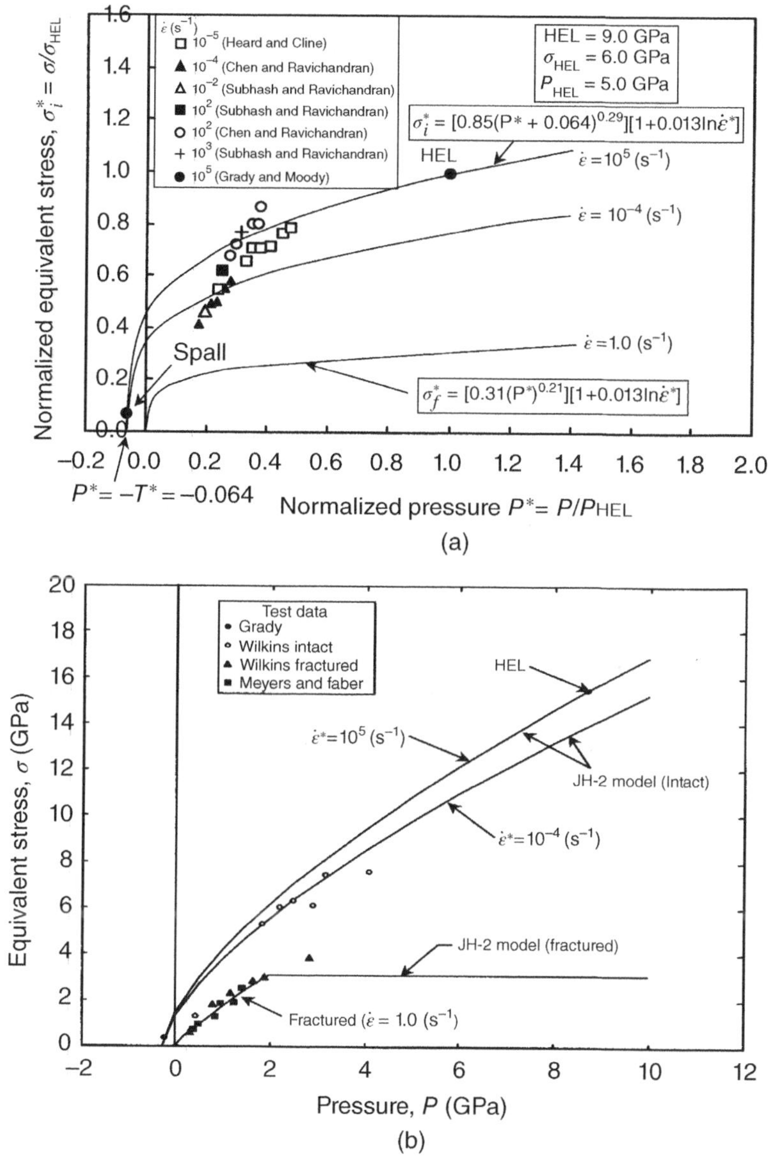

Figure 4.19 JH-2 model applied to the data on (a) AlN ceramics and (b) boron carbide. Source: (a) Holmquist et al. (2001). © 2001, Elsevier, (b) Johnson and Holmquist (1999), © 1999, AIP Publishing.

JHB Model

This model includes the same features as those of the JH-1 and JH-2 models but extends them to include phase changes that occur at high pressure in some ceramics, such as AlN. It has features similar to the JH-1 model except that the description of the strengths (intact and failed) and failure strain are described analytically, but they are not normalized as in the JH-2 model. The model is described in Holmquist and Johnson (2008), Johnson et al. (2003) and is named after the authors JHB.

In the JHB model, the strength is dependent on pressure, strain rate, and damage. The analytical description of strength goes from $\sigma = 0$ at a tensile pressure of $P = -T$ to a strength of $\sigma = \sigma_i$ at a confining pressure of $P = P_i$. Beyond this pressure, the strength is given by

$$\sigma = \sigma_i + (\sigma_{\max} - \sigma_i)\{1.0 - \exp[\alpha_i (P - P_i)]\} \tag{4.32}$$

The strength of the failed material follows a similar relationship except that it is lower than that of an intact ceramic and is given by

$$\sigma = \sigma_f + \left(\sigma_{\max}^f - \sigma_f\right)\{1.0 - \exp[-\alpha_f (P - P_f)]\} \tag{4.33}$$

Both these strengths are strain-rate dependent as per the relation

$$\sigma = \sigma_o + (1.0 + C \ln \dot{\varepsilon}^*) \tag{4.34}$$

The rest of the damage model is the same as the JH-1.

At high pressure, some ceramics exhibit a phase change, and this process is described schematically in Figure 4.20. When a material exhibiting a phase transformation is subjected to high pressure, it initially undergoes an elastic strain indicated by OA, and then a phase change occurs at a critical pressure level as indicated by the line AB. This phase change is accompanied by a large change in volume between states at A and at B. After the phase change, the behavior of the new phase is assumed to be linearly elastic (BC). This pattern mimics the behavior of AlN, which undergoes a phase change from a Wurtzite structure (hexagonal) to a rock salt (cubic) structure under high pressure (Johnson et al. 2003).

Motivated by similar linear trends observed in experiments for the two phases, the pressure (P) versus volumetric strain (μ) relationships can be described as follows:

$$\text{Phase I}: \quad P = K_1\mu + K_2\mu^2 + K_3\mu^3 \tag{4.35}$$

$$\text{Phase II}: \; P = \overline{K}_1\overline{\mu} + \overline{K}_2\overline{\mu}^2 + \overline{K}_3\overline{\mu}^3 \tag{4.36}$$

where $\overline{\mu} = \mu - \mu_o$ and $\overline{K}_1, \overline{K}_2, \overline{K}_3, \mu_o$ are constants. μ_o is experimentally determined by extrapolating the response of the new phase.

No bulking is allowed for these individual phases because there is no further phase change or fracture. Only elastic strains are allowed as described by the volumetric strain μ given by $(V_o/V-1)$. For simplicity, the transition region is assumed

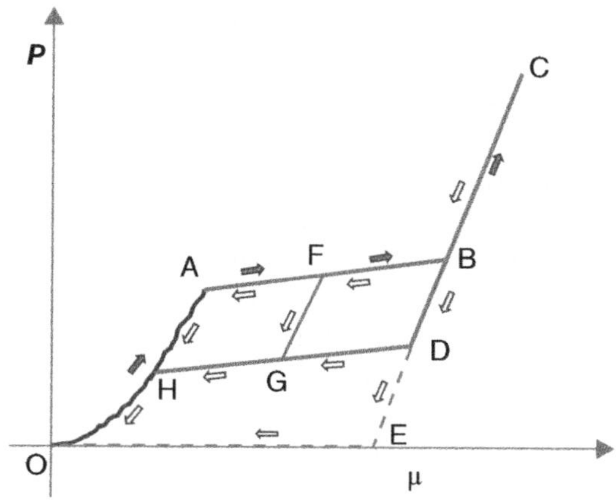

Figure 4.20 Schematic of the pressure volume relationship for ceramics that exhibit phase-transformations.

to be a linear function between Point A and Point B and may be chosen based on experimental data. The bulking pressure due to phase change is described in a similar manner for fracture except that a modified bulk modulus is used in the relationship as

$$\Delta P = -K_1^{\text{eff}}\mu_f^{\text{eff}} + \sqrt{\left(K_1^{\text{eff}}\mu_f^{\text{eff}}\right)^2 + 2\beta K_1^{\text{eff}}\Delta U} \tag{4.37}$$

where K_1^{eff} and μ_f^{eff} are interpolated values between K_1 and $\overline{K}_1$ and μ_f and μ_o, respectively, given by

$$K_1^{\text{eff}} = K_1\,(1-\lambda) + \overline{K}_1\lambda \tag{4.38}$$

$$\mu_f^{\text{eff}} = \mu_f - \lambda\mu_o \tag{4.39}$$

and λ is defined as

$$\lambda = \left(\mu_{\max} - \mu_1\right)\left(\mu_2 - \mu_1\right) \text{ for } \mu_1 \leq \mu_{\max} \leq \mu_2 \tag{4.40}$$

The phase change is assumed to influence only the change in pressure. Strength and damage features are assumed to be unaffected as there is no contrary test data available.

The JHB model is generalized to consider various issues associated with phase changes; the transformation can occur either at a constant pressure or under increasing pressure as indicated by the slope AB. The model also accounts for reverse phase transformations (indicated by unfilled arrows) when the pressure is decreased. This can be a partial or full reverse transformation, or no reverse transformation, as schematically represented by various paths in Figure 4.20. The level of hysteresis is defined by a parameter, h, where $h = 0$ represents no hysteresis and $h = 1$ represents full hysteresis. Thus, the above JH models were successfully able to model the behavior of a variety of ceramics.

4.5.1.3 Extended Mohr–Coulomb Model (Shafiq–Subhash Model)

In recent years, the suitability of the JH models to realistically capture the experimental data at ultrahigh pressures has been questioned (Bavdekar and Subhash 2018; Shafiq and Subhash 2016a). In the JH-2 model, the strength continues to grow with confinement pressure, whereas experimental data on ceramics reveal a strength saturation. In the JH models, strength is strain-rate dependent; however, the experimental data shown in Figure 4.9 from different experimental techniques and at different strain rates do not have strain-rate dependence. Also, the JH-1 and JH-2 models considered experimental data only up to the HEL, and the applicability of the model beyond the HEL is not considered. Due to these limitations, Shafiq and Subhash (2016a) proposed the extended MC model (here in after referred to as SS-EMC), described earlier in Eq. (4.7). The model can be rewritten in a simplified form as

$$\tau^* = a + be^{-kP^*} \tag{4.41}$$

where the superscript * indicates normalized quantities with the corresponding HEL values.

Comparing the proposed model to the earlier JH-2 model given in Eq. (4.29), we note that the JH-2 model has both pressure- and strain-rate-dependent fracture strength, whereas the SS-EMC in Eq. (4.41) has only pressure dependency of fracture strength. A comparison of these models with the experimental data was shown in Figure 4.17. It is seen that the SS-EMC model better captures the entire spectrum of data compared to the MC and JH models (Bavdekar and Subhash 2018). Note here that the gray band is the 99% confidence interval for the SS-EMC model for the experimental data. While the MC model fails to capture the transition between the linear dependence and saturation of failure strength, the JH-2 model predicts a continuous increase in strength with pressure beyond the HEL and does not saturate, contrary to the experimental data. In general, most ceramics show a saturation of strength beyond $10^4\ s^{-1}$, as was seen in Figure 4.3b (Grady 1998) which the SS-EMC model effectively captures.

The absence of a strain-rate term in the SS-EMC model in Eq. (4.41) should not come as a surprise because the experimental data shown in Figure 4.17 comes from a variety of materials, test methods, and a range of strain rates. Despite these variations, the trend in the failure strength is consistently the same. It has also been recognized by many researchers that it is the pressure that influences the low- and high-strain rate failure behavior of ceramics and that the contribution of strain rate is negligible (Grady 1998; Wilkins et al. 1969). This lack of strain rate sensitivity can also be noted from the weak strain rate coefficient C in the JH-2 model (Eq. (4.29)). While the constants A and N have values in the range of 0.3–0.95, the values of C are an order of magnitude lower and span in the range of

0.003–0.012, implying its weak influence on the ceramic fracture strength (Shafiq and Subhash 2016a).

In addition, the SS-EMC model can also be represented in a linear fashion for the low-pressure regime with slope $\alpha = -\frac{bk\tau_{\mathrm{HEL}}}{P_{\mathrm{HEL}}}$ as stated in Eq. (4.10) (Bavdekar and Subhash 2018; Shafiq and Subhash 2016a). Thus, if the HEL components of a material are known, the maximum shear strength ($\tau = a\tau_{\mathrm{HEL}}$), cohesive strength from Eq. (4.9), i.e. $\tau_0 = (a+b)\tau_{\mathrm{HEL}} = 0.09 \times \tau_{\mathrm{HEL}}$, and the pressure-shear slope α of a ceramic can all be determined, a unique advantage of SS-EMC not available in the JH-2 model. Chen and Ravichandran (2000) also showed by conducting confinement tests on AlN, that when describing the failure behavior of intact ceramics at low confinement pressures both the JH-2 model and MC converge to the same form.

4.5.2 Micromechanical Models (Wing-Crack Models)

Unlike phenomenological models that capture the overall experimental trends with a set of equations, micromechanics-based models for the damage behavior of ceramics focus on crack growth from preexisting defects and subsequent fragmentation behavior during compressive failure. With appropriate crack growth dynamics, these models have been able to successfully capture both the strain-rate- and pressure-dependent compressive failure behaviors in the low and intermediate strain rate regimes. Models for crack initiation from preexisting defects, such as the wing-crack growth models, originally proposed by Brace and Bombolakis (1963) and later extensively analyzed by Nemat-Nasser and Horii (Horii and Nemat-Nasser 1986; Nemat-Nasser and Horii 1982) and Ashby and Hallam (1986) have played a central role in understanding the mechanics of failure behavior of ceramics under compressive loading. Extension of this modeling approach to dynamic loading has been attempted by several researchers. Deng and Nemat-Nasser (1992) and Nemat-Nasser and Deng (1994)) considered an array of interacting and dynamically growing wing-cracks to investigate the rate-dependent dynamic damage evolution in brittle solids. Ravichandran and Subhash (1995) used a wing-crack model under biaxial compression to investigate strain-rate dependence of ceramic fracture strength. Huang and Subhash (2003) developed a dynamic damage growth model under lateral confinement and investigated the pressure dependence of fracture strength. Paliwal and Ramesh (2008) assumed a power law growth rate of wing-cracks to analyze the influence of various parameters and captured the failure growth and strain rate dependence of fracture strength under multiaxial compression. These models are described briefly in this section.

The central premise of micromechanics-based modeling approach is that a ceramic material contains a distribution of inherent flaws within its

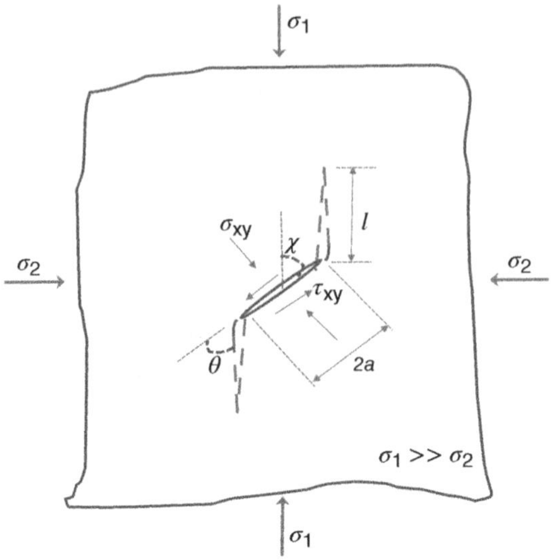

Figure 4.21 Schematic illustrating wing-crack growth from a preexisting inclined crack due to the applied biaxial compression.

microstructure and crack growth will occur from the preexisting defects upon application of a macroscopic compressive stress. A dilute distribution of cracks is initially assumed, which implies noninteracting cracks. The mechanism of axial crack growth from a favorably oriented microcrack under biaxial load is illustrated in Figure 4.21. The applied macroscopic load on the domain results in normal and shear stresses on the crack surfaces. While the normal compressive stress causes closure of crack surfaces, the shear stress, on the other hand, causes frictional sliding of crack surfaces resulting in tensile stress at the crack tips, producing the crack-tip opening. The tensile stress is perpendicular to the major principle stress orientation (after some initial crack growth), facilitating axial crack growth at the two ends of the crack as shown in the Figure 4.21. These tensile cracks are referred to as "wing-cracks." The growth of the tensile cracks is dictated by maximizing the mode *I* stress intensity factor, K_I, around the crack tip. Under biaxial loading, the cracks grow in the direction of maximum applied compression (in the direction of σ_1, since $\sigma_1 > \sigma_2$). Recall that under tensile loading, only a single crack dominates, whereas in compression, multiple favorably oriented preexisting cracks activate and grow into wing-cracks in a stable manner. Upon reaching a critical length, their stress fields interact and hence, the cracks coalesce to form large cracks and cause axial splitting of the ceramic. As a result, in compression, cracks grow approximately parallel to the direction of maximum applied compressive stress.

The wing-crack-growth model has been the foundation for modeling damage evolution in brittle materials during high-strain-rate loading. By defining the mechanism of crack growth through this model, one can understand the mechanics of damage evolution as a function of strain rate or confining pressure or stress state or even distribution of defects. The fundamental understanding of fracture behavior allows design of components by taking advantage of the knowledge of crack growth mechanics under a given loading scenario. For example, one may choose appropriate confinement levels to suppress crack growth during a specified loading condition.

In the analysis of wing-crack growth, the preexisting microcrack is assumed to have an initial length $2c$ and oriented at an angle θ to the vertical axis (see Figure 4.21). Upon application of load, damage initiates in the form of wing cracks from these microcracks, and the crack growth rate ($\dot{l}$) is described in various forms by different researchers. Freund (1998) presented a review of fracture criteria for brittle solids and defined $\dot{l}$ as

$$\dot{l} = C_R \frac{K_I - K_{IC}}{K_I - K_{IC}/2} \tag{4.42}$$

where C_R is the Raleigh wave velocity, and K_I and K_{IC} are the static stress intensity and static fracture toughness, respectively. It is often assumed that the Mode I dynamic stress intensity factor is related to the static stress intensity factor multiplied by a term that depends on the crack growth rate ($\dot{l}$) as (Freund 1972, 1998):

$$K_{Id} = k\left(\dot{l}\right) K_I \tag{4.43}$$

where

$$k\left(\dot{l}\right) = \left(1 - \frac{\dot{l}}{C_R}\right)\left(1 - \frac{\dot{l}}{2C_R}\right)^{-1} \tag{4.44}$$

For dynamic crack growth to occur, it is assumed that

$$K_{Id} = K_{IC} \tag{4.45}$$

Ravichandran and Subhash (1995) related the energy release rate G_c to the crack length (l) and crack growth rate ($\dot{l}$) as

$$\frac{G_c}{G_o} = h\left(l, \dot{l}\right) = \frac{c}{l} + \frac{c}{l_o} \tag{4.46}$$

where c is the half-crack length and l_o is a characteristic length defined by the product of the terminal velocity (~Rayleigh wave velocity) and time to reach the desired level of stress. G_o is the critical energy release rate at the crack initiation, and h is a function to be determined from experimental data. Based on the limited experimental data available Kobayashi et al. (1983) and Yang and Kobayashi (1990) postulated the form of h as shown above. Utilizing such a crack growth model and motivated by experimental data, Ravichandran and Subhash (1995) were able to capture both the QS and dynamic fracture strength response of ceramics along with the transition behavior between these two regimes. This transitional strain rate from the QS to the dynamic regime was determined in terms of material properties as

$$\dot{\varepsilon}_* = \frac{\sigma_0 C}{El_s} \tag{4.47}$$

where C is the characteristic speed where the inertial effects become significant. It was chosen to be 0.2 times the shear wave speed in the material. l_s is the specimen length, σ_0 is the QS compressive strength, and E is the Young's modulus.

Upon crack growth under dynamic loads, the failure of the specimen is expected to occur at the time t_c at which the crack grows to a critical length l_c. When the damage density reaches a critical value defined as $f_c = Nl_c^2$, where N is density of cracks, the cracks coalesce, and the material loses its load-bearing capacity. With this model, Ravichandran and Subhash (1995) were able to capture the failure strength of AlN as a function of strain rate.

Huang and Subhash (2003) utilized the wing-crack growth model for preexisting cracks, but defined damage in terms of wing-crack length through a damage parameter $D = NV$, where N is the number of flaws per unit volume, and V is a spherical volume influenced by a flaw of radius l. They related strain rate to a damage-growth rate under biaxial loading. By defining failure stress at the time when damage tends to a large value, they determined compressive failure strength as a function of strain rate. The uniaxial stress–strain response, damage growth, and reduction in modulus were determined as a function of applied confining pressure. Figure 4.22 shows stress–strain curves for various confinement levels (at a

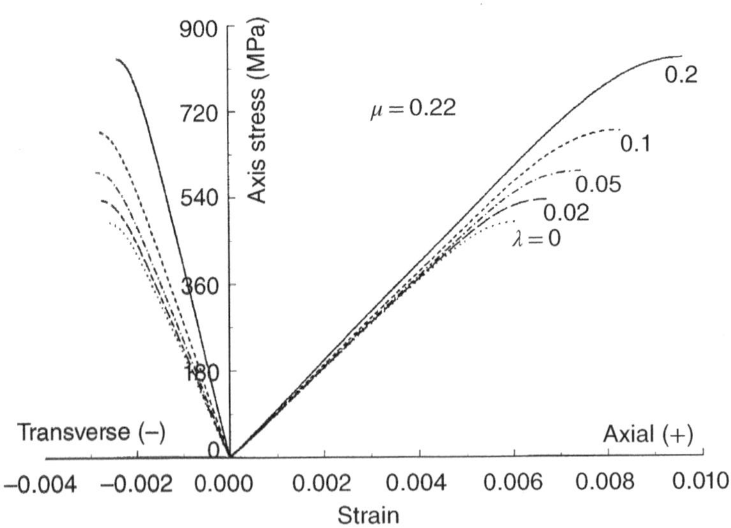

Figure 4.22 Stress–strain curves as a function of confining stress. Note the increase in strength and axial strain with increase in confinement ratio. Source: Huang and Subhash (2003). © 2003, Elsevier.

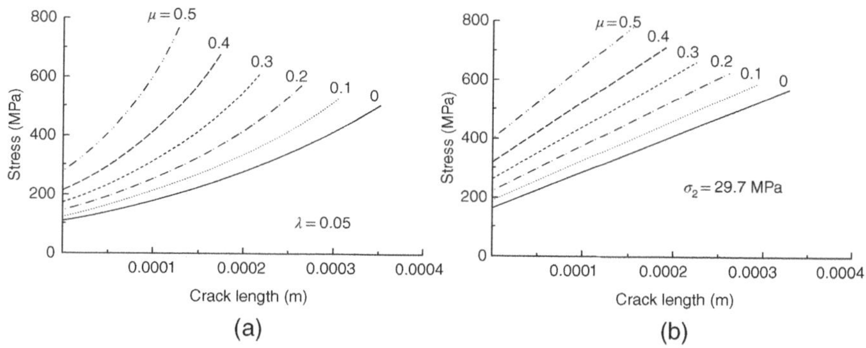

Figure 4.23 Influence of confinement state and friction coefficient on wing-crack-growth characteristics under (a) proportional confinement and (b) fixed confinement. Source: Huang and Subhash (2003). © 2003, Elsevier.

friction coefficient $\mu = 0.22$) which reveal that with an increase in confining stress ($\lambda = \frac{\sigma_2}{\sigma_1}$, where σ_2 is lateral stress and σ_1 is axial stress), a marked increase in compressive fracture strength and axial strain and moderate increase in stiffness of the specimen were observed. These results qualitatively capture the experimental trends under static and dynamic loads (see Figure 4.5). It is also interesting to see that at low confining pressures, the transverse strain increases initially and then decreases, reflecting the suppression of crack growth (closure of cracks) at higher confinement pressures.

The analysis by Huang and Subhash (2003) also revealed interesting crack growth phenomena depending on the type of confinement: "proportional confinement," where the confinement level (σ_2) increases in proportion to the increase in applied axial compressive load (σ_1) and "fixed confinement," where the applied confinement pressure is held fixed at a specified level. The wing crack grows in a nonlinear fashion under proportional confinement (Figure 4.23a) and in a linear fashion under fixed confinement of about 30 MPa (Figure 4.23b). It is also shown that it becomes increasingly more difficult for a crack to grow with a higher friction coefficient (μ), leading to a higher fracture strength. Also due to an initial higher level of fixed confinement level compared to the proportional confinement, the stress required for crack initiation is higher; see Figure 4.23b.

Finally, the influence of strain rate and confinement state on the failure strength of a ceramic is shown in Figure 4.24. In both confinement states, there is an increase in failure strength at higher confinement levels even within the QS strain-rate regime. For a given confinement stress, there is no increase in strength until a critical strain rate is reached, and, beyond that, the strength increases dramatically. Similar behavior is observed for all confinement pressures. For strain rates less than $10^4\,\text{s}^{-1}$, the failure strength is independent of strain rate,

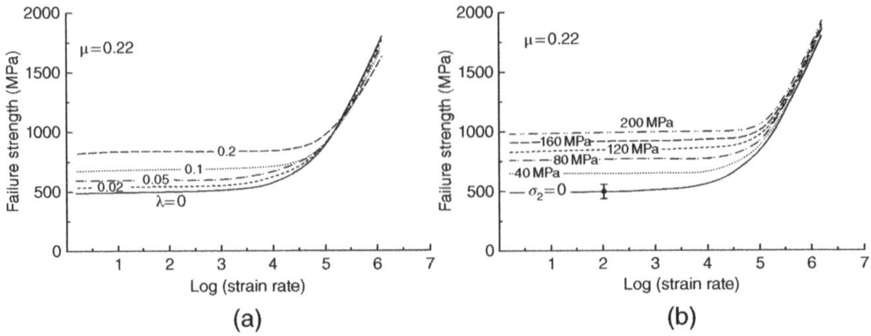

Figure 4.24 Influence of confinement pressure and strain rate on failure strength of brittle materials under (a) proportional and (b) constant confinement states. Source: Huang and Subhash (2003). © 2003, Elsevier.

but depend on the level of confinement. Beyond $10^4\ \mathrm{s}^{-1}$, the failure strength is a strong function of strain rate but becomes less dependent on the type of confining pressure and magnitude.

While most of the previous models assumed a dilute distribution of noninteracting microcracks, Paliwal and Ramesh (2008) considered interacting microcracks by assuming a power-law form for the wing-crack growth rate as

$$\dot{l} = c_{\max}\left(\frac{K_I - K_{IC}}{K_I - K_{IC}/2}\right)^{\gamma} \tag{4.48}$$

where

$$c_{\max} = C_R/\alpha \tag{4.49}$$

The parameters α and γ, respectively, were introduced to describe the strain-rate sensitivity of the damage-growth-rate process and the transitional strain rate at which the strength starts to increase dramatically. The constitutive law consisted of distributed damage evolution function in terms of crack growth law, given by

$$\dot{\sigma} = E\left(1 - \frac{\pi^2}{30}\left(1+\overline{v}\right)\left(5-4\overline{v}\right)\Omega\right)\dot{\varepsilon} - E\frac{\pi^2}{30}\left(1+\overline{v}\right)\left(5-4\overline{v}\right)\dot{\Omega}\varepsilon \tag{4.50}$$

where the damage evolution was described in terms of crack density parameter Ω, whose growth rate is defined by flaw distribution and crack growth dynamics as

$$\dot{\Omega} = 2\eta\sum_{j=1}^{N} g\left(s_j\right)\Delta s_j l_j \dot{l}_j \tag{4.51}$$

where the probability distribution function is defined by $g(s_j)$. For a family of flaws with flaw size $2s_j$, the associated wing-crack length is $2l_j$. Δs_j represents increments in the half flaw size, and η represents the flaw density per unit area.

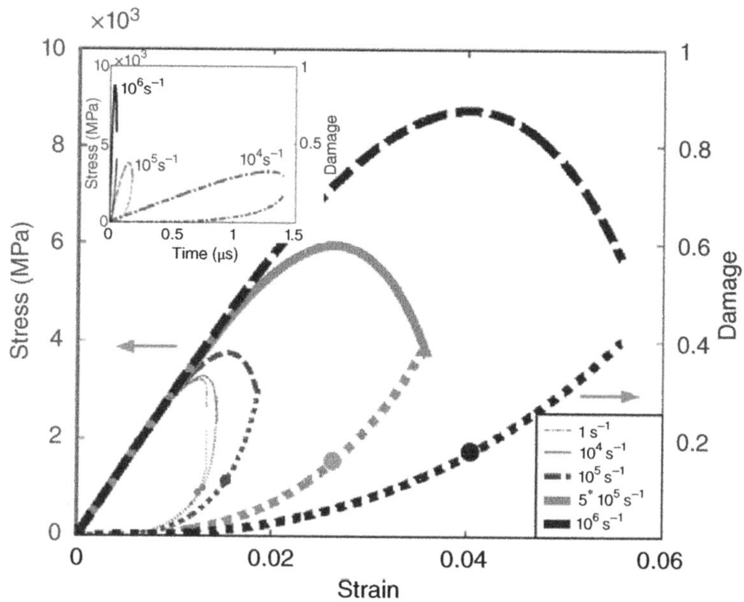

Figure 4.25 Compression stress–strain response and damage growth in ceramics at various strain rates. Source: Paliwal and Ramesh (2008). © 2008, Elsevier.

The appeal of the model comes from the realization that the strength of a ceramic is a function of confinement level, flaw size, and flaw distribution, which can be statistically represented. Since manufacturers cannot fully control the nature of flaws inadvertently introduced during processing, their influence can be reduced by precompressing the ceramic. The model was applied to uniaxial loading of a ceramic specimen under constant strain rate loading over a range of strain rates. The results, shown in Figure 4.25, reveal that an increase in strain rate yields not only a higher peak stress but also a larger inelastic strain due to a higher level of damage evolution. The peak stress also becomes increasingly sensitive to strain rate beyond $10^4\ s^{-1}$, similar to the work of Huang and Subhash (2003). The growth rate of damage is faster at lower strain rates, and hence, the fracture strength as well as strain to failure, are lower, as the crack growth reaches a critical value to cause macroscopic fracture.

To capture the influence of flaw distribution on fracture strength, Paliwal and Ramesh (2008) used Gaussian and Weibull (with skew parameter Γ) distributions of flaw density (see Figure 4.26a1) and noted that the distribution function does not have a significant influence on the stress–strain response at the two strain rates indicated (see Figure 4.26a2). On the other hand, for the two Gaussian distributions with the same mean value but different flaw size distribution shown in

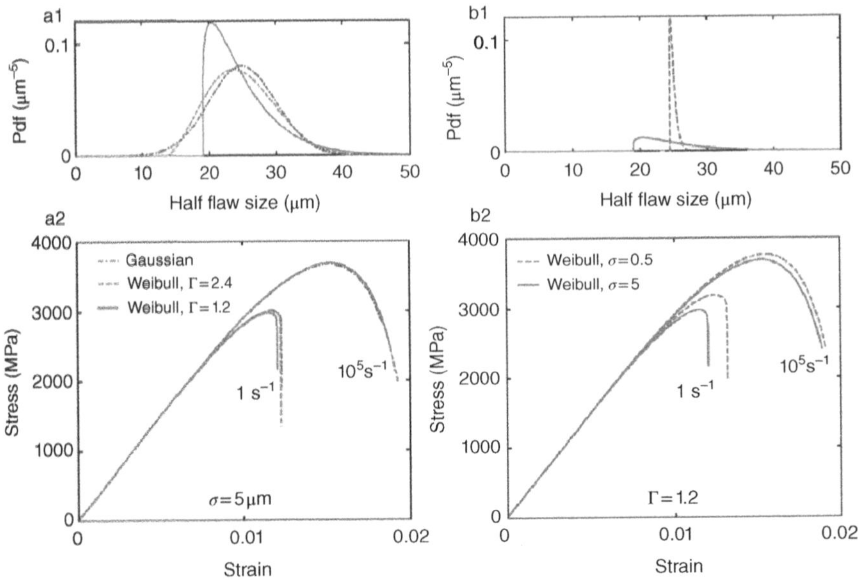

Figure 4.26 (a1) Assumed Weibull and Gaussian distribution of flaws and (a2) the corresponding stress–strain responses. It is seen that the assumed defect distribution does not significantly alter the resulting stress–strain response at the two strain rates. (b1) Assumed narrow and broad defect distributions and (b2) corresponding stress–strain curves at two strain rates. When the defect distribution is narrow, the strength is higher at both low and high strain rates than when the defect distribution is broad. Source: Paliwal and Ramesh (2008). © 2008, Elsevier.

Figure 4.26b1, the response is different (here σ is the standard deviation). When the flaw-size distribution is narrow, the strength is higher than when the flaw-size distribution is wide. This result implies that there is a higher probability of a bigger flaw size dominating the failure process; see Figure 4.26b2. Also, at lower strain rates, the difference in distribution of flaws yields a greater difference in strength than at higher strain rates. This is because at a higher strain rate there is less time for a single crack to dominate the behavior as the stress wave travels past the crack before the stress at the crack-tips reaches a critical value to grow.

Finally, the model by Paliwal and Ramesh (2008) also captured the effect of flaw density η on strain rate dependence of ceramic compressive strength. It was found that, for low flaw density, the material has a higher strain rate sensitivity of compressive strength as seen Figure 4.27. This is because when the flaw density is low, the cracks are farther apart, and the wing-crack growth dynamics become more dominant as these cracks have to grow longer to coalesce for macroscopic fracture. Thus, an earlier transitional value for the strain rate sensitivity of strength is noted with a lower flaw density. On the other hand, a higher flaw density means

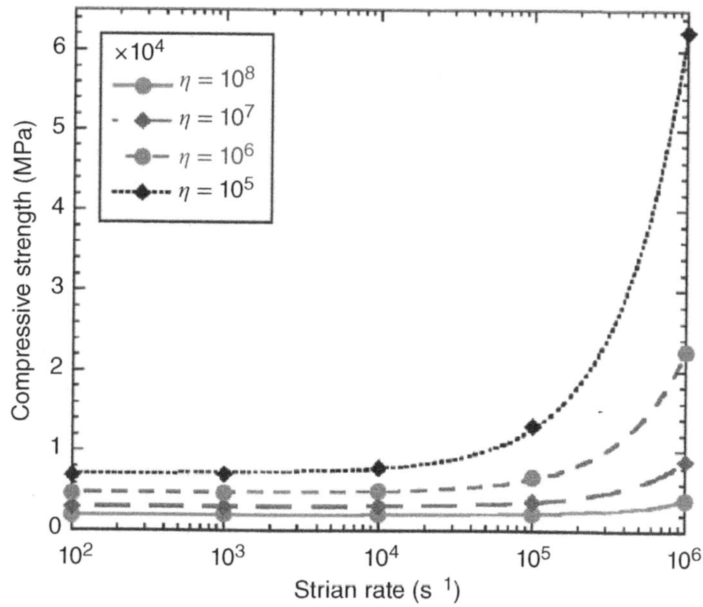

Figure 4.27 Compressive strength of a ceramic as a function of flaw density and strain rate. Source: Paliwal and Ramesh (2008). © 2008, Elsevier.

closer crack spacing, which does not permit longer wing-crack growth before they coalesce. Hence, the transitional strain rate shifts to a higher value. This behavior is consistent with the observations of Nemat-Nasser and Deng (1994). In a more recent micromechanical model developed by Hu et al. (2015), the competition between microcracking and inelasticity was also captured. At low confining stress levels, the wing-crack mechanism dominates, leading to the degradation of the modulus and peak strength of the material, whereas at high enough confining stresses, the cracking mechanism is completely shut down and dislocation mechanisms become dominant.

In summary, micromechanics-based wing-crack growth models capture not only the evolving damage in ceramics but also the strength dependence on the level of damage, distribution of damage, and strain rate.

4.5.3 Energy-based Models

In these models, the fracture process is controlled by the energy available for crack growth. An elastic solid fractures under an applied load because in doing so, it reduces its total energy. The total energy input during loading can be split into bulk elastic strain energy and surface energy due to the creation of cracks. Bhattacharya et al. (1998) considered axial splitting of specimens under far field uniaxial compression and combined this energetic approach with effective properties of the solid and the principle of minimum complementary energy. Similar to micromechanical models, a ceramic is considered to have a distribution of various microstructural heterogeneities, such as single crystal grains, grain boundaries, multiple phases, microcracks, and voids. This heterogeneity gives rise to either anisotropy in elastic properties at a given point or local variation of elastic properties resulting in a complex stress distribution in the intact solid. Upon application of uniaxial load, changes in the microstructural stresses will nucleate new cracks, and growth of favorably oriented cracks will cause axial splitting of the solid, thus reducing the total elastic energy. The elastic strain energy released by splitting is assumed to be the same as the fracture energy required for the creation of new surfaces. In this formulation, the specimen is assumed to fail in columnar fashion (parallel to the applied uniaxial load) with each column of width equal to the grain size. The philosophy of the failure process is schematically illustrated in Figure 4.28, where σ is the axial stress and σ_c is the applied confinement pressure.

Using the above approach, Bhattacharya et al. (1998) derived a critical condition for axial splitting in the form of as a function of material properties and applied stresses. The total energy of the unsplit specimen is equal to its stored elastic energy, which can be written in terms of its elastic modulus tensor (or compliance tensor) and the strain field (or stress field) over the volume of the specimen. For a cuboidal specimen with dimensions $h \times t \times t$, the elastic energy of the unsplit

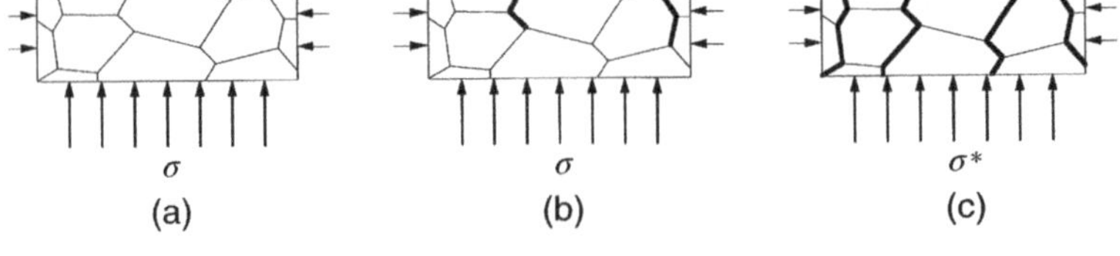

Figure 4.28 (a) Schematic of grains in a domain subjected to axial loading with lateral confinement, (b) nucleation of cracks along grain boundaries, and (c) axial cracks growth and creation of columnar fragments of width equal to grain size. Source: Bhattacharya et al. (1998). © 1998, Elsevier.

specimen can be written as

$$\varepsilon_u = -\frac{ht^2}{2E_u}\left(\sigma^2 + 2\sigma_c^2\left(1 - \nu_u\right) - 4\nu_u\sigma\cdot\sigma_c\right) \tag{4.52}$$

where E_u and ν_u are the modulus and Poisson's ratio of the unsplit specimen.

For the split specimen, both the elastic energy and surface energy should be considered. The elastic energy is given by

$$\varepsilon_s = -\frac{ht^2}{2E_s}\left(\sigma^2 + 2\sigma_c^2\left(1 - \nu_s\right) - 4\nu_s\sigma\sigma_c\right) \tag{4.53}$$

The material is isotropic with Young's modulus, E_s, and Poisson's ratio, ν_s. To calculate the surface energy, it was assumed that the splitting produces columns of one grain width and that the surface energy density, γ, is constant independent of orientation. Therefore, the number of surfaces after splitting is $2t/d$ (where d is the typical grain size), each with area, ht. Thus, the total surface energy is

$$\mathcal{S} = 2\gamma\frac{ht^2}{d} \tag{4.54}$$

The criterion for splitting is obtained from the principle of minimum complementary energy. Note that splitting (fragmentation) reduces elastic energy and hence, $\varepsilon_u - \varepsilon_s \geq 0$. Therefore, the fracture criterion is based on the fact that a reduction in elastic energy must overcome the surface energy associated with the newly created surfaces. Therefore, at the critical state, we must have

$$\varepsilon_u - \varepsilon_s = \mathcal{S} \tag{4.55}$$

Substituting relevant equations into the above, we can calculate the applied stress, σ, at which it reaches the critical value, σ^*, by

$$(\sigma^*)^2\left(\frac{1}{E_u} - \frac{1}{E_s}\right) + 2\sigma_c^2\left(\frac{1-\nu_u}{E_u} - \frac{1-\nu_s}{E_s}\right) - 4\sigma^*\sigma_c\left(\frac{\nu_u}{E_u} - \frac{\nu_s}{E_s}\right) = -4\frac{\gamma}{d} \tag{4.56}$$

This quadratic equation can be solved to obtain the failure strength. One can now solve for uniaxial strength (i.e. failure strength at zero confining pressure) as

$$\sigma^*|_{\sigma_c=0} = \sqrt{\frac{\frac{4\gamma}{d}}{-\Delta_1}} \tag{4.57}$$

and the effect of confinement in terms of material properties as

$$\left.\frac{d\sigma^*}{d\sigma_c}\right|_{\sigma_c=0} = 2\frac{\Delta_2}{\Delta_1} \tag{4.58}$$

where

$$\Delta_1 = \left(\frac{1}{E_u} - \frac{1}{E_s}\right) \text{ and } \Delta_2 = \left(\frac{\nu_u}{E_u} - \frac{\nu_s}{E_s}\right) \tag{4.59}$$

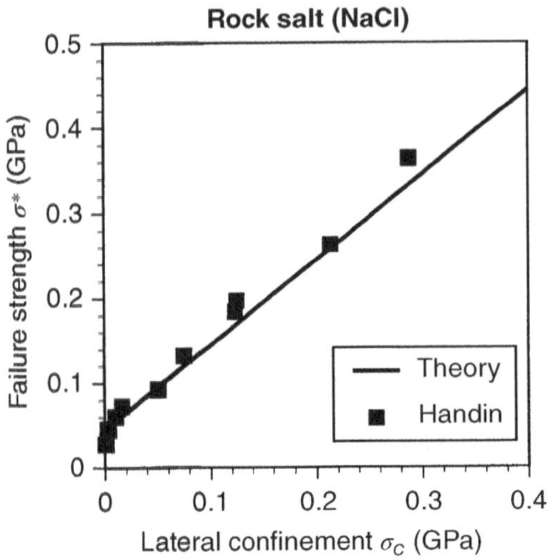

Figure 4.29 Comparison of theoretical predictions and reported experimental values of rock salt. Source: Bhattacharya et al. (1998). © 1998, Elsevier.

Using the above formulation, Bhattacharya et al. (1998) were able to calculate the uniaxial strength and biaxial strength of various brittle materials and a reasonable comparison with experiments were obtained on a few brittle materials such as rock salt as noted in Figure 4.29.

However, the model failed to capture the observed trends in strength of some materials such as limestone. Therefore, more sophisticated energy-based models that account for the distribution of defects, probabilistic nature of crack initiation, and strain-rate sensitivity are needed to further capture the experimental observations. The model also underestimates the failure strength with increasing lateral confinement. This may be because as the lateral confinement increases, the failure mode changes from splitting to shear faulting to crushing. The current model also does not account for the evolution of damage; instead, it simply assumes a final state with elastic parameters as E_s and ν_S. Additional weaknesses include lack of consideration of the strain rate sensitivity of strength and noninclusion of other relevant energy terms which account for motion of fragments as they dynamically move apart to create new surfaces.

4.5.4 Models Based on Hertzian Contact Mechanics

While the micromechanical models describe the evolution of damage in terms of wing-crack growth, the Hertzian contact-based indentation models help understand the sequence of evolving damage patterns during a penetration event by a point load on a semi-infinite surface (e.g. projectile impact on a flat surface) and relate these patterns to elastic-plastic stress fields that emerge at that location.

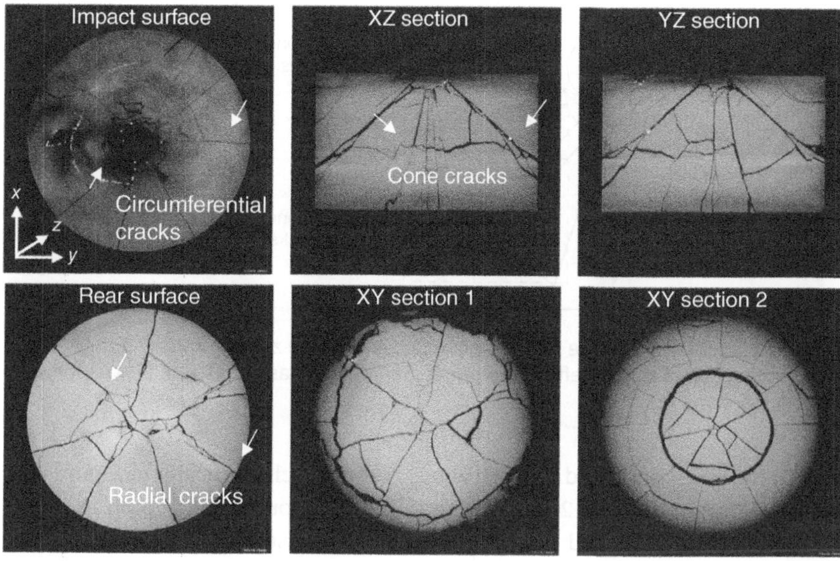

Figure 4.30 Ball-impact damage on SiC cylinder on front surface, back surface, and on radial cross sections.

Among numerous efforts available in the literature, we will selectively present the most relevant models.

Figure 4.30 shows images of damage induced by a low-velocity (<100 m/s) ball impact on a cylindrical SiC ceramic target. In addition to the severe damage at the impact location in the center of the target, radial cracks emanate from the center of the target both on the impact surface as well as on the rear surface (compare these images to those in Chapter 3). A few circumferential cracks connecting these radial cracks are also seen on these surfaces. In the interior of the cylinder in the radial sections (*XZ* and *YZ* planes), we notice a single dominant cone crack emerging from the impact location, appearing as a ring in the two *XY*-sections shown at different depths.

Ball-impact experiments at higher velocities showed more severe damage patterns (Iyer 2007). Multiple cone cracks and concentric ring cracks evolved from the impact site. These interior cone macrocracks remain isolated at lower velocities up to 160 m/s. With an increase in impact velocity, more cone cracks evolve and additional microcracks develop within the regions between the cone cracks, leading to fragmentation of the ceramic.

The stress state responsible for these crack systems was analyzed by Iyer (2007) utilizing the Hertzian contact theory. Based on observations of crack patterns and analysis of the elastic stress state in the target at the time of impact, the existence of four damage zones beneath the target were identified. The correspondence

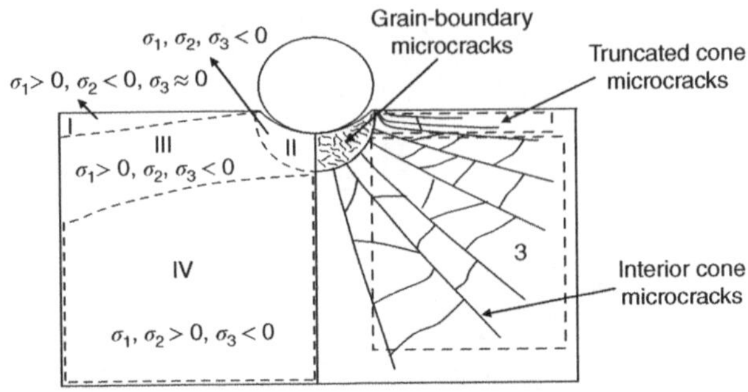

Figure 4.31 Schematic of the ball-impact induced damage zones (right) and the corresponding stress states (left) in the target. Source: Subhash et al. (2008). © 2008, John Wiley & Sons.

between the stress state and damage zones is shown schematically in Figure 4.31 (Iyer 2007; Subhash et al. 2008), where the illustration of damage on the right (Damage Regions 1–3) and the corresponding stress states (Zones I–IV) on the left are shown. Damage Region 1 is approximately triangular shape but shallow and close to the impact surface. It contains intense damage with concentric, truncated macrocracks or surface cone macrocracks. Initially, these cracks grow normal to the target surface and then deflect to grow parallel to the surface at a shallow angle as shown schematically in the right side of the figure. Region 2 is a spherical region representing the Mescall zone directly beneath the impact axis and contains an intense network of distributed three-dimensional intergranular and transgranular microcracks. The Mescall zone is fully contained within the bulk solid and consists of highly comminuted material. The size of this zone and severity of damage within the zone increase with impact velocity. During this growth phase, the penetrator is in the dwell state as was discussed in Chapter 3. As this Mescall region grows bigger, at higher velocities, its boundary reaches the impact surface allowing penetration of the impactor. The corresponding velocity is called the dwell-penetration transition velocity. Region 3 contains a few large and discrete high-angle interior cone macrocracks. They initiate and diverge from the impact site and propagate into the bulk solid. A network of additional cracks connecting the large cone cracks also evolve at higher impact velocities. This connecting network causes extensive fragmentation of the target ceramic.

Motivated by the above observations, Iyer (2007) developed a stress-based analytical approach utilizing Hertzian contact stress and linear elastic fracture mechanics concepts. The damage evolved in the target material can be related to the induced stress field upon application of a concentrated load. The stress-influenced zone can be divided into four multiaxial stress states in the four zones depending on the tensile and compressive principal stresses. These stress states are indicated on the left side of Figure 4.31. Zone I corresponds to Region 1 with shallow cracks. In this region, the three principal stresses, σ_1, σ_2, and σ_3, respectively, are tensile, compressive and approximately zero (i.e. $\sigma_1 > \sigma_2 > \sigma_3 \approx 0$). The magnitude of σ_1 is the highest as this region is closer to the point of the load application. The impact due to the projectile causes intense damage in this region compared to other regions. Slightly away from the impact site and closer to the surface, tensile stress prevails with its maximum principal direction perpendicular to the surface, and cracks evolve almost parallel to the impact surface. Zone II corresponds to Region 2 where all the principal stress components are compressive (σ_1, σ_2, $\sigma_3 < 0$), resulting in a large hydrostatic compressive stress. Microcracking is initially suppressed, and at high stress levels, plasticity initiates. However, the low dislocation mobility in ceramics limits the overall ductility (Lankford et al. 1998), and instead microcracking is initiated due to high stress concentrations at these locations. The failure process is thus dominated purely by shear stress ($(\sigma_1 - \sigma_2)/2$). With increasing velocity, severe microcracking is expected, leading to extensive fragmentation, comminution and even pulverization of the ceramic. In Zone III (adjacent to Zones I and II), σ_1 is tensile, and σ_2 and σ_3 are compressive. Finally, in Zone IV, σ_1 and σ_2 are tensile, whereas σ_3 is compressive. Both Zones III and IV correspond to Region 3 and include a large network of cone and interconnected cracks. Note from the figure that this region is adjacent to Zone I, where only one principal stress component is tensile. In the interior of the target, two tensile principal stress components exist, but their magnitude is comparatively lower than in Zone I as this region is farther away from the impact site. Therefore, macrocracking will be less intense in Zone III compared to Zone I.

Similarly, Zone IV is mostly influenced by the elastic stress state except that cracks initiated from the nearby regions propagate in a radial fashion from the impact site. The region between two cone cracks is under a state of compression due to the bulking associated with the creation of new fracture surfaces along the radial direction. As is well known, in such regions, cracks evolve in a direction parallel to the maximum compressive stress (Ashby and Hallam 1986; Horii and Nemat-Nasser 1986; Subhash and Nemat-Nasser 1993). These cracks eventually connect with the cone cracks, forming the complex interconnected network, leading to fragmentation of the ceramic target. Thus, Hertzian contact mechanics can be effectively used to explain the damage patterns evolved within a ceramic target due to a high-velocity ball impact.

4.5.5 Expanding Cavity Models

The second approach to analyzing the stress state while the penetrator is in contact with the ceramic target is called the expanding cavity model (ECM).

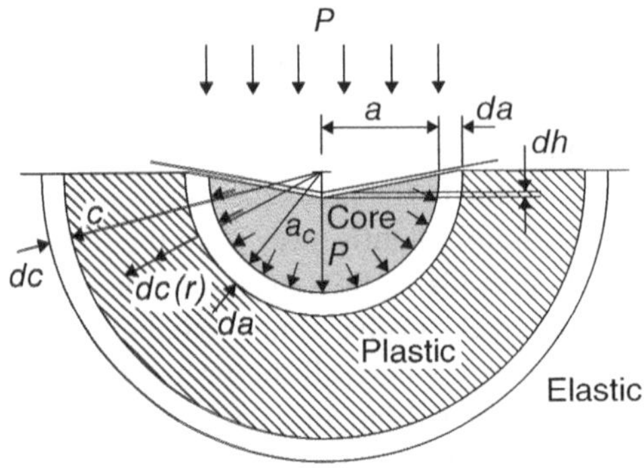

Figure 4.32 Schematic of the evolved elastic-plastic zones due to indentation in an expanding cavity model.

This model relates the pressure required to open a cavity from zero radius to the maximum size allowed by the target resistance. The ECMs are based on Hill's (1998) theory for the expansion of a spherical/cylindrical shell made of an elastic-perfectly-plastic material subjected to internal pressure. For elastic-plastic solids, Samuels and Mulhearn (1957) determined that the plastic zone beneath an indenter is hemispherical in shape, and it resembles the plastic region near an expanding spherical/cylindrical cavity. Johnson (1985) summarized these models, and a brief summary of relevant models is provided here (Johnson 1970, 1985).

The ECM assumes that the region of the specimen directly in contact with an indenter or a penetrator is under a large hydrostatic stress and has a hemispherical shape with radius a, as depicted in Figure 4.32. This region experiences both elastic and plastic responses with the application of a load. The hydrostatic core region is assumed to be surrounded by an incompressible hemispherical plastic zone with radius c, which is further constrained by the surrounding elastic zone. With an increase in penetration depth, both the radii of the core and the plastic zone increase. An increment of penetration depth dh causes an incremental radial displacement da of the core and the volume displaced is accommodated by a radial expansion dc of the plastic zone, which is eventually taken up by the elastic region. In the following, the ECMs as applied to both static and dynamic conditions are discussed.

4.5.5.1 Static Expanding Cavity Model

One of the main objectives of the ECMs is to estimate the hardness of a material based on its material properties. The relationship between hardness and yield strength ($H = 3Y$) originally developed by Tabor (1948) was improved by Johnson (1970) for conical indentations as follows:

$$\frac{H}{\sigma_y} = \frac{2}{3}\left[1 + \ln\left(\frac{1}{3}\frac{E}{\sigma_y}\cot\alpha\right)\right] \tag{4.60}$$

Equation (4.60) is based on Hill's analysis (Hill 1998) for elastic-perfectly-plastic materials, and it is known to break down when a material exhibits strong strain-hardening characteristics (e.g. AlN (Heard and Cline 1980)). Therefore, analytical models or finite element simulations have been popularly used to study indentation of strain-hardening materials (Fischer-Cripps 1997; Giannakopoulos and Suresh 1999; Mata and Alcala 2003; Mata et al. 2002; Mesarovic and Fleck 1999; Park and Pharr 2004; Zhang and Subhash 2001).

Gao et al. (2006) developed an ECM for strain-hardening materials subjected to indentation by conical and spherical indenters. They considered both power-law hardening and linear-hardening materials and developed explicit analytical expressions for hardness-yield strength relationships in terms of material properties, work-hardening characteristics, and indenter geometry. Utilizing a previously developed solution by Gao et al. (1991) and Gao (2003) for an internally pressurized spherical shell of an elastic, power-law-hardening material, Gao et al. (2006) derived the following relationship for conical indentations as:

$$\frac{H}{\sigma_y} = \frac{2}{3}\left\{\left(1-\frac{1}{n}\right)+\left(\frac{3}{4}+\frac{1}{n}\right)\left(\frac{1}{3}\frac{E}{\sigma_y}\cot\alpha\right)^n\right\} \tag{4.61}$$

and for spherical indentations as

$$\frac{H}{\sigma_y} = \frac{2}{3}\left\{\left(1-\frac{1}{n}\right)+\left(\frac{3}{4}+\frac{1}{n}\right)\left(\frac{1}{4}\frac{E}{\sigma_y}\frac{a}{R}\right)^n\right\} \tag{4.62}$$

where H is the hardness, E is the modulus, n is the power-law exponent, and σ_y is the yield stress, α is the half cone angle, a is the contact radius, and R is the radius of the spherical indenter. For elastic-perfectly plastic materials ($n = 0$), the above equations reduce to:

$$\frac{H}{\sigma_y} = \frac{2}{3}\left[\frac{7}{4}+\ln\left(\frac{1}{3}\frac{E}{\sigma_y}\cot\alpha\right)\right] \text{ for conical indentations} \tag{4.63}$$

and

$$\frac{H}{\sigma_y} = \frac{2}{3}\left[\frac{7}{4}+\ln\left(\frac{1}{4}\frac{E}{\sigma_y}\frac{a}{R}\right)\right] \text{ for spherical indentations.} \tag{4.64}$$

In both equations (4.63) and (4.64), the hardness is dependent upon the indenter geometry, and for a given indenter geometry, the indentation hardness depends on material properties such as E, $\sigma_{y,}$ and n. Conical indentations are self-similar, meaning that they are independent of any length parameter such as depth of indentation for the above material behaviors. On the other hand, spherical indentations are not self-similar; they depend on a and R for both types of elastic-plastic materials. Gao et al. (2006) utilized these models to describe the plastic-zone properties for materials with different strain-hardening characteristics. Parsard and Subhash (2017) also utilized the above equations to determine the yield strength of boron carbide and estimated the amorphization zone size beneath static and dynamic indentations.

The ECMs are also used for calculating the target resistance to penetration. Bishop et al. (1945) were the first to consider the static pressure required to open a cavity in a metal target. This effort was later expanded by Tate (1967) and by Forrestal et al. (1988) and Sternberg (1989). For ceramics and brittle materials, the concept was adopted by Forrestal and Longcope (1990) and Satapathy and Bless (1996). These models were motivated by the physical observation of evolved damage features in a ceramic target impacted by a projectile. As seen in numerous experimental results discussed in previous sections, upon impact, a cavity is formed in the target material, and this cavity is surrounded successively by comminuted material, a cracked material, and finally by an elastic region, as schematically depicted in Figure 4.33. To describe these features, steady-state penetration was considered by Satapathy and Bless (1996), and the resisting stress in the target was calculated by appropriate field equations in each region.

For the elastic region, the pressure and shear stresses were calculated by considering the traditional elasticity equations in spherical coordinates, i.e. equilibrium equations, strain–displacement relations, and the stress–strain relationships. Relevant boundary conditions were used at the elastic and cracked material interface, which are given by the equality of hoop stress and tensile strength, and continuity of radial stress. In the cracked region, hoop stress exceeds the tensile strength (σ_f)

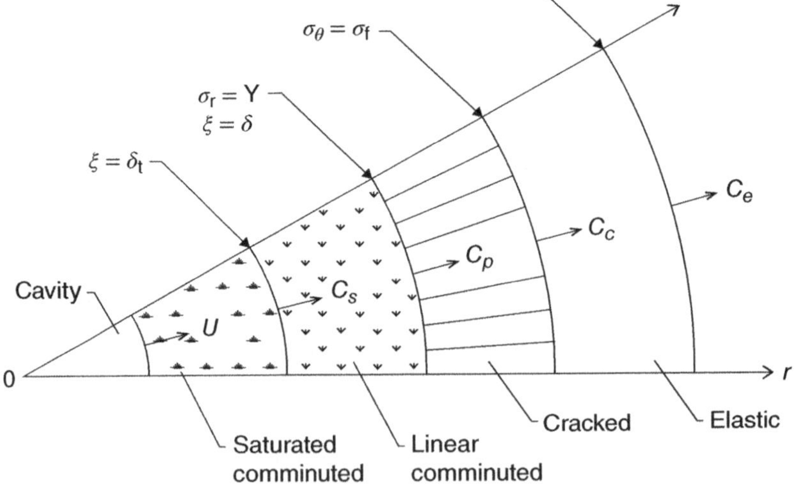

Figure 4.33 Schematic of the damage zones for modeling the response of a ceramic due to projectile impact. Source: Bavdekar et al. (2017). © 2017, Elsevier.

and radial cracks are formed. Hence, the hoop stress is relieved and assumed to be zero. The stress in each columnar fragment is considered uniaxial with a maximum compressive strength Y. When stress exceeds this value, comminution is assumed to occur. Finally, by considering the comminuted material response as a MC model with pressure-dependent shear strength, the target resistance (which is the radial stress at the cavity boundary), R_t, was derived by Satapathy and Bless (1996) as a function of material properties (E and ν) and confining pressure (or mean stress) as

$$R_t = Y\left[\frac{E}{3Y\left[1-(1-\upsilon)\sqrt{\sigma_f/2Y}\right]}\right]^p \tag{4.65}$$

where ν is Poisson's ratio, σ_f is tensile failure strength, p is related to the pressure-dependent constitutive behavior (assumed for the comminuted ceramic) and is given by $p = 4\,\alpha\,/\,(3+4\,\alpha)$, and α denotes the slope of the pressure (p)-shear strength (τ) curve for the comminuted ceramic region, where

$$\tau = \frac{(\sigma_r - \sigma_\theta)}{2} \tag{4.66}$$

$$p = \frac{(\sigma_r + 2\sigma_\theta)}{3} \tag{4.67}$$

$$a = \frac{\tau}{p} \tag{4.68}$$

where σ_r and σ_θ, are the radial and circumferential stresses. Using equations (4.66)–(4.68), the target resistance for AD995 alumina was calculated to fall between 7.7 and 8.6 GPa for a yield strength range of 2.62–3 GPa. The sizes of the pulverized and cracked zones were calculated, respectively, as 3.5 and 9 times the cavity size for a target of infinite thickness. Interestingly, the R_t value was found to be insensitive to the tensile strength of the ceramic.

Despite the ability of the model to capture the postimpact features such as cracked zone size and cavity size in the target, the model fails to depict the evolution of these features during the impact process. Hence, Satapathy (2001) developed a dynamic spherical cavity expansion model to better capture the strain-rate-sensitive behavior of the ceramic target.

4.5.5.2 Dynamic Expanding Cavity Model

The next logical extension to the ECM included strain-rate effects associated with high-velocity impact and the penetration process. While experiments that include dynamic effects on indentation hardness were pioneered by Koeppel and Subhash (1997, 1999) and Subhash et al. (1999), discussed in Chapter 2, Satapathy

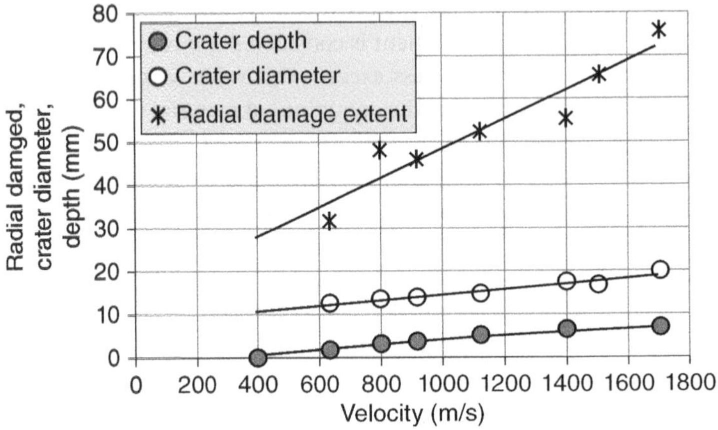

Figure 4.34 Plot of damage level as a function of impact velocity on SiC–N ceramics. Source: Normandia (2004). © 2004, John Wiley & Sons.

(2001) extended the ECM model to ballistic impact. Motivation for the model came from the sphere-impact experiments on confined SiC–N ceramics by Normandia (2004). Similar to the static ECM model described earlier, it was noted that a well-defined crater formed at the impact site at high velocities. Radial cracks and circular cone cracks surround this crater. The extent of the radial damage, crater radius, and depth of damage increased with impact velocity, as can be noted from Figure 4.34. There is also an increased level of fragmentation and comminution at higher velocities.

To capture these observations and model the target behavior, Satapathy (2001) considered a dynamic expanding cavity at a constant speed in an infinite medium to simulate impact by a projectile. The entire influenced zone is assumed to have spherical symmetry and consisted of five distinct zones growing at different velocities:

a. finite cavity expanding at a fixed velocity U,
b. comminuted zone growing at a speed of C_p,
c. radially cracked zone growing at a velocity of C_c,
d. elastic zone growing at a speed of dilatational elastic wave (C_e) and
e. an undisturbed zone.

A schematic of these zones and their growth rates are illustrated in Figure 4.33 (Bavdekar et al. 2017). Upon impact, the cavity is subjected to an instantaneous expansion velocity, U, which sends out radially expanding waves into the bulk solid. The dynamic cavity expansion causes propagation of comminuted, cracked, and elastic zones to spread into the undisturbed material at their characteristic velocities. At these propagation fronts, Hugoniot jump conditions for mass and momentum were applied. The comminuted zone closest to the cavity is assumed to exhibit a pressure-dependent strength (with zero strength in the absence of pressure) and has been modeled as a MC material by Satapathy (2001) and by an SS-EMC model by Bavdekar et al. (2017). The cracked region is assumed to have no hoop stress, only radial stress. The presence of only radial cracks also renders this zone anisotropic as far as elastic constants are considered. Shear saturation of comminuted material was also considered consistent with the experimental observations of Rosenberg et al. (1987). With these assumptions, dynamic cavity expansion equations were developed and then applied to model the penetration behavior of ceramic targets by Satapathy (2001). For a fully shear saturated comminuted region, the cavity expansion pressure P_c was calculated as

$$P_c = Y + 4\tau \ln \Delta + \frac{\rho V^2}{2}\left[3 + \frac{1}{\delta^4} - \frac{4}{\delta}\right] \tag{4.69}$$

where Y is the compressive strength, τ is the saturated shear strength of the comminuted zone, δ is the ratio of comminuted zone propagation speed to cavity expansion speed, and ρ is density. Satapathy (2001) showed that the cracked region disappears at high-cavity expansion speeds as the comminuted zone grows faster than the cracked zone. The rate of growth of the comminuted zone saturates and thereafter remains lower than the longitudinal wave speed. Most importantly, the target resistance term decreases with increasing penetrator velocity for steels, whereas for ceramics the target resistance term increases, suggesting a better penetration resistance at higher velocities.

Normandia (2004) utilized Satapathy's model (Satapathy 2001) to determine the target resistance due to a sphere impact on a hot pressed SiC–N ceramic. In this analysis, the depth of penetration is expressed in terms of normalized areal density, defined as the ratio of the product of target density and depth of penetration to the product of projectile density and its diameter. This normalized parameter was compared to the experimental data as shown in Figure 4.35, where it is seen that the normalized areal density penetrated increases with increasing impact velocity. The model found that as the velocity of the penetrator increased, the normalized penetration depth increased almost linearly. For a given impact velocity, the dynamic ECM model also predicts lower penetration with increasing target resistance. Also, the lower the inertial confinement, the lower is the integrated target resistance, and hence the greater the level of damage (Normandia 2004).

Most recently, Subhash and coworkers improved the above dynamic ECM model by incorporating the SS-EMC model (Bavdekar et al. 2017; Bavdekar and Subhash 2018). Recall that the comminuted material response was modeled using the linear-dependent M-C model in the earlier ECM by Satapathy (2001). This approach was moderately successful in capturing the experimental data on target resistance on a variety of ceramics. Recognizing the fact that the

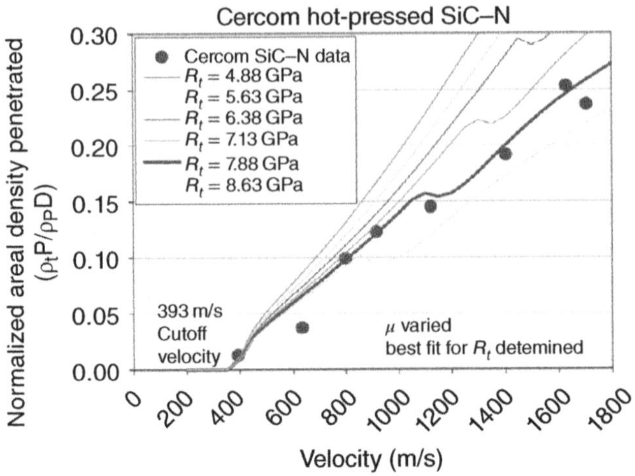

Figure 4.35 Normalized areal density of target as a function of impact velocity for various target resistances. Source: Normandia (2004). © 2004, John Wiley & Sons.

SS-EMC model from Eq. (4.7) (Shafiq and Subhash 2016a) with its exponential pressure dependence and strain-rate-independent strength, captures the strength saturation observed at high pressures in several intact ceramics, Bavdekar et al. (2017) modified the three parameter exponential form into a two-parameter exponential model, applicable to comminuted ceramics. For intact ceramics, the SS-EMC model was given earlier as

$$\tau^* = a + be^{-kP^*} \tag{4.41}$$

where the superscript * indicates normalized values with respect to the corresponding HEL values. Comminuted ceramics are assumed to have no strength when no confinement is applied, and hence, setting $\tau^* = 0$ at $P^* = 0$ in Eq. (4.41), yields $a = -b$ and the model is reduced to

$$\tau^* = a_1\left(1 - e^{-k_1 P}\right) \tag{4.70}$$

where a_1 is the saturated shear strength of the comminuted material, k_1 is the slope of curve interpreted as frictional sliding resistance similar to α in the traditional MC model and $\tau_s^* = a_1 = {}^{\tau_s}/_{\tau_{\mathrm{HEL}}}$. A comparison of the models for intact and comminuted materials is shown in Figure 4.36.

Recall that the constants in the EMC model were obtained from the curve fit for the experimental data on a variety of intact ceramics. Similarly, the constants in Eq. (4.70) for comminuted ceramics can also be found by plotting the shear strength of comminuted ceramics from Bourne (2008, 2012) and the corresponding HEL values for intact ceramics as seen in Figure 4.37, a linear relationship was noted. Interpolating values for other ceramics, it is found that the average value

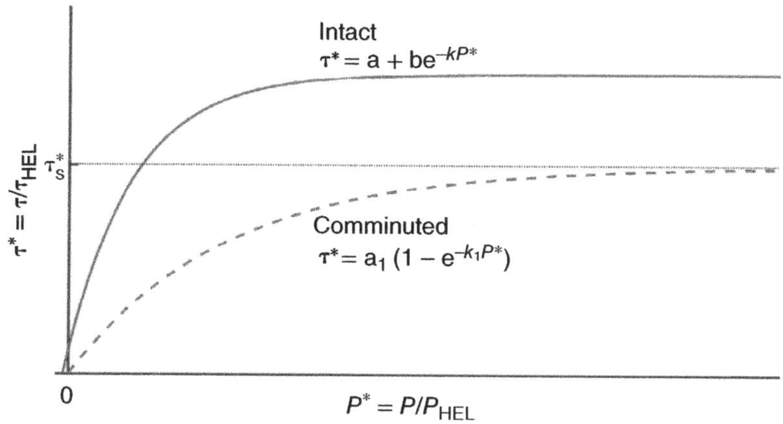

Figure 4.36 Comparison of the pressure-shear behavior of intact and comminuted ceramics in the SS-EMC model. Source: Bavdekar et al. (2017). © 2017, Elsevier.

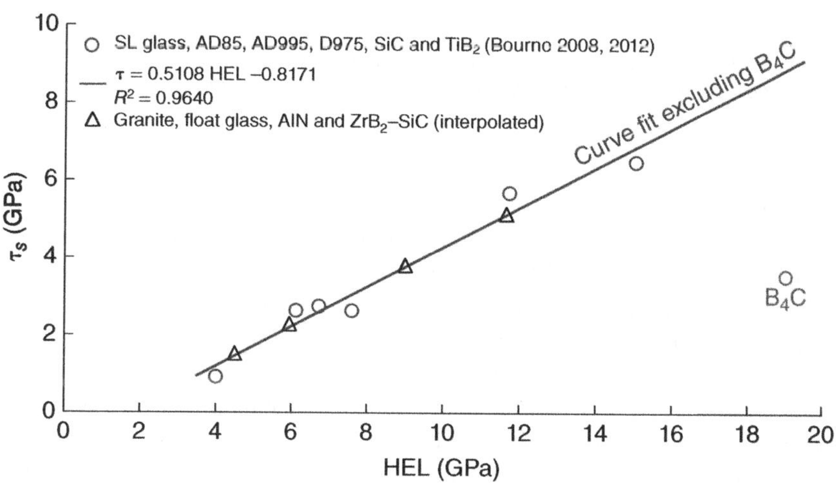

Figure 4.37 Plot of shear strength of comminuted ceramics versus HEL values for intact ceramics. The data for B_4C is excluded because of its anomalous failure behavior called amorphization. Source: Bavdekar et al. (2017). © 2017, Elsevier.

of a_1 was 1.11. Similarly, considering the penetration velocity and impact velocity relationship for many ceramics, they determined the average value of k_1 to be about 0.36. In this graph, the data from boron carbide is omitted because of its unique amorphization behavior, discussed in depth in Chapter 6.

Using the above SS-EMC model for comminuted ceramics, Bavdekar et al. (2017) calculated target resistance as a function of impact velocity following the

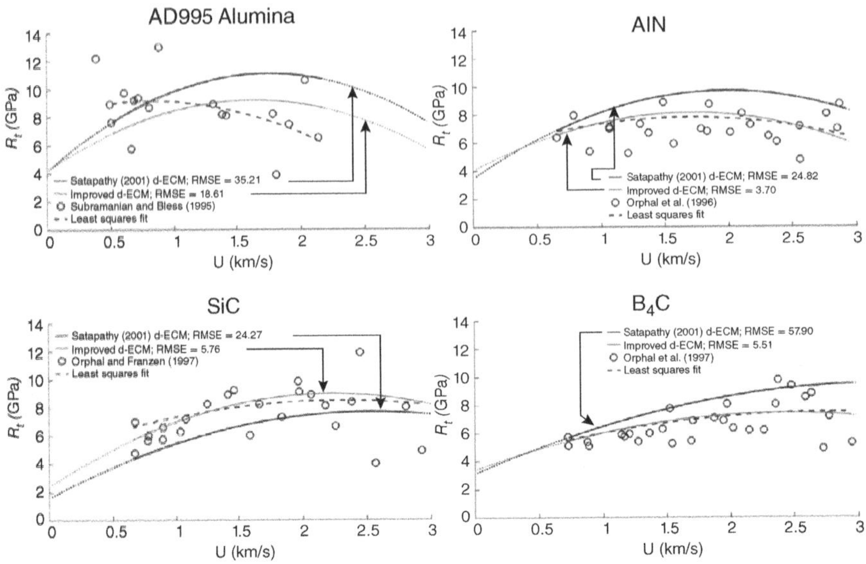

Figure 4.38 Comparison of experimental results and model predictions using the SS-EMC model for four structural ceramics. Source: Bavdekar et al. (2017). © 2017, Elsevier.

same analytical procedure as that of Satapathy (2001) and found a better match to the experimental data. The results for four ceramics are shown in Figure 4.38. The usefulness of such dynamic ECMs is clearly seen in these examples. The results indicate that the properties of comminuted ceramics have a bigger influence on the target resistance than those of intact ceramics. Sensitivity analysis by Bavdekar et al. (2017) also revealed that penetration resistance of ceramic targets is influenced in decreasing order: shear strength of comminuted material (a_1), frictional characteristics of comminuted material (k_1), and uniaxial compressive strength (Y) of the intact ceramic. Interestingly, the influence of bulking (volume increase due to damage) and compaction of the material on penetration depth are negligible.

4.5.6 Stochastic Models

The failure strength of ceramics is highly stochastic due to a large distribution of flaws of different shape and size. The propensities for activation and growth of cracks from these defects varies significantly even when the specimens are extracted from the same tile or a production batch. Despite the vast improvements in ceramic-processing technology, the ability to strictly control the microstructural defects and their distribution in a ceramic is still challenging. The random activation of cracks from these defects results in a large variability in the strength values. Despite considerable progress in processing technologies for stricter control of process parameters and reduced variability in defect type, size, and distribution, the inherent brittle nature of a ceramic is fundamentally a defect-driven response and hence statistical representation of the failure strength is possible. However, the sensitivity of ceramic strength to rate of loading and the level of confinement or precompression should also be considered in the stochastic modeling.

Traditionally, a Weibull-statistics-based analysis has been used to account for the statistical variation in properties (Weibull, 1939, 1951). The premise for such models is that ceramics have a large population of distributed intrinsic and extrinsic flaws whose characteristics and spatial distribution contribute to a large variation and uncertainty (scatter) in the strength values. Therefore, Weibull's "weakest link theory" based on the distribution of flaw strengths and associated macroscopic strength variations under QS loading has been extended to dynamic loading conditions as well. The philosophy of the weakest link assumes that a single flaw most sensitive to the applied loading conditions dominates the fracture process. Models based on the distribution of flaws and their rate dependency were first attempted by Kipp and Gardy (1980), which is described in a later section on fragmentation of ceramics. These statistical models do not account for defect geometry but instead consider only a distribution of a potential (strength or energy) associated with each defect. The current section is devoted exclusively to statistical models that focus on the influence of preexisting defects in the presence of confinement and high rate of loading.

4.5.6.1 Influence of Confinement Pressure

Confinement pressure resists crack opening and increases the strength of a ceramic. This behavior can be interpreted as reduced sensitivity of strength to preexisting flaws, or that the statistical distribution of strength is narrowed in the presence of confinement, i.e. a decrease in strength variability. With this philosophy, stochastic models have been developed to analyze the dynamic fracture of ceramics.

Significant progress has been made in understanding the statistical nature of strength under dynamic conditions by Zhou and Molinari (2004b, 2005). The approach involves finite element modeling with defects placed along the mesh facets and a cohesive element formulation for their growth (Zhou and Molinari 2004a). The failure mode is stochastically represented by giving each facet a strength value per specified distribution. The microcrack growth is initiated and the cohesive element formulation is invoked upon the effective shear stress exceeding the strength of the facet. Thus, defects are modeled as facets shared by two neighboring finite elements. The effective crack opening displacement for a cohesive element is given by Zhou and Molinari (2004b, 2005):

$$\delta_{\text{eff}} = \sqrt{\beta^2\delta_s^2 + \delta_n^2} \tag{4.71}$$

where δ_s and δ_n are tangential and normal crack-opening displacements, respectively, along the crack facets, and β is a parameter that assigns weight to these crack-opening displacements. The effective traction on cohesive elements is given by:

$$t_{\text{eff}} = \sqrt{\beta^{-2}\left|t_s\right|^2 + t_n^2} \tag{4.72}$$

where t_s and t_n are tractions along the tangential and normal directions, respectively. A simple linearly decaying (irreversible) cohesive law connecting the traction displacement was used.

To invoke the stochastic nature of the failure mode, each facet is given strength and fracture energy values in accordance with the Weibull distribution. These facets will initiate microcracks when the local effective shear stress exceeds the preassigned strength of the facet. Thus, the constitutive behavior of the ceramic consists of two parts: (i) homogeneous linear elastic solid and (ii) cohesive elements where microcracks initiate and grow along the element facets. While the former is routine, the latter part captures the statistical nature of the failure process in brittle ceramics by considering each element boundary as a potential crack nucleation site. Here all facets are assumed to have the same critical opening distance (δ_c), and the equivalent fracture energy (G_c) is given by

$$G_c = \sigma_s\Delta_c/2 \tag{4.73}$$

where σ_s is the scaling stress. Assuming a linearly decaying irreversible cohesive law, the probability of a facet failure under tensile loading is defined as (Zhou and Molinari 2004b)

$$P(\sigma) = 1 - \exp\left[-\left(\frac{A_{\text{facet}}}{A_s}\right)^{\frac{m_s}{m_a}}\left(\frac{\sigma}{\sigma_s}\right)^{m_s}\right] \tag{4.74}$$

where A_{facet} is the area of the facet, A_s is the scaling area, m_s is the Weibull strength distribution, and m_a is the Weibull modulus for the effective area. The distribution of local facet strength is given by

$$f\left(\sigma_{\text{facet}}\right) = \text{d}P/\text{d}\sigma_{\text{facet}} \tag{4.75}$$

Assuming a probability density function for various values of m (or m_s), i.e. statistically distributed strength values for microcracks as shown in Figure 4.39, Zhou and Molinari (2004b, 2005) described the failure response of brittle materials. For small m values (wide distribution of flaws), the failure strength of a ceramic is low (Zhou and Molinari 2004b), and with increasing m (narrow distribution), the

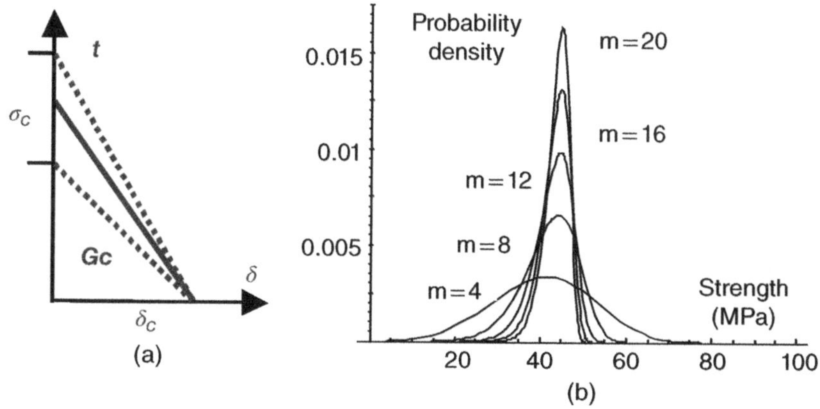

Figure 4.39 The representation of (a) cohesive law and (b) the assumed probability density distribution of microscopic strength. Source: Zhou and Molinari (2004b). © 2004, Elsevier.

strength of a ceramic increases. This result is anticipated because the wide distribution of flaw strengths (low *m*) implies a highly heterogeneous defect strength and greater probability of crack activation at low stress levels from sensitive flaws, eventually leading to the failure of the specimen. On the other hand, when the *m* value is high, the distribution of the flaw strengths is narrow, the material is more homogeneous, and there is lower probability of a single larger crack controlling the failure process.

4.5.6.2 Brittle Versus Ductile Behavior

Using a similar approach and assuming a distribution of fracture energy for the mesh facets, the analysis was extended to study the differences between the behaviors of ductile and brittle cracks on the tensile failure strength of a ceramic (Zhou and Molinari 2004b). The ductile and brittle cracks are classified based on the energy expended to open the crack. Activation of a ductile crack requires more energy and time (due to dislocation generation, motion, and accumulation leading to stress concentrations and eventual crack initiation) than a brittle crack, which is breakage of chemical bonds between atoms without the activation of dislocations. Under high strain rates, the applied stress amplitude rises rapidly compared to the crack growth rate, and ceramics exhibit higher strength than that under QS loading. Also, a ductile system (e.g. with high fracture energy $G_c = 50\,\text{N/m}$ in Figure 4.40) exhibits less scatter in the failure strength data because plasticity initiates at a fixed stress level, and the effects of any initial stress concentration are quickly subdued by plasticity. The lack of scatter in defect strength distribution results in the failure process being less stochastic. On the other hand, with a

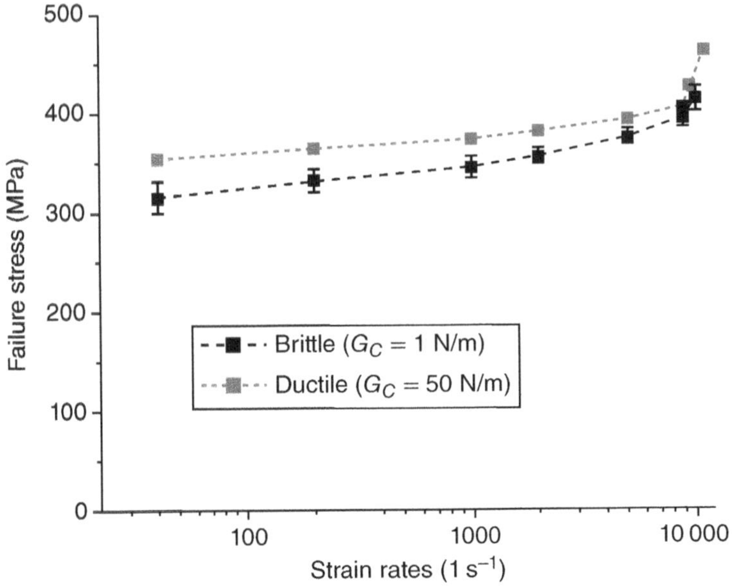

Figure 4.40 Influence of strain rate and fracture energy on failure strength for ductile and brittle cracks. Source: Zhou and Molinari (2004b). © 2004, Elsevier.

G_c of only 1 N/m and with a large distribution of defects, a brittle system exhibits greater scatter in the strength data. However, the level of scatter decreases with increases in strain rate as more flaws are activated throughout the material. The stress level rises quickly to activate many cracks of varying stress intensities at the same time, leading to an averaging effect (Zhou and Molinari 2004b). Thus, in the intermediate strain-rate range, the strain-rate sensitivity of the strength arises from the statistical distribution of flaws, whereas at much higher strain rates, the sensitivity to the flaw distribution decreases.

4.5.6.3 Loading Geometry (Uniaxial versus Biaxial Stress State)

Most of the available experimental data on the strength of brittle materials are based on uniaxial tension and compression tests and over a wide range of strain rates. Confinement experiments, on the other hand, impose a biaxial compression stress state but are difficult to implement experimentally. Computational studies are ideally suited to examine the effect of such loading configuration on the failure strength of ceramics. Zhao and Molinari (2005) compared failure strengths under both uniaxial and biaxial (equiaxial) tensile loading conditions and examined the influence of strain rate and defect distribution on local strength variations. A characteristic strength was defined as summation of local strength values as

$$\sum_{s} = \left(\frac{1}{N} \sum_{i=1}^{N} \sigma_i^{M_s} \right)^{1/M_s} \tag{4.76}$$

where the Weibull modulus, M_s, was defined as

$$\frac{\sum_{i=1}^{N} \sigma_i^{M_s} \ln \sigma_i}{\sum_{i=1}^{N} \sigma_i^{M_s}} - \frac{1}{N} \sum_{i=1}^{N} \ln \sigma_i - \frac{1}{M_s} = 0 \tag{4.77}$$

Using this formulation, Zhao and Molinari (2005) found that the QS strength values are greater under uniaxial loading than biaxial tensile loading. This behavior is rationalized as there is higher probability of activation of a more favorably oriented defect under biaxial loading than uniaxial loading. At increasing strain rates, a greater number of cracks activate during biaxial loading due to stress-wave loading, averaging the effect of the distribution and lowering the scatter.

In summary, an increase in strain rate reduces the scatter in the strength values because more cracks are activated within a given material volume and a decreased dominance of a single crack is noted as there is less time available to relax the stress in the surrounding region. Therefore, at high strain rates, stochasticity is reduced, resulting in an averaging effect; at higher strain rates, brittle failure is less dependent on "the weakest-link hypothesis." Compared to QS loading, activation of a greater number of cracks under dynamic loading dissipates more energy before macroscopic fracture, and thus a further increase in load is required for continued crack growth. As a result, the strength of brittle materials increases with strain rate and a similar trend is observed for both uniaxial and biaxial loading conditions. It was also found that a wider internal defect distribution (smaller m_s) resulted in a lower strength increase with strain rate.

An insightful way of relating defect size, defect distribution, and loading rate to the strength of a ceramic material was provided in a graphical form by Ramesh et al. (2015) where each of the relevant parameters are plotted on different axes as shown in Figure 4.41. The top right quadrant provides the relationship between the defect size and its distribution, where the two ends of the distribution show low populations. The bottom left quadrant of the figure depicts two loading rates representing low (QS) and high (dynamic) rates of stress rise. The square root singularity relationship between the stress and defect size is depicted in the bottom right quadrant. Following the dashed line at any given time, the low rate loading relates to the activation of a few large defects. On the contrary, the high-rate loading imposes a higher stress level (σ_2) and activates a large number of small defects of size greater than S_2. In essence, low-rate loading activates only a few critical

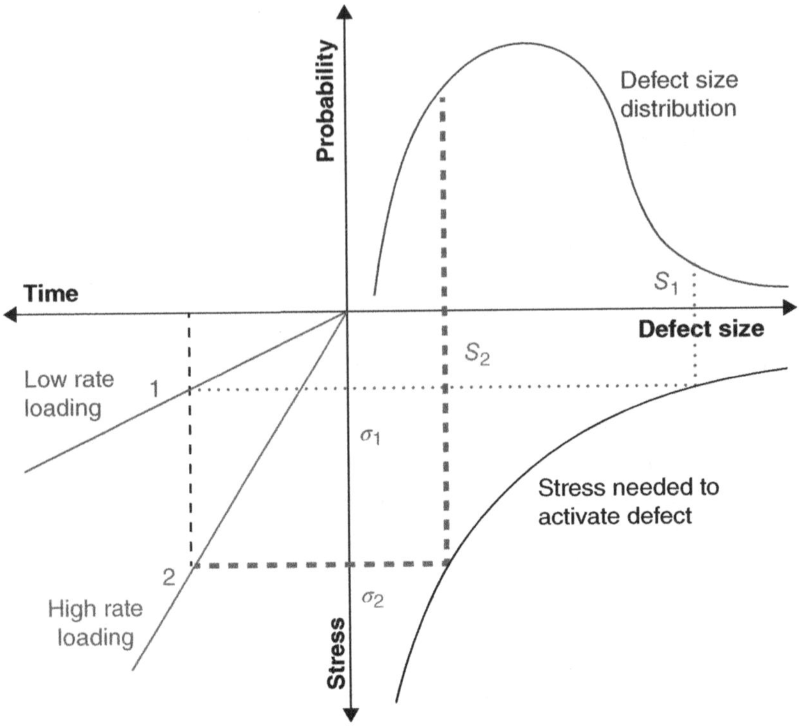

Figure 4.41 Graphical depiction of the relationship between defect size, defect distribution, and loading rate. Source: Ramesh et al. (2015). © 2015, Elsevier.

(large) flaws even at low stress levels leading to low strength, but the higher rate of loading not only activates the larger flaws, but also the entire distribution of flaw population because the rate of loading outpaces the rate of crack growth from the large cracks. This behavior leads to a higher strength of the ceramic. Thus, an intimate relationship between the rate sensitivity of fracture strength and inherent defect distributions in ceramics has been well captured through this depiction.

Clearly, stochastic models provide greater insight into the interactions among the applied stress rate, crack growth, and defect distribution. Such rationale can allow designers or process engineers to customize the defect distribution for enhanced performance and reduced stochasticity.

4.5.7 Modeling Fragmentation of Brittle Solids

Thus far, we have discussed the mechanics of crack growth and damage evolution and their influence on strength of a ceramic. However, the consequence of crack growth from multiple defects in a brittle material is severe fragmentation. The challenge for modeling the dynamic fragmentation process of brittle materials is the mathematical description of the constitutive properties of cracked and comminuted materials. In this chapter, we have thus far discussed both the experimental and computational efforts to evaluate the behavior of both an intact ceramic and a damaged ceramic. However, these models do not adequately describe the evolution of fragmentation size distribution as a function of strain rate. More importantly, the question of how fragmentation behavior influences the resulting homogenized dynamic response of the ceramic is not addressed.

Griffith (1943) was the first to suggest the correlation between strain rate and fragment size distribution. He utilized an energy balance approach and proposed that strain energy is entirely expended toward creating new surfaces due to fragmentation. Mott and Linfoot (1943) proposed a statistics-based theory of fragmentation built on the random creation of cracks and their interaction to explain fragment size distribution. However, unlike Griffith, they assumed that the energy requirement for actual fracture processes is negligible and fracture is instantaneous. Grady and colleagues (Grady 1982; Grady and Kipp 1985) refined the approach of Mott and Linfoot and suggested a linear distribution of fragment size, which showed improved predictive capability, and they calculated the time scale associated with the fracture event.

Brittle fracture progresses via nucleation of cracks from preexisting defects, and growth and coalescence of these cracks form fragments. However, crack growth characteristics are significantly different under QS and dynamic loading conditions as discussed earlier. During dynamic loading, a stress wave travels at longitudinal wave velocity, and when a crack is nucleated, it can travel at velocities up to the Rayleigh wave velocity. Since the longitudinal wave travels faster than the Rayleigh wave, the neighboring stress-free zone associated with any crack growth has little influence on the activation of new cracks because the stress wave has already traveled farther away from a crack and activated other defects. In other words, there is not enough time to unload a neighboring region. Hence, there is a revolution or mutiny under dynamic loading, where the critical stress is reached at multiple defects to simultaneously activate their growth in a given material volume. Under dynamic loading, higher magnitudes of incident stress lead to more neighboring cracks that activate. With the increasing number of growing cracks, the mean travel distance to coalesce decreases, and, as a result, fragment size decreases. Thus, the damage is more distributed within a material under dynamic loads, and this distribution of damage must be accounted for in modeling studies. This effect is often modeled as a reduction in material stiffness, especially under tensile loading.

In one of the earlier studies, Grady (1982) and Grady and Kipp (1985) derived a relationship between the fragment size (l) and strain rate ($\dot{\varepsilon}$) when a bar of length L

is held fixed at one end and pulled in tension at the other end at a constant velocity v_0:

$$l = 24^{1/3}\left(\frac{K_{Ic}}{\rho c\dot{\varepsilon}}\right)^{2/3} \tag{4.78}$$

Here, K_{IC} is the fracture toughness of the material, ρ is the density, and c is the longitudinal wave velocity. Later, Glenn et al. (1986) included the strain energy of fragments and specified a threshold stress ($\sigma_{\max}$) for fragmentation and determined the fragment size as

$$l = 4\sqrt{\frac{\alpha}{3}}\sin h\left(\frac{\phi}{3}\right) \tag{4.79}$$

where

$$\alpha = \frac{\sigma_{\max}^2}{\hat{E}\rho\dot{\varepsilon}^2} \tag{4.80}$$

and $\hat{E}$ is the effective tensile modulus and

$$\phi = \sin h\left\{\beta(3/\alpha)^{3/2}\right\} \tag{4.81}$$

Note that β is related to fracture toughness and longitudinal wave velocity as per

$$\beta = \frac{3}{2}\left(\frac{K_{IC}}{\rho c\dot{\varepsilon}^2}\right)^2 \tag{4.82}$$

In recent years, finite element modeling using the cohesive zone approach has become increasingly popular to investigate the dynamic fragmentation behavior of ceramics (Maiti and Geubelle 2004; Maiti et al. 2005; Zavattieri and Espinosa 2001). By choosing the appropriate traction separation law for finite element facets, these researchers have simulated spontaneous nucleation, rapid propagation, and coalescence of microcracks. The models capture the role of material microstructure (grain-level features) on fragment size and distribution. Numerical simulations revealed that fragmentation initiation time is independent of the microstructure. But the average grain size influences the average fragment size at higher loading rates. A comparison of predicted fragmentation size as a function of strain rate is shown in Figure 4.42 (Maiti and Geubelle 2004; Subhash et al. 2008). While Grady's model predicts a linear decrease in fragmentation size with strain rate (on a log–log plot), the other models predict a constant fragmentation size in the low strain rate regime and then a rapid decrease in size in the high strain rate regime above 10^5 s^{-1}.

Assuming a bilinear traction separation law, the fragmentation size in the QS strain rate range was derived by Maiti et al. (2005), as

$$l_{\text{stat}} = \frac{2(1-v)^2}{1-2v}\left(\frac{K_{IC}}{\sigma_{\max}}\right)^2 - \frac{\hat{E}}{b} \tag{4.83}$$

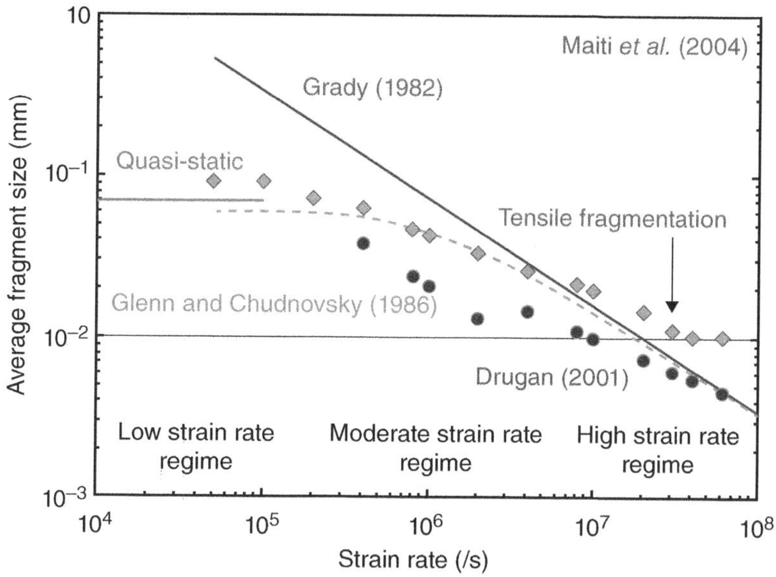

Figure 4.42 Comparison of fragmentation size as function of strain rate from various models. Source: Subhash et al. (2008). © 2008, John Wiley & Sons.

where b is the initial slope of the bilinear traction-separation law. Under shock wave loading Grady (1998) predicted the fragmentation size as

$$l = \frac{2E\gamma}{P_c^2} \tag{4.84}$$

where γ is the cohesive surface energy and P_c is the effective stress or confinement pressure in the shock wave experiments. Assuming $E = 400$ GPa for a typical structural ceramic, $\gamma = 5\ \mathrm{J/m^2}$ and a confinement pressure of 5–10 GPa in a shock wave plate impact experiment, the fragmentation size is 40–160 nm.

A more in-depth study of dynamic fragmentation considering nucleation, growth, and possible closing of cracks, as well as their interaction was conducted by Zhou et al. (2006c), where an average fragmentation size was calculated for a wide range of strain rates and material properties. Cohesive models developed by Maiti et al. (2005), analyzed the differences in crack growth dynamics under uniform tensile versus compressive loading. It was found that cracks grow over longer distances in tensile stress than in compressive stress under uniaxial loading for similar stress magnitudes.

The analytical and computational approaches adopted for fragmentation characteristics of ceramics have been able to capture the realistic behavior of brittle ceramics at a range of strain rates. Most of the stochastic models presented here apply tensile loading to the test domain while the majority of tests on brittle materials are conducted under uniaxial or biaxial compression. While ultimately crack growth is dominated by tensile component of stress, the role of shear and confinement in crack growth and eventual fragmentation at high strain rate loading are not yet fully established. Future efforts should focus on these challenging issues.

4.6 Summary and Conclusions

Modeling the dynamic response of ceramics continues to be a challenging task due to the complexities associated with several phenomena, such as crack nucleation, crack growth and interactions, microstructural dependence, stress state, level of confinement, rate of loading, and fragmentation characteristics. In this chapter, we have briefly attempted to provide a description of various modeling approaches and their utility in various applications. While models that describe fundamental dynamic behavior of brittle material focus on fracture strength dependence on strain rate or confinement pressure, models based on an expanding cavity are more suitable for understanding the penetration behavior of a projectile on a target. However, the emphasis in this chapter is on familiarizing a reader to various approaches available for modeling the fracture behavior of brittle materials subjected to high strain rate loads and point to some future directions for research.

References

Ashby, M. and Hallam, S. (1986). The failure of brittle solids containing small cracks under compressive stress states. *Acta Metallurgica* 34: 497–510.

Awasthi, A. and Subhash, G. (2019). High-pressure deformation and amorphization in boron carbide. *Journal of Applied Physics* 125: 215901.

Bavdekar, S. and Subhash, G. (2018). Comparison of pressure-sensitive strength models for ceramics under ultrahigh confinement. *International Journal of Impact Engineering* 118: 60–66.

Bavdekar, S., Parsard, G., Subhash, G., and Satapathy, S. (2017). An improved dynamic expanding cavity model for high-pressure and high-strain rate response of ceramics. *International Journal of Solids and Structures* 125: 77–88.

Bavdekar, S., Subhash, G., and Satapathy, S. (2019). A unified model for dwell and penetration during long rod impact on thick ceramic targets. *International Journal of Impact Engineering* 131: 304–316.

Bhattacharya, K., Ortiz, M., and Ravichandran, G. (1998). Energy-based model of compressive splitting in heterogeneous brittle solids. *Journal of the Mechanics and Physics of Solids* 46: 2171–2181.

Bishop, R., Hill, R., and Mott, N. (1945). The theory of indentation and hardness tests. *Proceedings of the Physical Society* 57: 147.

Bourne, N. (2008). The relation of failure under 1D shock to the ballistic performance of brittle materials. *International Journal of Impact Engineering* 35: 674–683.

Bourne, N. (2012). On the failure and dynamic performance of materials. *Experimental Mechanics* 52: 153–159.

Brace, W. and Bombolakis, E. (1963). A note on brittle crack growth in compression. *Journal of Geophysical Research* 68: 3709–3713.

Brar, N. and Bless, S. (1992). Failure waves in glass under dynamic compression. *International Journal of High Pressure Research* 10: 773–784.

Brar, N., Rosenberg, Z., and Bless, S. (1991). Spall strength and failure waves in glass. *Journal De Physique III* 1: 639–644.

Chen, W. and Luo, H. (2004). Dynamic compressive responses of intact and damaged ceramics from a single split Hopkinson pressure bar experiment. *Experimental Mechanics* 44: 295–299.

Chen, W. and Ravichandran, G. (1996a). An experimental technique for imposing dynamic multiaxial-compression with mechanical confinement. *Experimental Mechanics* 36: 155–158.

Chen, W. and Ravichandran, G. (1996b). Static and dynamic compressive behavior of aluminum nitride under moderate confinement. *Journal of the American Ceramic Society* 79: 579–584.

Chen, W. and Ravichandran, G. (2000). Failure mode transition in ceramics under dynamic multiaxial compression. *International Journal of Fracture* 101: 141–159.

Chen, W.W., Rajendran, A.M., Song, B., and Nie, X. (2007). Dynamic fracture of ceramics in armor applications. *Journal of the American Ceramic Society* 90: 1005–1018.

Chocron, S., Dannemann, K., Walker, J. et al. (2007). Constitutive model for damaged borosilicate glass under confinement. *Journal of the American Ceramic Society* 90: 2549–2555.

Chocron, S., Anderson, C., Dannemann, K., and Nicholls, A. (2013a). Pressure effects on the compressive response of confined intact and damaged soda-lime glass. *Experimental Mechanics* 53: 77–89.

Chocron, S., Anderson, C., and Nicholls, A. (2013b). Response of 3-D and plain-weave S-2 glass composites in out-of-plane tension and shear. *Experimental Mechanics* 53: 989–1000.

Coulomb, C. (1776). An attempt to apply the rules of maxima and minima to several problems of stability related to architecture. *Mémoires de l'Académie Royale des Sciences* 7: 343–382.

Deng, H. and Nemat-Nasser, S. (1992). Dynamic damage evolution in brittle solids. *Mechanics of Materials* 14: 83–103.

Drucker, D. and Prager, W. (1952). Soil mechanics and plastic analysis or limit design. *Quarterly of Applied Mathematics* 10: 157–165.
Fischer-Cripps, A. (1997). Predicting Hertzian fracture. *Journal of Materials Science* 32: 1277–1285.
Forrestal, M. and Longcope, D. (1990). Target strength of ceramic materials for high-velocity penetration. *Journal of Applied Physics* 67: 3669–3672.
Forrestal, M., Luk, V., and Watts, H. (1988). Penetration of reinforced concrete with ogive-nose penetrators. *International Journal of Solids and Structures* 24: 77–87.
Freund, L. (1972). Crack propagation in an elastic solid subjected to general loading—II. Non-uniform rate of extension. *Journal of the Mechanics and Physics of Solids* 20: 141–152.
Freund, L.B. (1998). *Dynamic Fracture Mechanics*. Cambridge University Press.
Gao, X. (2003). Strain gradient plasticity solution for an internally pressurized thick-walled spherical shell of an elastic-plastic material. *Mechanics Research Communications* 30: 411–420.
Gao, X., Wei, X., and Wang, Z. (1991). A general analytical solution of a strain-hardening elastoplastic plate containing a circular hole subjected to biaxial loading – with applications in pressure-vessels. *International Journal of Pressure Vessels and Piping* 47: 35–55.
Gao, X., Jing, X., and Subhash, G. (2006). Two new expanding cavity models for indentation deformations of elastic strain-hardening materials. *International Journal of Solids and Structures* 43: 2193–2208.
Ghosh, D., Subhash, G., and Orlovskaya, N. (2010). Slip-line spacing in ZrB_2-based ultrahigh-temperature ceramics. *Scripta Materialia* 62: 839–842.
Giannakopoulos, A. and Suresh, S. (1999). Theory of indentation of piezoelectric materials. *Acta Materialia* 47: 2153–2164.
Glenn, L., Gommerstadt, B., and Chudnovsky, A. (1986). A fracture-mechanics model of fragmentation. *Journal of Applied Physics* 60: 1224–1226.
Grady, D. (1982). Local inertial effects in dynamic fragmentation. *Journal of Applied Physics* 53: 322–325.
Grady, D. (1998). Shock-wave compression of brittle solids. *Mechanics of Materials* 29: 181–203.
Grady, D. and Kipp, M. (1985). Geometric statistics and dynamic fragmentation. *Journal of Applied Physics* 58: 1210–1222.
Grady, D. and Lipkin, J. (1980). Criteria for impulsive rock fracture. *Geophysical Research Letters* 7: 255–258.
Griffith, L. (1943). A theory of the size distribution of particles in a comminuted system. *Canadian Journal of Research* 21: 57–64.
Heard, H. and Cline, C. (1980). Mechanical behaviour of polycrystalline BeO, Al_2O_3 and AlN at high pressure. *Journal of Materials Science* 15: 1889–1897.
Hill, R. (1998). *The Mathematical Theory of Plasticity*. Oxford University Press.

Holmquist, T. and Johnson, G. (2002). Response of silicon carbide to high velocity impact. *Journal of Applied Physics* 91: 5858–5866.

Holmquist, T. and Johnson, G. (2008). Response of boron carbide subjected to high-velocity impact. *International Journal of Impact Engineering* 35: 742–752.

Holmquist, T.J., Johnson, G.R., Lopatin, C. et al. (1995). *High Strain Rate Properties and Constitutive Modeling of Glass*. Albuquerque, NM: Sandia National Labs.

Holmquist, T., Templeton, D., and Bishnoi, K. (2001). Constitutive modeling of aluminum nitride for large strain, high-strain rate, and high-pressure applications. *International Journal of Impact Engineering* 25: 211–231.

Horii, H. and Nemat-Nasser, S. (1986). Brittle failure in compression – splitting, faulting and brittle-ductile transition. *Philosophical Transactions of the Royal Society A-Mathematical Physical and Engineering Sciences* 319: 337–374.

Horii, H. and Nemat-Nasser, S. (1985). Compression-induced microcrack growth in brittle solids: axial splitting and shear failure. *Journal of Geophysical Research: Solid Earth* 90: 3105–3125.

Hu, G., Liu, J., Graham-Brady, L., and Ramesh, K. (2015). A 3D mechanistic model for brittle materials containing evolving flaw distributions under dynamic multiaxial loading. *Journal of the Mechanics and Physics of Solids* 78: 269–297.

Huang, C. and Subhash, G. (2003). Influence of lateral confinement on dynamic damage evolution during uniaxial compressive response of brittle solids. *Journal of the Mechanics and Physics of Solids* 51: 1089–1105.

Iyer, K. (2007). Relationships between multiaxial stress states and internal fracture patterns in sphere-impacted silicon carbide. *International Journal of Fracture* 146: 1–18.

Jiang, H. and Xie, Y. (2011). A note on the Mohr–Coulomb and Drucker–Prager strength criteria. *Mechanics Research Communications* 38: 309–314.

Johnson, K. (1970). Correlation of indentation experiments. *Journal of the Mechanics and Physics of Solids* 18: 115–126.

Johnson, K.L. (1985). *Contact Mechanics*. Cambridge Cambridgeshire, NY: Cambridge University Press.

Johnson, G. and Holmquist, T. (1992). A computational constitutive model for brittle materials subjected to large strains, high strain rates and high pressures. In: *Shock Wave and High-Strain-Rate Phenomena in Materials* (eds. M.A. Meyers, L.E. Muir and K.P. Staudhammer), 1075–1081. New York: Marcel Dekker.

Johnson, G.R. and Holmquist, T.J. (1994). An improved computational constitutive model for brittle materials. In: *AIP Conference Proceedings*, 981–984. American Institute of Physics.

Johnson, G. and Holmquist, T. (1999). Response of boron carbide subjected to large strains, high strain rates, and high pressures. *Journal of Applied Physics* 85: 8060–8073.

Johnson, G., Stryk, R., Beissel, S., and Holmquist, T. (2002). An algorithm to automatically convert distorted finite elements into meshless particles during dynamic deformation. *International Journal of Impact Engineering* 27: 997–1013.

Johnson, G., Holmquist, T., and Beissel, S. (2003). Response of aluminum nitride (including a phase change) to large strains, high strain rates, and high pressures. *Journal of Applied Physics* 94: 1639–1646.

Kanel, G.I., Razorenov, S.V., Utkin, A.V. et al. (1998). Influence of the load conditions on the failure wave in glasses. *High Pressure Research* 16: 27–44.

Kimberley, J., Ramesh, K., and Daphalapurkar, N. (2013). A scaling law for the dynamic strength of brittle solids. *Acta Materialia* 61: 3509–3521.

Kipp, M.E. and Grady, D.E. (1980). *Numerical Studies of Rock Fragmentation*, 721–735. Sandia National Laboratories.

Kipp, M. and Grady, D. (1994). Shock phase transformation and release properties of aluminum nitride. *Le Journal de Physique IV* 4: C8-249–C248-256.

Kobayashi, A., Emery, A., and Liaw, B. (1983). Dynamic fracture toughnesses of reaction-bonded silicon-nitride. *Journal of the American Ceramic Society* 66: 151–155.

Koeppel, B. and Subhash, G. (1997). An experimental technique to investigate the dynamic indentation hardness of materials. *Experimental Techniques* 21: 16–18.

Koeppel, B. and Subhash, G. (1999). Characteristics of residual plastic zone under static and dynamic Vickers indentations. *Wear* 224: 56–67.

Lankford, J. (1981). Mechanisms responsible for strain-rate-dependent compressive strength in ceramic materials. *Journal of the American Ceramic Society* 64: C33–C34.

Lankford, J., Predebon, W., Staehler, J. et al. (1998). The role of plasticity as a limiting factor in the compressive failure of high strength ceramics. *Mechanics of Materials* 29: 205–218.

Luo, H., Chen, W., and Rajendran, A. (2006). Dynamic compressive response of damaged and interlocked SiC–N ceramics. *Journal of the American Ceramic Society* 89: 266–273.

Maiti, S. and Geubelle, P. (2004). Mesoscale modeling of dynamic fracture of ceramic materials. *CMES-Computer Modeling in Engineering & Sciences* 5: 91–101.

Maiti, S., Rangaswamy, K., and Geubelle, P. (2005). Mesoscale analysis of dynamic fragmentation of ceramics under tension. *Acta Materialia* 53: 823–834.

Mata, M. and Alcala, J. (2003). Mechanical property evaluation through sharp indentations in elastoplastic and fully plastic contact regimes. *Journal of Materials Research* 18: 1705–1709.

Mata, M., Anglada, M., and Alcala, J. (2002). A hardness equation for sharp indentation of elastic-power-law strain-hardening materials. *Philosophical Magazine A-Physics of Condensed Matter Structure Defects and Mechanical Properties* 82: 1831–1839.

Mesarovic, S. and Fleck, N. (1999). Spherical indentation of elastic-plastic solids. *Proceedings of the Royal Society A-Mathematical Physical and Engineering Sciences* 455: 2707–2728.

Millett, J. and Bourne, N. (2000). The effect of an internal interface on the shear strength of a shock-loaded, high density glass. *Scripta Materialia* 43: 77–82.

Mohr, O. (1900). Welche Umstände bedingen die Elastizitätsgrenze und den Bruch eines Materials. *Zeitschrift des Vereins Deutscher Ingenieure* 46: 1572–1577.

Molinari, J., Gazonas, G., Raghupathy, R. et al. (2007). The cohesive element approach to dynamic fragmentation: the question of energy convergence. *International Journal For Numerical Methods in Engineering* 69: 484–503.

Mott, N. and Linfoot, E. (1943). A Theory of Fragmentation. Extra-Mural Research No. F72/80. *Ministry of Supply, AC 3348.*

Nabarro, F. (1997). Fifty-year study of the Peierls–Nabarro stress. *Materials Science and Engineering: A* 234: 67–76.

Nemat-Nasser, S. and Deng, H. (1994). Strain-rate effect on brittle failure in compression. *Acta Metallurgica Et Materialia* 42: 1013–1024.

Nemat-Nasser, S. and Horii, H. (1982). Compression-induced nonplanar crack extension with application to splitting, exfoliation, and rockburst. *Journal of Geophysical Research* 87: 6805–6821.

Normandia, M. (2004). Impact response and analysis of several silicon carbides. *International Journal of Applied Ceramic Technology* 1: 226–234.

Orphal, D.L. and Franzen, R.R. (1997). Penetration of confined silicon carbide targets by tungsten long rods at impact velocities from 1.5 to 4.6 km/s. *International Journal of Impact Engineering* 19: 1–13.

Orphal, D.L., Franzen, R.R., Piekutowski, A.J., and Forrestal, M.J. (1996). Penetration of confined aluminum nitride targets by tungsten long rods at 1.5–4.5 km/s. *International Journal of Impact Engineering* 18: 355–368.

Orphal, D.L., Franzen, R.R., Charters, A.C. et al. (1997). Penetration of confined boron carbide targets by tungsten long rods at impact velocities from 1.5 to 5.0 km/s. *International Journal of Impact Engineering* 19: 15–29.

Paliwal, B. and Ramesh, K. (2008). An interacting micro-crack damage model for failure of brittle materials under compression. *Journal of the Mechanics and Physics of Solids* 56: 896–923.

Paliwal, B., Ramesh, K., and McCauley, J. (2006). Direct observation of the dynamic compressive failure of a transparent polycrystalline ceramic (AlON). *Journal of the American Ceramic Society* 89: 2128–2133.

Park, Y. and Pharr, G. (2004). Nanoindentation with spherical indenters: finite element studies of deformation in the elastic-plastic transition regime. *Thin Solid Films* 447: 246–250.

Parsard, G. and Subhash, G. (2017). Raman spectroscopy mapping of amorphized zones beneath static and dynamic Vickers indentations on boron carbide. *Journal of the European Ceramic Society* 37: 1945–1953.

Parsard, G., Subhash, G., and Jannotti, P. (2018). Amorphization-induced volume change and residual stresses in boron carbide. *Journal of the American Ceramic Society* 101: 2606–2615.

Ramesh, K., Hogan, J.D., Kimberley, J., and Stickle, A. (2015). A review of mechanisms and models for dynamic failure, strength, and fragmentation. *Planetary and Space Science* 107: 10–23.

Rani, R.S., Prasad, K., and Krishna, T. (2014). Applicability of Mohr–Coulomb & Drucker–Prager models for assessment of undrained shear behaviour of clayey soils. *International Journal of Civil Engineering and Technology* 5: 104–123.

Ravichandran, G. and Subhash, G. (1995). A micromechanical model for high-strain rate behavior of ceramics. *International Journal of Solids and Structures* 32: 2627–2646.

Rosenberg, Z., Yaziv, D., Yeshurun, Y., and Bless, S. (1987). Shear-strength of shock-loaded alumina as determined with longitudinal and transverse manganin gauges. *Journal of Applied Physics* 62: 1120–1122.

Rosenberg, Z., Brar, N., and Bless, S. (1991). Dynamic high-pressure properties of aln ceramic as determined by flyer plate impact. *Journal of Applied Physics* 70: 167–169.

Sammis, C. and Ashby, M. (1986). The failure of brittle porous solids under compressive stress states. *Acta Metallurgica* 34: 511–526.

Samuels, L. and Mulhearn, T. (1957). An experimental investigation of the deformed zone associated with indentation hardness impressions. *Journal of the Mechanics and Physics of Solids* 5: 125–134.

Satapathy, S. (2001). Dynamic spherical cavity expansion in brittle ceramics. *International Journal of Solids and Structures* 38: 5833–5845.

Satapathy, S. and Bless, S. (1996). Calculation of penetration resistance of brittle materials using spherical cavity expansion analysis. *Mechanics of Materials* 23: 323–330.

Shafiq, M. and Subhash, G. (2014). A novel technique for the determination of surface biaxial stress under external confinement using Raman spectroscopy. *Experimental Mechanics* 54: 763–774.

Shafiq, M. and Subhash, G. (2016a). An extended Mohr–Coulomb model for fracture strength of intact brittle materials under ultrahigh pressures. *Journal of the American Ceramic Society* 99: 627–630.

Shafiq, M. and Subhash, G. (2016b). Dynamic deformation characteristics of zirconium diboride–silicon carbide under multi-axial confinement. *International Journal of Impact Engineering* 91: 158–169.

Staehler, J.M., Predebon, W.W., Pletka, B.J., and Subhash, G. (2000). Micromechanisms of deformation in high-purity hot-pressed alumina. *Materials Science and Engineering: A* 291: 37–45.

Sternberg, J. (1989). Material properties determining the resistance of ceramics to high velocity penetration. *Journal of Applied Physics* 65: 3417–3424.

Subhash, G. and Nemat-Nasser, S. (1993). Dynamic stress-induced transformation and texture formation in uniaxial compression of zirconia ceramics. *Journal of the American Ceramic Society* 76: 153–165.

Subhash, G., Koeppel, B., and Chandra, A. (1999). Dynamic indentation hardness and rate sensitivity in metals. *Journal of Engineering Materials and Technology-Transactions of the Asme* 121: 257–263.

Subhash, G., Maiti, S., Geubelle, P., and Ghosh, D. (2008). Recent advances in dynamic indentation fracture, impact damage and fragmentation of ceramics. *Journal of the American Ceramic Society* 91: 2777–2791.

Subhash, G., Awasthi, A., Kunka, C. et al. (2016). In search of amorphization-resistant boron carbide. *Scripta Materialia* 123: 158–162.

Subramanian, R. and Bless, S.J. (1995). Penetration of semi-infinite AD995 alumina targets by tungsten long rod penetrators from 1.5 to 3.5 km/s. *International Journal of Impact Engineering* 17: 807–816.

Tabor, D. (1948). A simple theory of static and dynamic hardness. *Proceedings of the Royal Society of London Series a-Mathematical and Physical Sciences* 192: 247–274.

Tate, A. (1967). A theory for the deceleration of long rods after impact. *Journal of the Mechanics and Physics of Solids* 15: 387–399.

Vogler, T., Reinhart, W., and Chhabildas, L. (2004). Dynamic behavior of boron carbide. *Journal of Applied Physics* 95: 4173–4183.

Weibull, W. (1939). A statistical theory of the strength of materials. *Proceedings of the Royal Swedish Institute of Engineering Research* 151: 1–45.

Weibull, W. (1951). A statistical distribution function of wide applicability. *Journal of Applied Mechanics-Transactions of the Asme* 18: 293–297.

Westerling, L., Lundberg, P., and Lundberg, B. (2001). Tungsten long-rod penetration into confined cylinders of boron carbide at and above ordnance velocities. *International Journal of Impact Engineering* 25: 703–714.

Wilkins, M., Cline, C., and Honodel, C. (1969). Fourth Progress Report of Light Armor Program. California Univ., Livermore. Lawrence Radiation Lab.

Yang, K. and Kobayashi, A. (1990). Dynamic fracture responses of alumina and 2 ceramic composites. *Journal of the American Ceramic Society* 73: 2309–2315.

Zavattieri, P. and Espinosa, H. (2001). Grain level analysis of crack initiation and propagation in brittle materials. *Acta Materialia* 49: 4291–4311.

Zhang, W. and Subhash, G. (2001). An elastic-plastic-cracking model for finite element analysis of indentation cracking in brittle materials. *International Journal of Solids and Structures* 38: 5893–5913.

Zhou, F. and Molinari, J. (2004a). Dynamic crack propagation with cohesive elements: a methodology to address mesh dependency. *International Journal For Numerical Methods in Engineering* 59: 1–24.

Zhou, F. and Molinari, J. (2004b). Stochastic fracture of ceramics under dynamic tensile loading. *International Journal of Solids and Structures* 41: 6573–6596.

Zhou, F. and Molinari, J. (2005). On the rate-dependency of dynamic tensile strength of a model ceramic system. *Computer Methods in Applied Mechanics and Engineering* 194: 1693–1709.

Zhou, F., Molinari, J., and Li, Y. (2004). Three-dimensional numerical simulations of dynamic fracture in silicon carbide reinforced aluminum. *Engineering Fracture Mechanics* 71: 1357–1378.

Zhou, F., Molinari, J., and Ramesh, K. (2005). A cohesive model based fragmentation analysis: effects of strain rate and initial defects distribution. *International Journal of Solids and Structures* 42: 5181–5207.

Zhou, F., Molinari, J., and Ramesh, K. (2006a). Analysis of the brittle fragmentation of an expanding ring. *Computational Materials Science* 37: 74–85.

Zhou, F., Molinari, J., and Ramesh, K. (2006b). Characteristic fragment size distributions in dynamic fragmentation. *Applied Physics Letters* 88: 261918.

Zhou, F., Molinari, J., and Ramesh, K. (2006c). Effects of material properties on the fragmentation of brittle materials. *International Journal of Fracture* 139: 169–196.

5

Shock Response of Brittle Solids

5.1 Introduction

Chapters 2 and 4 discussed the behavior of ceramics at quasi-static and intermediate strain rate regimes between 10^{-5} and $10^3\ s^{-1}$. In this chapter, we extend our discussion to ceramic behavior under extremely rapid strain rates, above $10^5\ s^{-1}$ (i.e. shock loading). As already discussed in Chapter 2, test methods in the low or intermediate strain-rate range such as Split-Hopkinson Pressure Bar (SHPB) use specimens that are longer in the direction of loading (e.g. x-axis) and experience uniaxial stress (i.e. $\sigma_x \neq 0$, $\sigma_y = \sigma_z = 0$). Under this loading, the maximum stress a ceramic specimen can withstand is limited by its failure strength. On the other hand, the specimen geometry in ultrahigh-strain-rate experiments (e.g. flyer plate impact) is plate-like, where the lateral dimension is significantly longer than the axial (loading) direction, and the specimen experiences uniaxial strain (i.e. $\varepsilon_x \neq 0$, $\varepsilon_y = \varepsilon_z = 0$) conditions. The lateral confinement allows the specimen to be subjected to any magnitude of stress. Here, we present the theory of shock wave propagation followed by the behavior of a wide variety of ceramics from both experimental and theoretical perspectives. Finally, a summary of molecular dynamic simulations of high pressure and shock response of ceramics that capture atomistic level deformation mechanisms and thermal effects are presented.

Various features and failure mechanisms that evolve in materials at different length and time scales are presented in Figure 5.1 (Ramesh et al. 2015). This chart was initially composed for a massive planetary impact, but it is also applicable to projectile impact and shock. Due to the large mass and high impact velocities of these celestial bodies, extremely high pressures, strain rates, and temperatures are developed. The "time" scale is the top axis of the figure, and the units are based on τ, the ratio of diameter of the impacting object and its impact velocity. It is not a true measurement of impact duration. Parameters of interest such as pressure, strain rate, and temperature as well as relevant deformation processes, failure mechanisms, and modeling approaches used for four different locations

Dynamic Response of Advanced Ceramics, First Edition.
Ghatu Subhash, Amnaya Awasthi, Dipankar Ghosh.

V ↓ a		~5τ	~10–100τ	Time →
Domain	**Source**	**Shock**	**Strength/flow**	**Structure/ interior**
Pressure, *Pa*	10^{11}	10^{9}	10^{7}	10^{6}–10^{7}
Strain rate, s^{-1}	10^{5}	10^{4}–10^{2}	10^{1}–10^{-2}	10^{-6}
Temperature, *K*	10^{4}–10^{3}	10^{2}	10^{2}	10^{2}–10^{1}
Critical processes	Strong shocks, jetting	Propagating shocks, excavation flow	Collapse mechanisms, post-shock strength	Interactions of cracks and waves with interal structure
Failure mechanisms	Melting, vaporization, phase transitions	Massive cracking massive shear banding massive void growth	Shear zones collapsed bands comminuted flow	Cracking
Kinds of models	Equation of state	Plasticity dynamic failure mechanics	Plasticity, flow of disordered media	Structural deformation and failure

Figure 5.1 Domains, mechanisms, variables, and their ranges during high velocity impact. Source: Ramesh et al. (2015). © 2015, Elsevier.

(domains) relative to the distance from the blast (shock or impact) site are listed in the first column. Given in the top horizontal row are various domains classified based on the severity of deformation or nature of the impact-influence felt by these regions. They are classified as the source (site of the impact), the shock region, followed by a zone where inelastic processes that influence structural strength occur, and finally, a domain where microstructural changes are possible which may not affect the structure level properties. The four domains occur in the above order with distance from the impact location. Closest to the source of impact, the material experiences strong shocks with several hundred GPa of pressure, deformation strain rates in excess of 10^5 s^{-1}, and temperatures several times above the melting point of the material. Away from the impact site, these numbers decrease gradually. To accommodate the massive input of energy, the material activates various energy dissipation mechanisms including melting and vaporization of material, phase transformation, plasticity, microcracking, and fragmentation. Of interest to the current chapter are the first two columns in the figure which deal with shocks where the material description is given by the equation of state (EOS) that relates to pressure–volume relationship and models for dynamic plasticity and fracture.

5.2 Principles of Shock Loading

In Chapter 2, the fundamentals of elastic stress wave propagation were discussed. In this chapter, we will present concepts of shock wave propagation, shock profiles, and various deformation mechanisms as applied to ceramics from experimental, theoretical, and computational perspectives. In-depth discussions on shock physics related issues in solids and fluids are available in a recent two-volume monograph by Dennis Grady titled *Physics of Shock and Impact* (Grady 2017a,b).

5.2.1 High-amplitude Compressive Waves: Plastic Waves

For one-dimensional longitudinal stress wave propagation, the elastic wave velocity in the specimen is given by $C_L = \sqrt{\frac{E}{\rho}}$, (Eq. (2.2)). At low impact velocities on linearly elastic materials, the amplitude of the applied stress limits the specimen response to the elastic regime, and the wave velocity remains independent of stress and strain. When the stress is applied rapidly, i.e. when the loading device moves at a rate comparable to the stress wave velocity in the material, lateral inertial effects come into play. This state of confinement can be achieved in laboratory experiments by designing the specimen with a large lateral dimension and small axial dimension. Now the wave velocity is given by Eq. (2.1). Upon application of the rapid load, the particles move at certain rate along the loading direction, but the material motion in the lateral direction is restrained due to the confinement offered by the surrounding material. The ability to resist applied stress can exceed the elastic limit of the material (under uniaxial stress) as the specimen is laterally confined. This stress state suppresses the early tensile cracking and allows initiation of plastic (inelastic) flow in the material. In the plastic regime, the wave velocity is a function of the slope of the stress–strain curve at that strain level as shown in Figure 5.2, meaning that the wave velocity varies depending on the magnitude of the imposed maximum plastic strain (and stress). Unlike the elastic wave velocity, the plastic wave velocity is not a constant value due to continuously varying strain level and hardening response at each material point.

To derive the plastic wave velocity, consider the propagation of a uniaxial stress wave of a sufficiently high amplitude (greater than the elastic limit) in a long rod. The axial stress in the rod can be written as a function of axial strain as:

$$\sigma_x = \sigma_x\left(\varepsilon_x\right) \tag{5.1}$$

Here $\varepsilon_x = \frac{\partial u}{\partial x}$, where u is the displacement in axial (along x-axis) direction. Differentiating Eq. (5.1), we get

$$\frac{\partial \sigma_x}{\partial x} = \frac{\partial \sigma_x}{\partial \varepsilon_x}\frac{\partial \varepsilon_x}{\partial x} = T\frac{\partial^2 u}{\partial x^2} \tag{5.2}$$

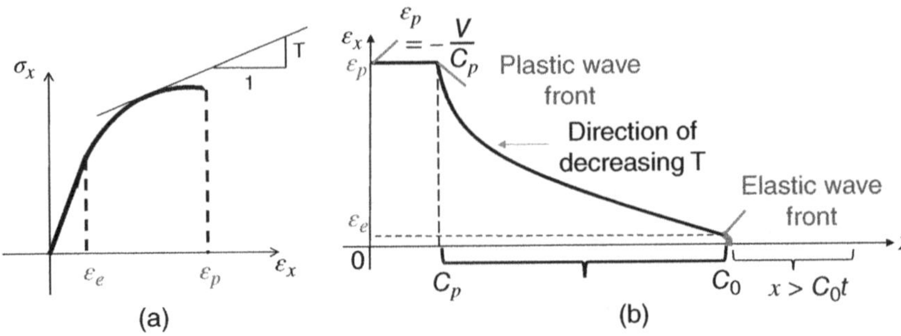

Figure 5.2 (a) Typical stress–strain curve showing the changes in slope with increasing strain. (b) Regions of a high-amplitude stress wave traveling down a long rod. The plastic wave velocity, which is related to the slope of the stress–strain curve, varies depending on the maximum strain level.

where $\frac{d\sigma_x}{d\varepsilon_x}$ is the slope, T, of the stress–strain curve (see Figure 5.2). Using Newton's second law of motion ($F = ma = \rho A dx \frac{\partial^2 u}{\partial t^2}$ and hence $\frac{\partial \sigma_x}{\partial x} = \rho \frac{\partial^2 u}{\partial t^2}$) and neglecting body forces, Eq. (5.2) can be written as $\rho \frac{\partial^2 u}{\partial t^2}$ (Achenbach 1973) and

$$\frac{T}{\rho}\frac{\partial^2 u}{\partial x^2} = \frac{\partial^2 u}{\partial t^2} \tag{5.3}$$

Equation (5.3) is the wave equation, where, the plastic wave velocity is given by:

$$C_P = \sqrt{\frac{T}{\rho}} \tag{5.4}$$

Note that C_P is a function of strain because T is the slope of the stress–strain curve in the plastic regime; see Figure 5.2. For small strains (i.e. in the elastic regime), $T = E$ and hence $C_P = C_L$, the elastic wave velocity from Eq. (2.2). Beyond the yield point, $T < E$ and $C_P < C_L$. Thus, the plastic wave velocity is lower than the elastic wave velocity. Therefore, when a rod is impacted at one end with a stress amplitude greater than the yield strength of the rod material, the material initially experiences elastic deformation followed by plastic deformation. Thus, both elastic and plastic waves are generated, each traveling at their characteristic wave velocities.

Assume that the impact generates a maximum strain, ε_P, in the rod. In general, the elastic strain is negligible for metals when compared to the plastic strain amplitude, and so the amplitude of the elastic strain pulse will be much smaller than the plastic pulse amplitude. Figure 5.2b illustrates the strain profile in such a rod with a propagating elastic–plastic wave. The x-axis is shown as x/t (velocity) and the ordinate is strain. Three regions are formed in the rod:

a. At a distance $x > C_0 t$, there is no strain ($\varepsilon_x = 0$) because the wave has not arrived at this point of the rod yet.
b. At distances $x < C_0 t$, the strain amplitude is greater than the elastic strain but less than the maximum plastic strain imposed due to impact. Thus, the strain gradually increases until a magnitude of ε_p is reached at a distance $C_p t$. Note that larger strains slow the wave propagation due to the reducing slope T of the material response. It can be shown that (Meyers 1994):

$$T = \rho \left(\frac{x^2}{t^2} \right) \tag{5.5}$$

c. For $0 < x < C_p t$, the strain (ε_p) is constant with a magnitude equal to that at the maximum stress, and is given by

$$\varepsilon_p = -\frac{V}{C_p} \tag{5.6}$$

where V is the impact velocity. The different strain regions along the length of the rod are shown in Figure 5.2b. Note that the plastic wave front travels at a lower velocity than the elastic wave front, falling behind the elastic front with time. As time progresses, the distance between the elastic and plastic fronts increases, and the wave profile spreads in the material.

5.2.2 Beyond Plastic Waves: Development of Shock Waves

In the previous example, the medium of wave propagation was a slender rod subjected to an applied load that produced stress in excess of its yield strength, resulting in plastic deformation progressing along the bar and the material flowing in a lateral (radial) direction (Poisson's effect). With time, the energy dissipates, and the amplitude of plastic wave decreases. However, upon application of high amplitude shock loading, the material point is compressed axially in one direction and undergoes significant volume compression. A high pressure (compressive) wave now travels through the material depending on the stress amplitude. This means that the plastic front will not spread as depicted in Figure 5.2b, but with increasing pressure, it will become steep and sharp, rising as much as 50–100 times the flow stress of the material depending on the impact velocity. Such a situation often occurs in ballistic impact or high-energy explosive loading where the material is subjected to high-amplitude shock loading.

At these high pressures, metals do not deform by the traditional dislocation-based (shear driven) plasticity but flow without any resistance to shear, similar to a Newtonian fluid. The stress state can thus be deemed hydrostatic, and the shear stress is assumed negligible. Shock loading can therefore be analyzed using

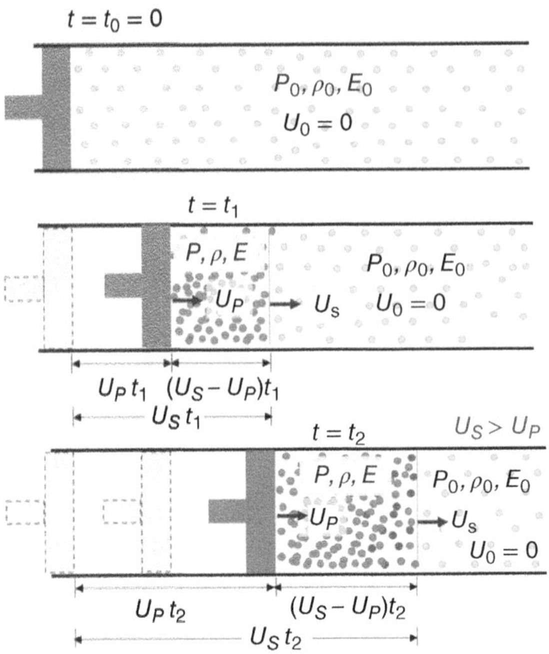

Figure 5.3 Illustration of a piston rapidly pushing on the ideal gas molecules to create a shock front. Source: Meyers (1994). © 1994, John Wiley & Sons.

a hydrodynamic treatment. The derivation of shock equations often begin with the ideal gas law. To illustrate the process, consider a piston rapidly compressing gas molecules inside a cylinder as illustrated in Figure 5.3. The ideal gas law for an isentropic process (adiabatic and reversible) is given by:

$$PV^{\gamma} = k \tag{5.7}$$

where P is pressure, V is volume, and γ and k are constants. For an ideal gas, γ is the ratio of specific heat at constant pressure to specific heat at constant volume. Differentiating Eq. (5.7), we get:

$$\frac{dP}{dV} = -\gamma \frac{P}{V} \tag{5.8}$$

Note that, mathematically, both sides of the equation increase with P, but the compressibility (dP/dV) of a material decreases with increasing P, which is captured by the negative sign. This relationship is explained by visualizing the behavior of gas molecules under increasing pressure, depicted in Figure 5.3. As the pressure increases, the molecules in a finite volume are pushed closer to each other, and it becomes increasingly difficult for the molecules to move any closer.

Thus, the compressibility decreases as the pressure increases. Therefore, the velocity of the disturbance (pressure wave) in the shock regime, also called the shock front velocity, is given by:

$$U = \sqrt{\frac{dP/dV}{\rho}} \tag{5.9}$$

At high stress amplitudes, the disturbance (shock front) travels faster (due to increased local density from molecules being compressed) than in the low amplitude region. As the shock amplitude increases, there will be a large discontinuity between the undeformed (stationary) molecules ahead of the shock front and the molecules under high pressure behind the shock front. Therefore, discontinuities in pressure, temperature, and density evolve in the medium at the shock front. Because the material particles get closer with pressure (increase in density), the velocity U of the shock wave will continue to increase with pressure. Therefore, the fundamental requirement for shock is given by $\left(\frac{\partial^2 P}{\partial U^2}\right) > 0$.

5.2.3 Equations of State (Shock Hugoniot Relationships)

At lower strain rates, the material response is described in terms of the stress–strain behavior. However, shock loading does not result in traditional plastic behavior. The material response is now expressed in terms of a pressure-specific volume relationship, referred to as the EOS or Hugoniot. Let us once again consider the cylinder with a rapidly moving piston, as shown in Figure 5.3, compressing a gas of initial volume, V_0, density, ρ_0, and pressure, P_0, at time $t_0 = 0$, and internal energy of E_0. As the piston moves with a velocity U_P, it compresses a finite volume of gas ahead of it. With time, the overall volume of compressed molecules increases. Assume that the front of the compressed volume moves with a velocity U_S imparting a particle velocity u_P in that volume. Applying the three conservation laws (mass, momentum, and energy), we obtain the following equations (Meyers 1994):

conservation of mass:

$$U_S \rho_0 = \rho \left(U_S - u_P\right) \tag{5.10}$$

conservation of momentum:

$$\rho \left(U_S - u_P\right) = P - P_0 \tag{5.11}$$

conservation of energy:

$$E_1 - E_0 = \frac{1}{2} u_P^2 \tag{5.12}$$

The last equation can be written in terms of pressure and volume as

$$E_1 - E_0 = \frac{1}{2}\left(P + P_0\right)\left(V_0 - V\right) \tag{5.13}$$

Equations (5.10), (5.11), and (5.13) are collectively called the Rankine–Hugoniot (R-H) equations. Note that there are five variables (P, u_P, U_S, ρ or V, and E) but only three equations. If we had an additional independent equation, we could determine all parameters as a function of the unknown parameter. This is achieved by measuring the shock velocity U_S and particle velocity u_P during an experiment and fitting a polynomial curve through the data to obtain an equation of the type

$$U_S = C_o + S_1 u_P + S_2 u_P^2 \tag{5.14}$$

where C_0 is the bulk sound wave velocity, and S_1 and S_2 are coefficients from the polynomial fitting process for the experimental data. In general, a linear instead of polynomial fit is assumed, especially when no phase transformations are present. Thus, Eq. (5.14) reduces to

$$U_S = C_o + S_1 u_P \tag{5.15}$$

From the R–H relations and experimental data, we can fit curves for various relationships such as: P versus U_S, P versus u_P, P versus ρ, P versus V/V_0, and E versus U_S. In general, the $(P-\rho)$ or $(P-V)$ relationship is also referred to as the R-H relationship or EOS. This line has a slope of $(\rho_0 U_S)^2 = \frac{P_1 - P_0}{V_1 - V_0}$. An example is shown in Figure 5.4 where the two straight lines are the Rayleigh lines at two different particle velocities. Each line describes the shock state at a maximum pressure of P_1, i.e. the path of the P–V data when a maximum pressure of P_1 (corresponding to impact velocity V_1) is imposed.

The Hugoniot is defined as the locus of all final shocked states (pressure–volume relationship) in a material for various maximum pressure values. As can be seen from the Hugoniot (curved) line, the slope of the curve increases with increasing pressures. It becomes increasingly difficult for molecules to come closer at higher pressures, and thus the velocity of shock wave increases with increasing

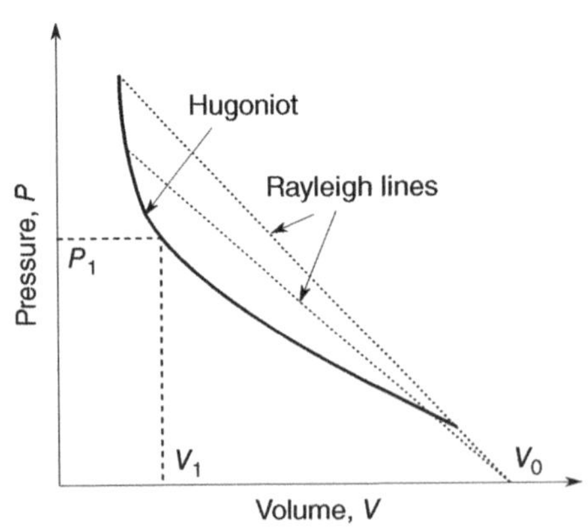

Figure 5.4 Pressure versus volume relationship during shock loading. The Hugoniot and Rayleigh lines are labeled. Each Rayleigh line corresponds to P–V relationship during one specific impact velocity. The Hugoniot is the locus of the end-states of P–V relationships at all these velocities.

pressure as the molecules come closer and the density of the medium increases. This increasing shock velocity with pressure is captured in Eq. (5.11) (the conservation of momentum), which by rearranging, we obtain

$$\frac{P - P_0}{u_P} = \rho_0 U_S \tag{5.16}$$

If the experiments are conducted at ambient conditions ($P_0 = 0$), the pressure induced by the shock is given by $P = \rho_o U_s u_p$. In addition, the maximum velocity of the shock front is given by the longitudinal elastic wave velocity C_o, and hence the stress (pressure) induced by the shock loading is given by $P = \rho_o C_o u_p$.

5.2.4 Plate Impact Experiments

To obtain EOS, we must experimentally measure shock and particle velocities, and this is accomplished through flyer plate impact experiments, illustrated in Figure 2.16. Consider two plates, a flyer plate (or projectile) impacting a stationary target plate at a velocity v, as shown in Figure 5.5. Upon impact, the particles along the impact interface experience shock stress in both the materials.

The equations corresponding to the pressure induced (Meyers 1994) in the projectile and target material, respectively, are:

$$P_1 = \rho_{0_1} U_{S_1} u_{P_1} \tag{5.17a}$$

$$P_2 = \rho_{0_2} U_{S_2} u_{P_2} \tag{5.17b}$$

The EOS for the two materials can be written as $U_{S_1} = C_1 + S_1 u_{P_1}$ and $U_{S_2} = C_2 + S_2 u_{P_2}$, respectively. If the target and projectile are made of the same material (or when both the impedances are the same), it is called symmetric impact, and half of the incident momentum is transferred to the target. Therefore, the maximum particle velocity in the target is half the impact velocity:

$$u_P = \frac{1}{2} V_1 \tag{5.18}$$

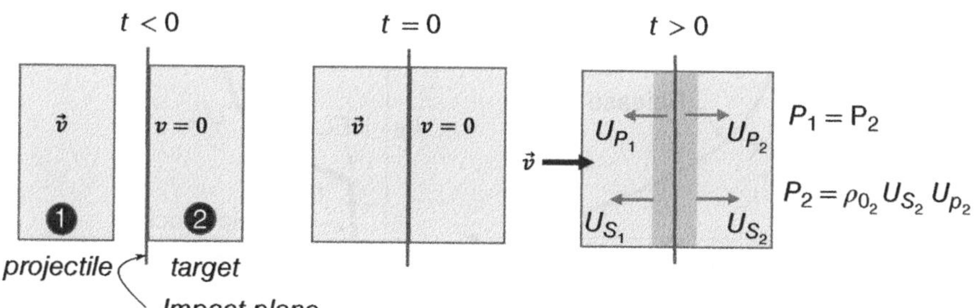

Figure 5.5 Plate impact experiment showing the particle and wave velocities. Source: Meyers (1994).

The pressure can be calculated from Eqs. (5.15, 5.17a, 5.17b) as

$$P = \rho_0 \left(C_2 + S_2 u_P\right) u_P \tag{5.19}$$

If the flyer plate and target are not made of the same materials, then for the projectile, the pressure is given from Eq. (5.19) as

$$P_1 = \rho_{0_1} C_1 u_{P_1} + \rho_{0_1} S_1 u_{P_1}^2 \tag{5.20}$$

And for the target

$$P_2 = \rho_{0_2} \left(C_2 + S_2 u_{P_2}\right) u_{P_2} \tag{5.21}$$

where u_{P_2} is the solution of the two equations.

5.2.5 Calculation of Shock Stress and Strain

A schematic of an idealized shock wave profile in terms of particle velocity versus time profile, as measured by an interferometer, is shown in Figure 5.6a. It consists of a steep shock front followed by a plateau and then release. Most materials remain elastic up to a certain pressure, and this limit is called the Hugoniot elastic limit (HEL), commonly referred to as the "dynamic yield strength" of a material. Unlike the idealized response in Figure 5.6a, real materials exhibit a transition between the elastic shock regime and final Hugoniot (the highest shock pressure applied) state. This transitional regime refers to the elastic–plastic deformation. In the elastic range, the stress rise is steeper, and this part of the wave is called the "elastic precursor." Table 5.1 provides a comparison of typical HEL values for some metals and ceramics (Meyers 1994).

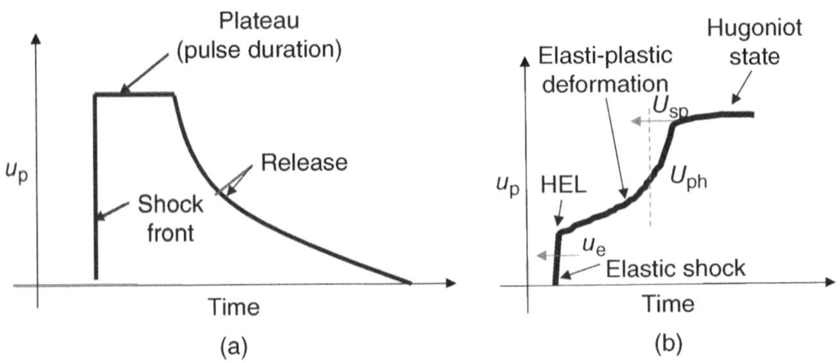

Figure 5.6 Anatomy of (a) an idealized shock wave profile. Source: Meyers (1994). (b) Realistic shock front profile. Source: Based on (Grady 1998).

Table 5.1 HEL values for some common materials.

Material	HEL (GPa)
Aluminum (Al)	0.6
Nickel (Ni)	1.0
Iron (Fe)	1–1.5
Titanium (Ti)	2
Aluminum oxide (Al_2O_3)	8–12
Sapphire	20
Silicon carbide (SiC)	16–18
Boron carbide (B_4C)	18–20

The midpoint of the shock front, labeled in Figure 5.6b, is often used to calculate the Hugoniot stress, σ_{ph}, and strain, ε_{ph}, as (Reinhart and Chhabildas 2003):

$$\sigma_{ph} = \sigma_e + \rho_o U_{sp} \left(u_{ph} - u_e\right) \tag{5.22}$$

$$\varepsilon_{ph} = \varepsilon_e + \left(u_{ph} - u_e\right) / U_{sp} \tag{5.23}$$

where u_{ph} is one-half of the impact velocity for symmetric impact, and σ_e and ε_e are the stress and strain at the elastic limit (HEL). u_e is the in-material particle velocity measurement prior to transitioning to a plastic wave, and U_{sp} is the shock wave velocity in the plastic front that is taken at the center of the plastic shock (Reinhart and Chhabildas 2003). In general, U_{sp} is significantly lower than the elastic shock velocity, u_e, but it is quite possible that U_{sp} can be equal to u_e at extremely high shock pressures. For example, the plastic shock front velocity is almost the same as that of the elastic shock front for an alumina ceramic at shock pressures above 100 GPa (Reinhart and Chhabildas 2003). The in-material particle velocity, u_e, is determined using impedance-matching relations at the interface between the ceramic and the LiF window and is given by:

$$u_e = u_w \left(Z_w + Z_m\right) / \left(2Z_m\right), \tag{5.24}$$

where u_w, Z_w and Z_m are the velocities measured in the window material and the shock impedances of the window and ceramic, respectively.

These conditions result in an almost hydrostatic loading in the center of the specimen. Assuming no resistance to shear, the P-V relationship will follow as shown in Figure 5.4. However, in real materials, non-hydrodynamic (shear) components dominate, especially for ceramics with a significant HEL. Thus, for real materials, as shown in Figure 5.7a, the Hugoniot curve is displaced from the R-H curve

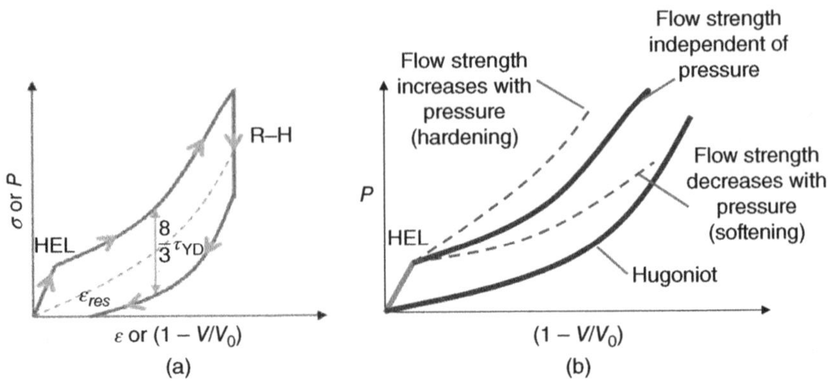

Figure 5.7 (a) Idealized pressure-specific volume relationship for real materials. The Hugoniot is displaced due to the shear resistance offered by real materials. (b) Change in Hugoniot response when the flow strength increases or decreases with pressure.

by $\frac{4}{3}\tau_{YD}$ where τ_{YD} is the dynamic shear yield strength. In this figure, volumetric strain is plotted on x-axis rather than volume as in Figure 5.4. The magnitude $\frac{4}{3}\tau_{YD}$is determined as follows. Uniaxial strain conditions ($\varepsilon_1 > 0$ and $\varepsilon_2 = \varepsilon_3 = 0$) prevail in plate-impact experiments, and the pressure and maximum shear stress at any material point are given by (assuming axisymmetric conditions, i.e. $\sigma_y = \sigma_z$):

$$P = tr\ (\sigma) = \frac{1}{3}\left(\sigma_x + 2\sigma_y\right) \tag{5.25}$$

$$\tau_{max} = \pm\frac{1}{2}\left(\sigma_x - \sigma_y\right) \tag{5.26}$$

where the stress in the lateral direction, σ_y, is given by $\sigma_y = \frac{v}{1-v}\sigma_x$.

From the above two equations, we get

$$P = \sigma_x \pm \frac{4}{3}\tau_{max} \tag{5.27}$$

or

$$\sigma_x = P \pm \frac{4}{3}\tau_{max} \tag{5.28}$$

The stress release path is different from the loading path because of hysteresis (irreversible) effects in real materials, and, hence, there is always residual strain (ε_{res}) in the material. Beyond the HEL the ideal pressure–volumetric strain relationship for a ceramic is parallel to the R-H curve (the dashed line in Figure 5.7a). This response implies that the flow strength of the material is independent of pressure. However, ceramics can also exhibit either hardening or softening behavior beyond HEL as shown in Figure 5.7b. As will be seen later, ceramics such as silicon carbide (SiC) and alumina show hardening behavior, whereas, boron carbide exhibits softening behavior. As the applied pressure increases, the elastic effects below the HEL can be deemed negligible (especially in metals whose HEL values are well below 2 GPs as seen in Table 5.1), and the three curves collapse into a single curve originating from the origin as in Figure 5.7b. The maximum stress attained is only limited by the gas gun's capacity to launch a flyer plate at a given velocity. Some ceramics have been subjected to Hugoniot states as high 100 GPa (Reinhart and Chhabildas 2003).

5.2.6 Propagation of the Shock Wave

To illustrate shock wave propagation through the material during a flyer plate impact experiment, we now plot distance (x) versus pressure amplitude (P), as shown schematically in Figure 5.8. Upon impact at time t_1, the generated ideal shock pulse consists of an elastic front traveling at C_o, the sound velocity under a stress-free condition. Initially, the front travels with a characteristic longitudinal wave velocity (C_o) of the material calculated from Eq. (2.1) at ambient conditions and is faster than the shock front velocity (U_S), which is a function of impact velocity. The response of the material is elastic until the stress amplitude reaches the HEL. This is followed by the shock front where the state of stress is hydrostatic ($\sigma_x = \sigma_y = \sigma_z = -P$), and the hydrodynamic theory applies in metals. When the pressure increases above the HEL, the shock front emerges and travels at a velocity U_s, which is a function of impact velocity. The shock front velocity increases with an increase in pressure (impact velocity) due to material densification as explained in relation to Figure 5.3. It is then followed by a plateau whose duration, ideally, is twice the travel time in the flyer plate.

The plateau is followed by a shock release in the target after the flyer plate is unloaded. The velocity of the release wave depends on the current pressure level

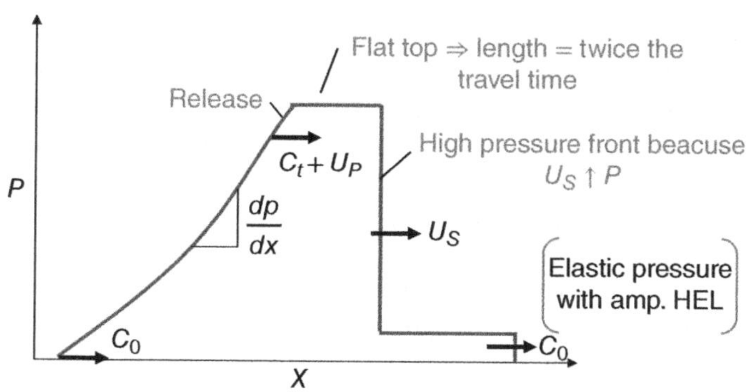

Figure 5.8 Features in a typical shock wave. Note the differences in wave velocities of the shock front and release wave. Source: Meyers (1994).

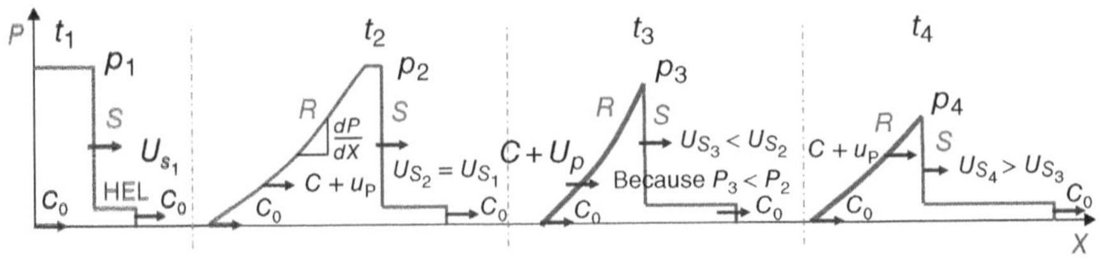

Figure 5.9 Illustration of shock wave profile dissipation with propagation distance. Source: Meyers (1994).

because the density of the material is higher at this state. Higher pressures produce faster release wave velocities. Because the release occurs from the peak pressure, the release wave travels at a velocity of $\left(C = \sqrt{\frac{dP/dV}{\rho}}\right)$, faster than the shock front, and it chases the shock front. Along similar lines, the tail of the release wave travels at a lower velocity than the release wave front due to lower pressure (and lower density of the material) at the tail-end. In addition, the material particles in the "shock-state" (i.e. at peak pressure) move with velocity u_P, and therefore, the release wave velocity is $(C + u_P)$ at the top of the release front as shown in Figure 5.8. As the time progresses, the slope of the release wave decreases due to the reducing pressure during release. This decreasing slope represented as $\left.\frac{dP}{dX}\right|_2 > \left.\frac{dP}{dX}\right|_3 > \left.\frac{dP}{dX}\right|_4$ in Figure 5.9, where the subscript denotes the snapshots at various times during the shock wave propagation. Recall that the stress release from any high-pressure state is elastic, and due to the release front (also called the rarefaction wave) being faster than the shock front, it overtakes the shock front over time. This process causes a reduction in the peak duration initially (e.g. $t = t_2$), and eventually reduces the peak pressure (e.g. at $t = t_3$, and $t = t_4$) as illustrated in Figure 5.9.

The above concepts can be better illustrated with an x-t diagram shown in Figure 5.10. In this figure, the slope of each line is given by the inverse of wave velocity, and so the greater the slope of a line, the slower the wave velocity and vice versa. Note from the figure that while the shock front travels at a velocity U_s, the release (rarefaction) wave travels faster and catches up with it, thus reducing the shock amplitude after a sufficiently long time. In metallic materials, the stress amplitude decay is often related to energy dissipation due to metallurgical changes (dislocation generation and their motion or defect generation and growth). However, in high shear strength materials such as ceramics, the shock wave amplitude continually decreases with time (propagation distance) because the high shear deformation causes energy dissipation under these non-hydrodynamic conditions. This dissipation causes heat generation, defect development, and possibly high-pressure phase transformations.

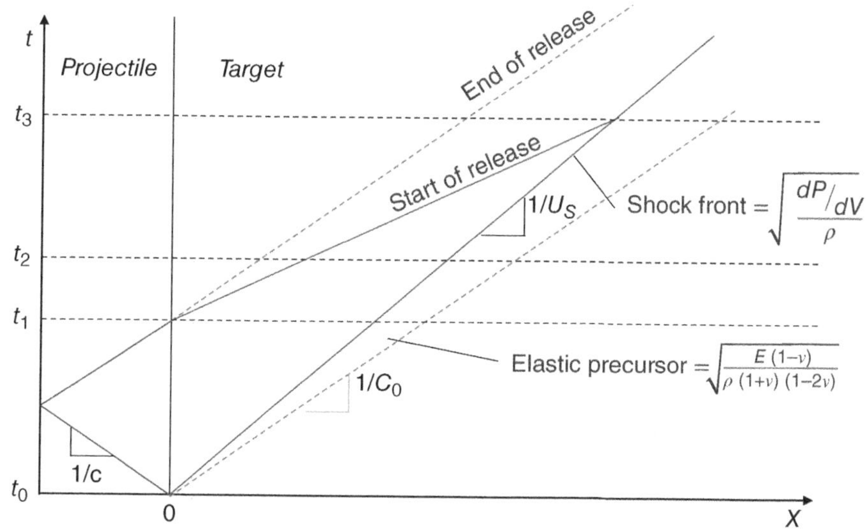

Figure 5.10 $X-t$ diagram illustrating the shock front and shock release states at various times corresponding to Figure 5.10. Source: Meyers (1994).

5.2.7 Shock Wave Profile in Structural Ceramics

The hydrodynamic treatment of shock waves is well suited for metals, which deform plastically and cannot sustain large shear stress at high pressures. However, the behavior of ceramics under shock loading is considerably different due to their limited number of slip systems. Their strong covalent bonding can support large shear stresses within the compressive shock pulse without macroscopic failure. The high confinement pressure prevents macroscopic cracking but may allow ductile deformation mechanisms (e.g. microplasticity, phase transformation, etc.) and defect generation before brittle fracture is eventually initiated.

The typical features of a shock profile for a ceramic are shown in Figure 5.11. This profile includes all possible deformation features during the shock loading in a high strength structural ceramic, but not all mechanisms may activate in all ceramics. The elastic shock (i.e. stress until the HEL) is followed by a plastic ramp, which is then followed by a plastic shock front before reaching the high-pressure state (Hugoniot). The plastic ramp is an indication of activation of yielding in various grains in a polycrystalline material. The heterogeneous yielding at various stress levels of differently oriented grains results in a gradual stress rise until all the grains are plastically deformed (Grady 1998). Then the plastic shock ensues. If phase transformation does occur in a material, it will be reflected as another steep rise (referred to as the phase transformation wave) before the final Hugoniot

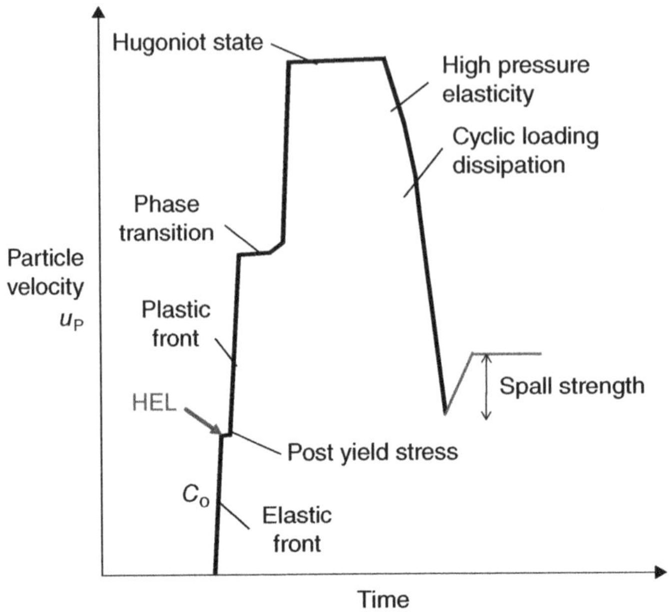

Figure 5.11 Anatomy of a shock profile in high strength structural ceramics. Source: Meyers (1994).

state is achieved. Upon release, the material unloads elastically from high pressure. Further metallurgical changes (e.g. tensile cracking or reverse phase transformation or plastic recovery) are possible during this elastic unloading, which dissipate energy. Finally, the tensile stress developed could cause spalling of the material. Spalling causes an elastic pullback signal whose magnitude can be used to calculate the spall strength of the material (to be discussed later).

5.3 Experimental Results of Shock Testing of Ceramics

5.3.1 Typical Shock Response of A Ceramic

Having discussed the basic theory of shock wave propagation in solids, we now focus on the experimental details. A schematic of the plate impact experiments to generate uniaxial strain shock waves and relevant principle of operation were earlier discussed in Chapter 2. Figure 5.12a shows the measured particle velocity–time profiles at various impact velocities for AD-995 alumina ceramic (Grady 1998). At low-impact velocities (0.56 km/s), the induced pressure is just sufficient to exceed the HEL of the material. The shock precursor (elastic portion)

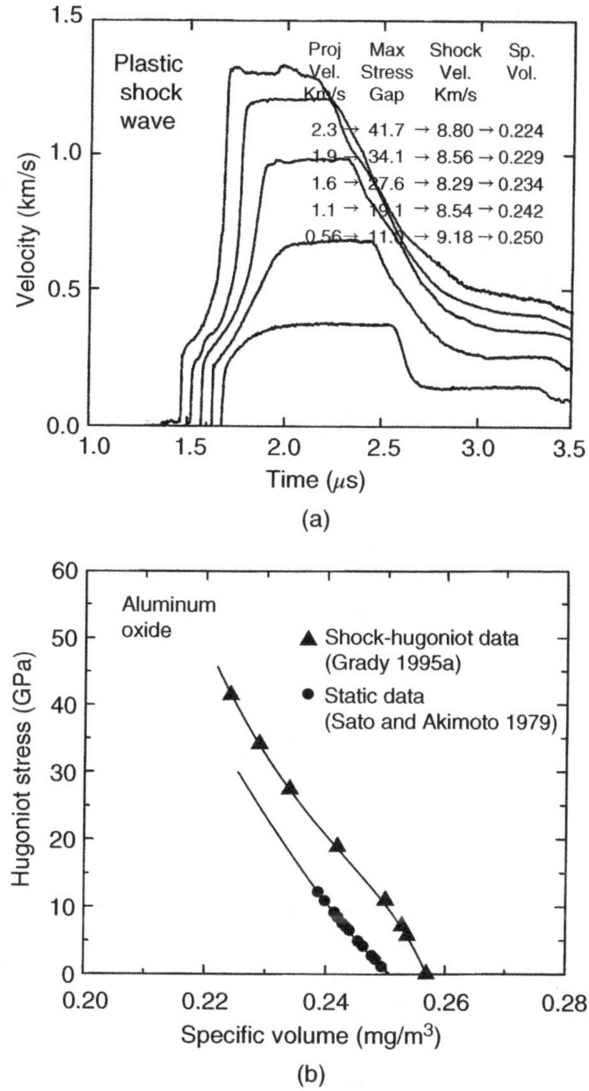

Figure 5.12 (a) Shock profiles of alumina at various velocities and (b) the relationship between the Hugoniot stress and the specific volume. The difference between the static and shock data indicates the shear resistance offered by the ceramic under shock. Source: Grady (1998). © 1998, Elsevier.

has a velocity of 10.7 km/s, the elastic wave speed at ambient conditions in the material. With increasing flyer plate velocities, the amplitude of shock, particle velocity, and slope of the plastic front increase. At higher pressures, the plastic shock wave becomes steeper and follows almost right behind the elastic front. Due to a greater compressed state at higher velocities, the release wave travels faster than the plastic shock front, reducing the duration of the Hugoniot (the high-pressure) state, as was discussed earlier in Figure 5.10. This behavior is also reflected in the decreasing slope of the release wave at higher projectile velocities. The average HEL where the onset of shear failure occurs is determined as 6.2 ± 0.4 GPa. From the above profiles and the theory presented earlier, the Hugoniot (peak stress versus specific volume relationship) can be determined for the alumina as shown in Figure 5.12b. The static data (called hydrostat) under hydrostatic conditions at high pressures from Sato and Akimoto (Sato and Akimoto 1979) is also shown for comparison. The offset between static and shock data reflects that the imposed shock compression produces a sustained non-hydrostatic (shear) stress state in the material. Also, with an increase in pressure, an upward convex shape is evident in the plot.

5.3.2 Post-HEL and Pre-Hugoniot Behavior

During shock compression, a range of deformation mechanisms are expected to arise in ceramics. Beyond the HEL, most ceramics exhibit a strain-hardening behavior, similar to the one shown for alumina in Figure 5.12a and for SiC in Figure 5.13a. However, ceramics such as boron carbide (B_4C) exhibit a strain-softening response, where a loss of strength is seen beyond the HEL, as illustrated in Figure 5.13b (Grady 1998; Reinhart and Chhabildas 2003). This softening behavior in boron carbide has been attributed to the collapse of the crystal structure, referred to as amorphization, when the stress exceeds HEL. For metals, the HEL is low, and any post-HEL softening/hardening effects are not detectable. Thus, the ideal Hugoniot curve (starting from the origin), shown in Figure 5.7b, is a good approximation. But, ceramics have an order of magnitude greater HEL with a strong post-HEL response. This response is then followed by the Hugoniot material response.

The data on shock wave velocity and normalized particle velocity for several structural ceramics is presented in Figure 5.14 (Grady 1998). The particle velocity is normalized by the quantity ($C_L - C_B$), where C_L is the longitudinal wave velocity and C_B is the ultrasonic bulk velocity. Assuming a first-order EOS relationship between the shock velocity and particle velocity given in Eq. (5.15) and realizing that $C = C_B$ for ceramics, we note from the data that ceramics such as TiB_2 and ZrO_2 have slope S values greater than 1 and shock velocities significantly higher than C_B. Such high values are interpreted as having propensity for yielding or a material phase transition (Grady 1998). On the other hand, ceramics such as B_4C and silicon nitride (Si_3N_4) have lower C values than C_B. This behavior is interpreted as loss of strength or lattice collapse at high shock pressures. Indeed, as shown in Figure 5.13b, B_4C does show a loss of strength above HEL, and it is now well accepted that B_4C exhibits lattice collapse at high pressures, a phenomenon commonly referred to as amorphization (Chen et al. 2003; Ghosh et al. 2012).

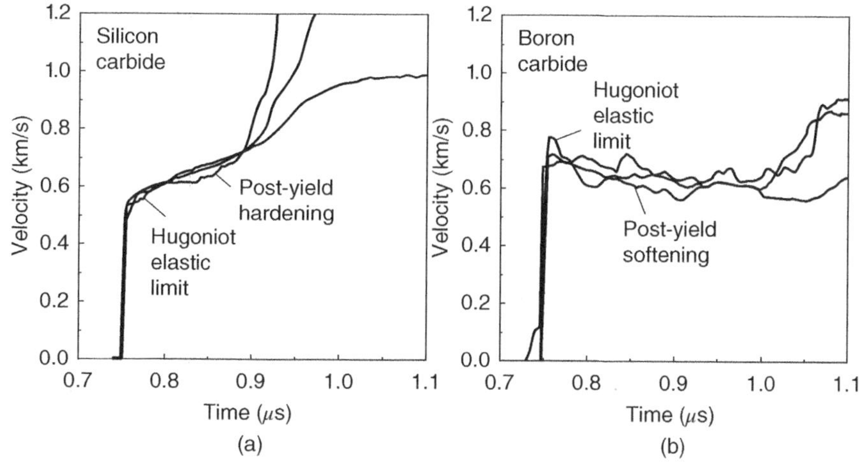

Figure 5.13 Post-HEL response of (a) SiC revealing strain-hardening and (b) boron carbide revealing strain-softening behavior. Source: Grady (1994). © 1994, EDP Sciences.

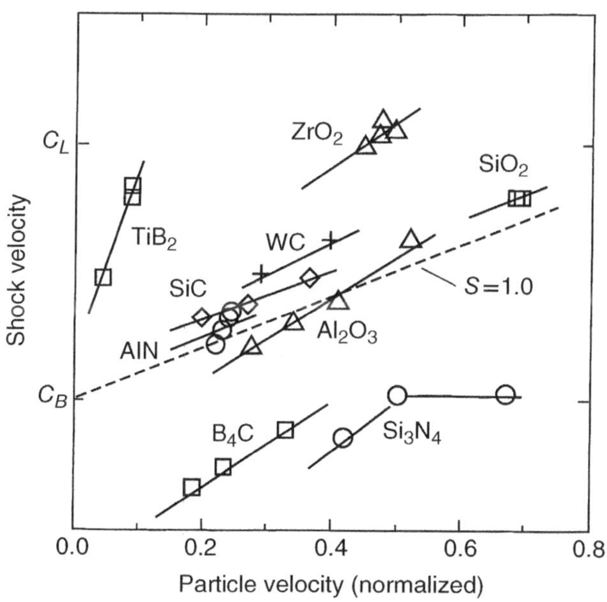

Figure 5.14 Shock wave velocity versus particle velocity of various structural ceramics. Those with slope above $S = 1$ have phase transitions and those below $S = 1$ exhibit lattice collapse. Source: Grady (1998). © 1998, Elsevier.

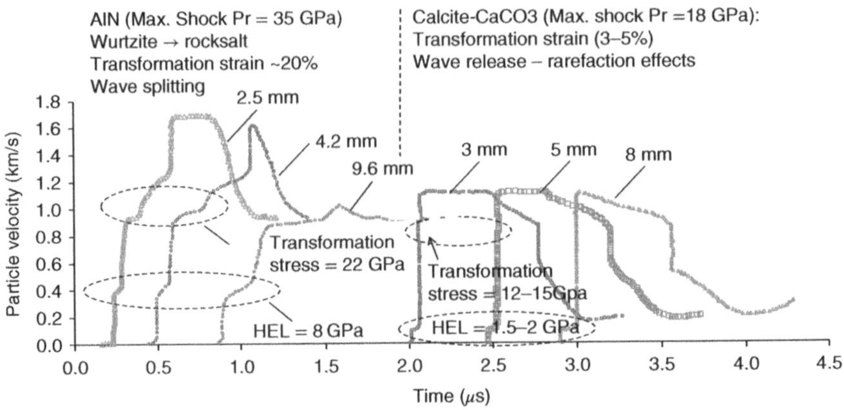

Figure 5.15 VISAR measurements of AlN (left) and calcite (right). Source: Grady (1998). © 1998, Elsevier.

A more in-depth discussion about the amorphization behavior in B_4C is presented in Chapter 6.

Figure 5.15 illustrates two examples in which the volume change due to phase transformation during shock loading is shown (Grady 1998). It also illustrates the wave propagation characteristics through specimens of various thicknesses. Aluminum nitride (AlN) undergoes a phase transformation from wurtzite to the rocksalt crystal structure at high pressures with a change in volumetric strain as large as 20%. Therefore, as the shock wave propagates through a thicker AlN plate, it not only spreads in the time domain (referred to as wave splitting in some articles) but also attenuates in amplitude due to energy expended in the phase transformation. In these experimental profiles, the phase transformation is reflected as a second discontinuity (HEL being the first one) in the wave profile. The HEL is about 8 GPa, and the phase transformation pressure occurs between 19 and 22 GPa. The maximum shock pressure reached in all three experiments is about 35 GPa. However, as the shock wave traverses through a thicker plate, the wave splitting indicates a slowing down of the transformation wave and hence the rarefaction wave catches up with this wave to decrease the maximum shock stress. On the other hand, crystalline calcite ($CaCO_3$) has a significantly lower HEL of only 1.5–2 GPa. The low-pressure metastable calcite transition also occurs around this pressure, and another transition occurs at a high pressure of about 12–15 GPa. However, unlike AlN, the volumetric strain for these transitions is fairly small (3–5%), and hence, the particle velocity profile displays no wave splitting as it propagates through the thicker sample. Regardless, the release wave catches up with the wave front in the thicker sample although it has not yet decreased the peak stress. A steep drop in the wave release is also indicative of the reverse phase transformation in calcite whereas the gradual release in AlN indicates the phase transformation is nonreversible.

Shock loading in some ceramics leads to a high-pressure phase transformation from a low-density crystal structure to a high-density crystal structure, such as in AlN and Calcite. As seen in Figure 5.15, the shock signature during the phase transformation appears as a strain-hardening response reflecting the energy required for continued phase transformation and the concomitant volume change (decrease). In all these cases, "phase transformation" refers to a change in one deterministic crystal structure to another deterministic structure. For SiC, the low pressure zinc blend structure transforms to a high pressure rock-salt structure with a concomitant 9% decrease in volume at a shock pressure of 105 GPa (Branicio et al. 2018a,b). The lower density zinc blend phase has a Si–C–Si angle of 109°, which reduces to 90° in the high-density rock-salt structure at the higher pressure. On the other hand, the post-yield softening response in boron carbide evident in Figure 5.13b is reflective of amorphization, a mechanism that refers to change in crystal structure from a deterministic structure to a nondeterministic amorphous structure. The volume change is difficult to determine due to a lack of ordered structure. In accordance with the conventional wisdom of high-pressure transformations, it is commonly assumed that the associated volume change in B_4C is also negative, where the volume decreases or the density increases. This view, however, is not consistent with the observed post yield softening of B_4C compared to the post-yield hardening response of SiC. This issue will be discussed in-depth in Chapter 6.

5.3.3 Damage Induced During Shock Loading

How does the damage evolve during shock loading? With an increase in the shock pressure in one direction (σ_x), the shear stress ($(\sigma_x - \sigma_y)/2$) continues to rise in the material, which will eventually exceed the critical strength of the material at preexisting defects. Cracking and dislocation motion may initiate early at some preferentially oriented defects to relieve the stress rise, and this behavior is reflected as the failure ramp beyond the HEL, marked "elastic–plastic deformation" in Figure 5.6b. This process is not catastrophic because only a few defects are activated, and these defects are under a high level of lateral confinement from the surrounding material in addition to the applied compressive shock loading. Only fracture nucleation occurs in the immediate post-HEL shock regime, which is indicated as failure ramp in Figure 5.16 (Grady 1998). With an increase in pressure, cracks also initiate at other defects, and the previously initiated cracks start to grow under combined shear and compressive loading. This phase is referred to as deformation shock. In the final stages, these cracks have grown to sufficiently large lengths and coalesce to form fragments.

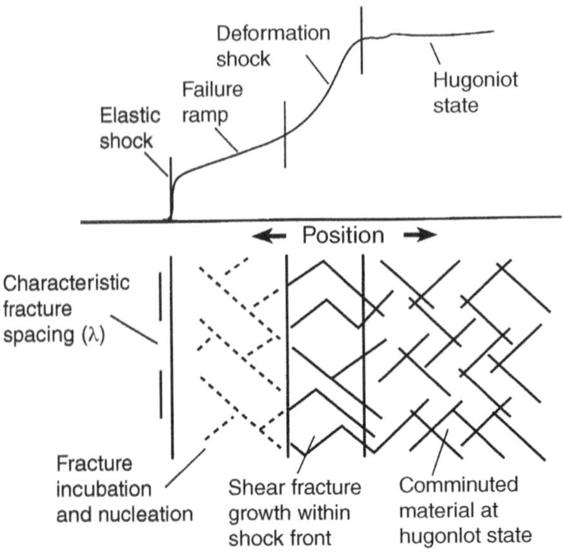

Figure 5.16 Illustration of damage evolution during various phases of the shock loading. Source: Grady (1998). © 1998, Elsevier.

The higher the strain rate, the more cracks nucleate and grow, producing a smaller fragment size. Grady (1998) estimated the upper bound for characteristic debris size (l) from Eq. (4.56). Using appropriate values, the equation yields $l = 40$–160 nm. Thus, fragments of nanometer size evolve during the failure ramp. Due to the high rate of loading, there is also a significant incubation time (inertia) for the cracks to grow. With time, these fractures grow and the characteristic spacing (λ) of fracture in the process zone has been estimated by Grady (1998) as:

$$\lambda \sim \left(\frac{48\gamma}{p\dot{\varepsilon}^2}\right)^{1/3} \tag{5.29}$$

and the strain rate is calculated from

$$\dot{\varepsilon} \sim \frac{3\sigma}{2\eta} \tag{5.30}$$

where, η is viscosity coefficient and σ is the jump in the stress amplitude. For an alumina ceramic subjected to a shock state of 19 GPa and strain rate of $1.7 \times 10^5\ \text{s}^{-1}$, the fracture spacing is 130 μm. A higher shock pressure of 42 GPa and a strain rate of $10^7\ \text{s}^{-1}$, the fracture spacing reduces to less than 8 μm. Therefore, at higher strain rates, comminution is expected with the debris sizes in the nanometer range.

5.3.4 Unloading and Spall

A release wave develops upon unloading of shock wave from the high pressure. This unloading process or release of compressive stress is akin to tensile loading,

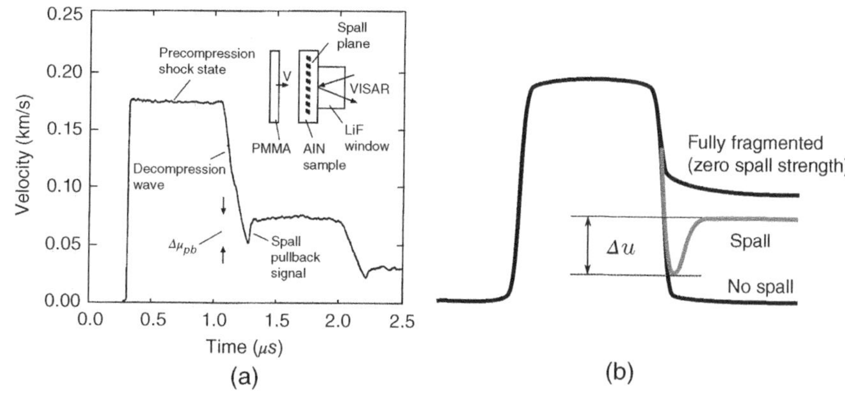

Figure 5.17 (a) Typical spall signature in AlN. Source: Grady (1998). © 1998, Elsevier. (b) Three possible scenarios in release wave characteristics to detect whether spall occurred or not.

and many dissipative mechanisms such as microcracking and even fracture are activated during this phase. The tensile fracture during this unloading phase is called "spalling," and the stress at which this fracture occurs is called the "spall strength," which can be defined as the residual cohesion of the material under dynamic tensile load. This stress is also referred to as the dynamic tensile hydrostatic strength of a material. Thus, shock compression experiments can also be effectively used to determine the spall strength of a ceramic. If the shock amplitude is low, the ceramic may remain undamaged during the loading phase of the shock, and the ensuing release wave causes fracture of the ceramic. The stress at which the spall occurs can be inferred from the shock profile; an example is presented in Figure 5.17a. There are three possible scenarios of material response as depicted in Figure 5.17b. At high particle velocities (high shock pressures) the release is immediately followed by a flat wave profile. This means that the material is fully pulverized during the loading phase, and the material cannot sustain a release wave. Therefore, the spall strength in this case is zero (thick black curve). At low shock pressures, the material may withstand the shock and is fully unloaded from the Hugoniot state. In this situation, the release wave profile consists of a flat response at a low amplitude over a short duration and may be followed by an increase at a later time. Since the material has fully unloaded (thin gray curve), the material has not yet spalled, and no spall information can be gathered. Finally, the release signal can suddenly jump up and then flatten (thick gray curve). The jump in particle velocity occurs because during the release, the resistance for material separation requires additional force, and this is reflected as a "spall pull-back" signal as shown.

The magnitude of the pull back is then used to calculate the spall strength of the ceramic as:

$$\sigma_{sp} = \frac{1}{2}\left(Z_s + Z_w\right)\Delta u_{pb} \tag{5.31}$$

where Z_s and Z_w are the impedances of the specimen and the LiF window, respectively, and Δu_{pb} is the amplitude of spall pull-back signal (jump) as shown in Figure 5.17. Grady (1998) also proposed another equation for the spall strength at a specific strain rate as

$$\sigma_{sp} = \left(6\gamma \rho^2 c^3 \dot{\varepsilon}\right)^{1/3} \tag{5.32}$$

where γ is the surface energy during the crack extension. Clearly, the spall strength is a function of strain rate. At high strain rates (and sufficiently high shock loads), the spall strength may be zero. Table 5.2 summarizes various properties of common ceramics derived from shock experiments (Grady, 1998). The measured γ values (γ_{meas}) were calculated from the spall strength and other properties in Eq. (5.32), whereas the theoretical γ values (γ_{th}) were estimated from the intermolecular cohesion length, b_c, as per Lawn and Wilshaw (1993), and is given by (Table 5.2)

$$\gamma_{th} = Eb_c/\pi^2 \tag{5.33}$$

Note from the table that for most materials, the spall strength is less than 1/10*th* of the HEL value. However, the spall strength of a ceramic is a strong function of microstructure. When the microstructure contains a high volume-fraction of defects (such as the open structure of glass or extensive microporosity), significant

Table 5.2 Elastic properties and strength values of certain ceramics.

Material	ρ (kg/m^3)	E (GPa)	υ	b_c (nm)	σ_{th} (GPa)	σ_{hel} (GPa)	σ_{sp} (GPa)	γ_{th} (J/m^2)	γ_{meas} (J/m^2)
AlN	3250	321	0.24	0.19	44.9	7–9	0.5–0.6	1.2/6.2	28–48
Al_2O_3	3970	401	0.24	0.19	59.5	11–12	0.5–0.8	2.1/7.7	15–61
B_4C	2510	462	0.17	0.16	64.5	17–19	0.4–0.5	1.4/7.5	23–40
SiC	3120	434	0.16	0.19	55.9	15–16	0.3–0.4	1.4/8.4	7–17
SiO_2	2630	90	0.08	0.16	24.0	7–9	0.8–0.11	1.0/1.5	5–14
TiB_2	4510	523	0.05	0.24	42.6	14–17	0.4–0.5	0.8/12.7	20–34
WC	14 930	627	0.21	0.25	52.2	4–6	2.5–3.0	1.0/15.9	490–845
ZrO_2	6030	218	0.31	0.20	69.4	13–16	1.5–1.9	4.4/4.4	640–1300

Source: Grady (1998). © 1998, Elsevier.

damage accumulation can occur even when the shock amplitude is below the elastic limit. In some materials, no significant damage occurs when shocked beneath the elastic limit, but the material can still spall due to its low tensile strength. For example, Coors AD-86 alumina has an HEL of 6 GPa but exhibits decreasing spall strength from 300 MPa to zero when the shock pressure is increased from 4 GPa to the HEL value (Rosenberg et al. 1985, 1987, 1996; Rosenberg and Yeshurun 1985, 1986). As the shock pressure is increased close to the HEL value, damage develops behind the compression wave, and upon the tensile wave arrival, the spall strength of the ceramic can be low depending on the state of damage in the material at this time. Similar testing revealed that Lucalox alumina (Munson and Lawrence 1979) had negligible spall strength, which implies that the material was fully damaged under shock loads at these amplitudes, and the tensile wave may have arrived after the damage state was fully developed in the material. Cagnoux and Longy (1988) also found that alumina has a negligible spall strength even at loads below HEL. However, they noticed that the spall strength increased at higher Hugoniot stress values. This result was attributed to microplasticity at defects such as grain boundaries and glassy second phases along the grain boundaries. In a pure alumina material, significant microplasticity has been observed in the grains under dynamic loads (Cagnoux and Longy 1988; Lankford et al. 1998; Staehler et al. 2000). Alumina with a smaller grain size reveals a higher HEL value than a large grain ceramic, and higher porosity leads to lower HEL (Cagnoux and Longy 1988). More discussions on spall strength as a function of strain rate and crystallographic orientation will be provided in the section on molecular dynamic simulations of shock loading on various ceramics.

5.4 Rationalizing Ultrahigh Strain-rate-dependent Behavior of Ceramics

Although the idealized shock response is illustrated with a sharp discontinuity at the HEL, the P–V relationship shows a smooth increase. This relationship can be converted to a stress versus strain-rate plot. Due to the vast range in the strain rate and stress values, it is common practice to plot them on a log–log scale, and, in general, this relationship is linear. For the majority of materials, the stress rise (or strain rise) rate is proportional to the peak stress to the fourth power (Grady 1998). Recall that the stress imposed on the material can reach to any level depending on the impact velocity. This is different than the HEL strength values plotted earlier in Chapter 4 which remain constant beyond a strain rate. The question then is why does the material continue to sustain a level of high stress at these high strain rates? The simple answer is viscous effects, discussed in the next section.

5.4.1 Viscous Effects

The inertia-dominated material response in the strain rate range of $10^2–10^4\ s^{-1}$, discussed in Chapter 4, is responsible for the increased strength in materials and is a nonviscous effect. At strain rates well beyond the above range, the stress rise is attributed to viscosity in the form of dislocation drag effects. Viscosity is a material constant that relates the viscous stress component to the velocity gradient (strain rate). In the steady-state shock process, the viscosity is viewed as the diffusion of momentum along the axes of wave propagation (Grady 1998). For a steady shock wave, the viscosity coefficient η is given by:

$$\eta \cong \frac{1}{4} S\sigma\,(\Delta t) \tag{5.34}$$

where σ is the jump in the stress amplitude, Δt is the rise time of the wave, and $S = 1$. The viscosity coefficient is related to material parameters (elastic and viscous components) that are responsible both for the upward concavity of the Hugoniot curve seen in Figure 5.4 as well as the geometry of the Rayleigh line. More specifically, the viscous coefficient is related to the pressure derivative of the bulk modulus, S_1 in Eq. (5.15), which relates the shock velocity, U_s, and particle velocity, u_p. All these parameters provide a dimensionless measure of the concavity of the Hugoniot line, which is essentially a quantification of viscosity. Clearly, one can notice from Eq. (5.34) that the stress is directly proportional to viscosity and inversely proportional to the rise time (Δt) of the wave. Therefore, with an increase in viscosity and decrease in time, the stress continues to rise. Microstructurally, the increase in stress with strain rate is related to the momentum diffusion through acoustic scatter and acoustic emission. The wave energy is scattered as the shock wave interacts with microstructural features such as defects, dislocations, grain boundaries, pores, and vacancies. Acoustic emission occurs due to dynamic fracture, unstable crack propagation and adiabatic slip. Thus, plasticity, slip, and microcracking are all possible under shock loading and contribute to increasing stress with strain rate.

5.4.2 HEL and Hardness

It is informative to compare the dynamic yield strength (HEL) with the static yield strength or hardness. Recall that HEL represents the yield strength under dynamic uniaxial strain conditions, while the static yield strength (or fracture strength in case of ceramics) values are obtained under uniaxial stress conditions. For metals, the Tabor's rule connects static hardness (H) to static yield strength (Y) through the relationship: $H = 3Y$ (Tabor 1948). This relationship was also found to be valid in the moderate strain rate regime of $10^3\ s^{-1}$ for metals (Koeppel and Subhash 1999; Subhash 2000). Thus, the dynamic yield strength could be approximated from dynamic hardness. However, for ceramics such a relationship was not

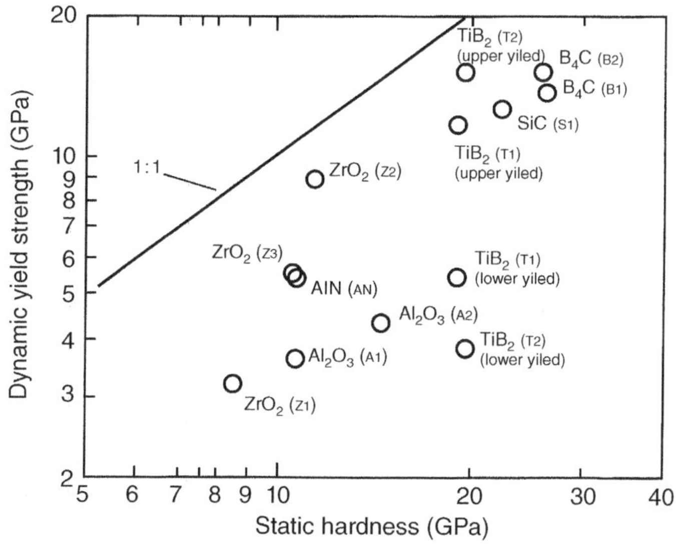

Figure 5.18 Static hardness values versus the calculated dynamic yield strength. Source: Grady (1998). © 1998, Elsevier.

explored due to a lack of yield strength values in the quasi-static regime. With the availability of dynamic yield strength (HEL), we can now try to relate to static hardness. The dynamic yield strength (Y_d) is calculated (Bourne et al. 1998; Bourne et al. 1995) as

$$Y_d = \frac{(1-2\nu)}{1-\nu}\sigma_{HEL} \tag{5.35}$$

where ν is Poisson's ratio and σ_{HEL} is the stress at HEL. The results are shown in Figure 5.18. For ceramics with perfect close-packed structures (e.g. ZrO_2, Al_2O_3, AlN), the dynamic yield strength values typically range between $G/5$ and $G/20$, where G is the shear modulus. The values derived from the above equation range between $G/25$ and $G/40$. The lower dynamic yield strength values under shock loading point to the dominance of shear in creating damage at high strain rates (i.e. the damage evolves at significantly lower yield values than would occur under static loads).

To understand the influence of plasticity and microcracking on the failure strength of ceramics and to elucidate the relationship between hardness and yield strength, Lankford et al. (Lankford 2004; Lankford et al. 1998) compared the uniaxial failure stress values of an alumina hot-pressed from pure elemental powders to those of a sintered alumina. Note that sintered ceramics in general have more microstructural inhomogeneities due to the addition of a second phase to aid in sintering. The expected yield strength values for sintered alumina were calculated from hardness, $Y = H/3$ as per Tabor's rule. In the pure alumina samples, dislocation-initiated fracture resulted in failure strength values close to the yield strength of the ceramic, whereas in the sintered alumina, the uniaxial failure strength values are far less than $H/3$, confirming the dominance of fracture originating from the microstructural defects in the sintered ceramics. Thus, in such ceramics where plasticity initiates the fracture process (instead of pure, microcracking-dominated fracture), the dynamic yield strength values are close to those calculated from hardness values. It is also worthwhile to mention that in these plasticity-dominated ceramics, although dislocations evolve well before fracture is initiated, the dislocations themselves are eventually responsible for the initiation of fracture due to their inability to dislocation glide and consequent stress to build up (Lankford et al. 1998).

5.5 Effect of Initial Porosity on Hugoniot

Porosity is a dominant microstructural defect in many ceramics. Therefore, it is worthwhile to discuss progress in the theoretical development of analyses of the Hugoniot response of ceramics with parametric dependence on initial porosity. We inform the reader that porous ceramic materials have also been termed "distended" ceramics, and levels of porosity termed "distention." Grady et al. (2013) investigated shock Hugoniot data for full density and porous compounds of boron carbide, silicon dioxide, tantalum pentoxide, uranium dioxide, and playa alluvium, with the aim of representing them by equations-of-state of shock compression. In this work, the Rice–Walsh equation of state (R–W EOS) (Rice and Walsh 1957) was constructed to describe the shock Hugoniot properties of ceramic materials with different levels of porosity. Such development of EOS representation focuses on the thermodynamic behavior of the material along the Hugoniot, constructed to account for a shear-enhanced phase transformation occurring at high loads. R–W EOS (Rice and Walsh 1957) follows from an isobaric Taylor series expansion of the specific volume truncated to the linear term in the specific enthalpy, given as:

$$v(P, H) = v(P) + \eta(P)\left(H - H_r(P)\right) \tag{5.36}$$

To construct R-W EOS, the requirements include a reference EOS that provides the specific volume, $v(P, H)$, and specific enthalpy, $H_r(P)$ on some identified pressure path. Another requirement is a thermal EOS specified by a pressure-dependent relation for the Rice–Walsh factor $\eta(P) = \frac{\tau e^{-P/P_0}}{K_s(P)}$, where K_s is the bulk modulus. The reference state of the material is represented by the principal Hugoniot of the full-density solid as:

$$v_{h0}(P) = v_0\left[1 - \frac{1}{K_0' + 1}\ln\left(1 + \frac{K_0' + 1}{K_0}P\right)\right] \tag{5.37}$$

where $v_{h0}(P)$ represents the principal Hugoniot of the system, v_0 and K_0 are respectively the specific volume and bulk modulus for un-deformed material, and the primed variable represents a derivative with respect to P. When phase transformation occurs, the reference EOS is developed as follows, using addition of compression relations for both the low-pressure as well as the high-pressure phases of the material:

$$v_{h0}(P) = (1 - \mu(P, \alpha))\, v_{h1}(P) + \mu(P, \alpha)\, v_{h2}(P) \tag{5.38}$$

where functional forms for v_{h1} and v_{h2} can be obtained from Eq. (5.37), and $\mu(P, \alpha)$ is the Hugoniot transformation function and the mass fraction of the high-pressure phase, selected to be of the statistical Weibull type:

$$\mu(P, \alpha) = 1 - e^{-(p/\sigma)^n} \tag{5.39}$$

where the scale parameter, σ, represents transformation pressure, while the shape parameter, n, controls the spread in pressure over which transformation occurs. Also, the parameter α represents initial distention (initial porosity) of the material, and hence, the shape and scale parameters are essentially functions of the initial distention, and thus better represented as $\sigma(\alpha)$ and $n(\alpha)$, respectively. This model argues that phase transformation at high pressures is accelerated by shear deformation and that presence of porosity in test specimens enhances shear deformation. Hugoniot energy conservation equations for principal Hugoniot and distended media Hugoniot are given as:

$$H_{h0} = \frac{1}{2}P\left(v_0 - v_{h0}(P)\right), H_h(P) = \frac{1}{2}P\left(v_{00} + v_h(P)\right) \tag{5.40}$$

where v_0 and v_{00} are initial specific volumes for full volume and distended media, respectively. Introducing Eq. (5.40) into EOS expression of Eq. (5.36) and solving, we have:

$$v_h(P) = v_{h0}P + \frac{\eta}{1 - \eta P/2}\left(\frac{1}{2}P\left(v_{00} - v_0\right)\right) \tag{5.41}$$

Shock and particle velocities pertaining to the corresponding Hugoniot state are respectively given as:

$$U_h(P) = \sqrt{\frac{P}{\rho_{00}}\left(1 - \frac{v_h(P)}{v_{00}}\right)^{-1}}, \quad u_h(P) = \sqrt{\frac{P}{\rho_{00}}\left(1 - \frac{v_h(P)}{v_{00}}\right)} \tag{5.42}$$

For illustration, we discuss the results of this theoretical development when applied to boron carbide for fully dense and two distention levels, 1.05 and 1.33 (Grady et al. 2013). As observed from Figure 5.19, Hugoniot data and respective EOS curves for low- and high-pressure boron carbide appear to be bounded by compression relations for low and high shock pressure phases of the material on either side (Grady et al. 2013). Using experimental data for low (Manghnani et al.

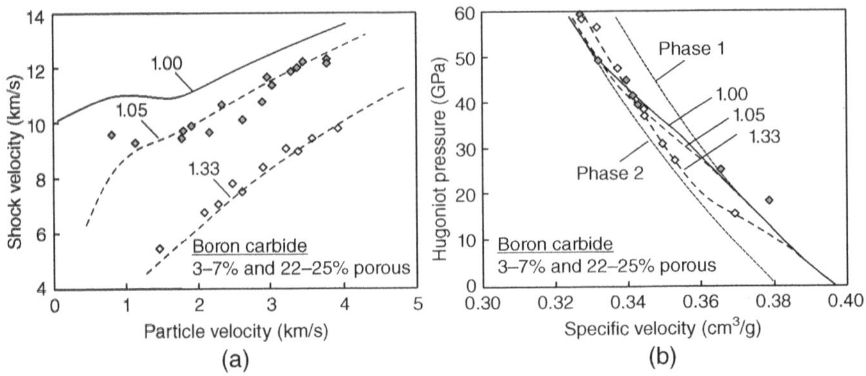

Figure 5.19 (a) Shock Hugoniot data for boron carbide from McQueen et al. (1970) and parameterization of Rice-Welsh EOS by Grady et al. (2013), identifying different levels of distention. (b) Compression curves for low- and high-pressure phases. Source: Grady et al. (2013). © 2013, Elsevier.

2000) and high (Yan et al. 2009) pressure phases, transformation pressure for fully dense boron carbide is found to be 41 GPa while that for 1.05 and 1.33 distention cases is 36 GPa and 16 GPa, respectively. Hence, porosity plays a role in reducing phase transformation stress of ceramic materials.

5.6 Temperature Rise in Shock Loading

Among the myriad of phenomena driven by shock loading, perhaps the least investigated for ceramic materials is temperature rise. For metals and materials in elemental form, temperature rise due to shock has been studied to obtain phenomenological data to create phase diagrams (Carter 1973) and for understanding the thermodynamic link in the cause of stress-induced localized melting, amorphization and phase transformation (Li et al. 2011; Meyers et al. 2003; Xu et al. 2008). Temperature rise for ceramics was briefly alluded to by Dandekar (Dandekar 2001) and Zhao (Zhao et al. 2016) for boron carbide under shock. In recent years, due to the advancement of molecular dynamics simulations, interest in temperature rise at the nanoscale owing to shock loading has increased. Some examples of temperature rise in shock, as computed via molecular dynamics for shock loading are covered in Section 5.7 and Chapter 6.

While assessing the temperature rise during shock, it is important to note the presence of two distinct length and time scales: (i) global, distinguished by a gradual temperature rise, equilibrium processes and near-adiabatic behavior, and (ii) local, earmarked by complex, nonequilibrium and rapid temperature rise. While temperature rise at the local scale would likely be resolved through molecular dynamics calculations, the global temperature rise has been calculated using equilibrium thermodynamics (Grady 2017c). To theoretically derive the global temperature rise in shock, first consider the differential analogue of R–H in Eq. (5.13) given by:

$$dE = \frac{1}{2}\left(V_0 - V\right) dP - \frac{1}{2} PdV \tag{5.43}$$

In addition, the differential form of second law of thermodynamics for the same process for isentropic conditions (Grady 2017c) is:

$$c_v dT = -c_v \gamma T \frac{dV}{V} + PdV + dE \tag{5.44}$$

Joining the above two equations gives:

$$c_v \frac{dT}{dV} + c_v \frac{\gamma\,(V)}{V} T = \frac{1}{2}\left(V_0 - V\right) P'\,(\varepsilon) + \frac{1}{2} P\,(V) \tag{5.45}$$

where $P'\,(\varepsilon) = \frac{dP}{d\varepsilon}$. Assuming the specific heat, c_v, to be constant and the Gruneisan parameter, γ, to follow $\rho\gamma = \rho_0\gamma_0$, the above equation can be integrated to predict temperature for a given strain, $\varepsilon\ (= 1 - V/V_0)$ (Carter 1973):

$$c_v T\,(\varepsilon) = c_v T_0 e^{\gamma_0 \varepsilon} + \frac{1}{2} V_0 \int_0^{\varepsilon} \left(\xi P'\,(\xi) - P\,(\xi)\right) e^{\gamma_0(\varepsilon - \xi)} d\xi \tag{5.46}$$

In the above equation, the first term on the right represents isentropic temperature rise due to strain ε, while the second term provides excess temperature rise due to shock dissipation resulting from compression (Grady 2017c). Specializing this equation for a material whose Hugoniot is described by a linear shock–velocity versus particle–velocity relation:

$$P\,(\varepsilon) = \frac{\rho_0 C_0^2 \varepsilon}{\left(1 - S_1 \varepsilon\right)^2} \tag{5.47}$$

where C_0 is speed of sound and S_1 is constant of proportionality defined in Eq. 5.14. Upon simplification, Eq. 5.46 yields:

$$c_v T\,(\varepsilon) = c_v T_0 e^{\gamma_0 \varepsilon} + C_0^2 S_1 \int_0^{\varepsilon} \frac{\xi^2 e^{\gamma_0(\varepsilon - \xi)}}{\left(1 - S_1 \xi\right)^3} d\xi \tag{5.48}$$

For a material with linear shock-velocity versus particle-velocity behavior, the above equation simplifies to (Grady 2017c):

$$c_v T\,(\varepsilon) = c_v T_0 e^{\gamma_0 \varepsilon} + \frac{1}{3} C_0 S_1 \left(\varepsilon^3 + \frac{1}{4} \gamma_0 \varepsilon^4 + \ldots\right) \tag{5.49}$$

Equation (5.49) represents the global temperature rise in shock compression over a Hugoniot volume strain, ε. It is important to note that local temperature rise in typical shock conditions can approach or even cross the melting point, such as in the case of boron carbide (Dandekar 2001; DeVries et al. 2020).

5.7 Mechanics-based Modeling of Shock Response

Mechanistic models for the shock response of ceramics consider the shock wave interaction with the initial microstructure, which invariably contains preexisting, randomly distributed flaws such as grain boundaries, voids, impurities, and/or secondary phases. Many of these models are discussed in the context of glasses and transparent materials in Chapter 6. One of the well-received models was developed by Rajendran and Grove (1996) because of its versatility and comprehensive description of major microstructural inelastic deformation mechanisms during impact and/or high-strain-rate loading. In this model, the inelastic flow is assumed to occur when the ceramic is shocked above the HEL, but it does not occur during tensile loading, which may happen due to unloading or reflections from boundaries. Microcracking (e.g. wing crack growth from favorably oriented cracks or defects) is expected to occur during both tensile and compression loading. The model treats the consequence of these evolving microcracks as the degradation of elastic moduli. However, the number of flaws is assumed to be constant, and no additional flaws are nucleated during loading. These cracks grow to cause fragmentation and pulverization of the ceramic but only under compressive loading. Plasticity is not permitted under tensile loading but elastic deformation is assumed to occur. With continued loading these deformation mechanisms evolve, and, hence, a time-dependent crack-growth law is assumed for microcracking, which introduces strain-rate dependency. Finally, a strain-rate-dependent strength relationship is assumed during plastic deformation. In addition, the model also considers pore collapse, plastic strain due to viscoelastic flow and the strength of the intact and damaged ceramics.

Considering the above deformation features, Rajendran and Grove (1996) developed a mechanistic framework and performed numerical simulations of plate impact experiments over a wide range of impact velocities. The effects of viscoplastic flow, microcracking, and porosity on the shapes and amplitudes of the measured velocity profile were determined. These results revealed that damage occurs only under tension, and there is no crack growth under compression. But upon unloading and subsequent tensile loading, the damage evolves rapidly despite smaller tensile stresses. The model successfully captured the measured particle velocity profile in plate impact experiment. The microcrack growth during tensile loading in the model causes the bulk and shear moduli to degrade rapidly, and, as a consequence, the tensile stresses relax and unload to almost zero at both low and high velocities. Thus, the model was able to capture the essential features of ceramic behavior under rapid loading reasonably well. More detailed models of crack nucleation and growth, and development of failure waves during plate impact experiments will be presented in Chapter 7.

5.8 Molecular Dynamics Simulation of Shock Response of Ceramics

Shock experiments are expensive and time-consuming. Additionally, the diagnostic tools for monitoring material behavior at the desired resolution of nanometer length scale and picosecond time scale are limited. For example, many recent studies suggest temperature rise during shock loading to be well above the melting point of the ceramic (Branicio et al. 2018b; DeVries et al. 2020; Li et al. 2019). These results are in agreement with theoretical estimates of temperature rise derived as early as 1973 (Carter 1973) (see (Dandekar 2001)). However, there is no currently available experimental technique to measure temperature at the length and time scales of the simulations. Hence, researchers have resorted to atomic level simulations of shock experiments. Molecular dynamics (MD) simulations provide an ideal platform to provide insight into shock-induced structural phase transformation (SPT), presence, and character of plastic deformation mechanisms (e.g. twinning, dislocation plasticity, grain boundary rotation, and sliding), and the effects of crystal orientation, defects and grain boundaries (Branicio et al. 2018a). The MD simulations also help bridge the gap between the macroscopic experimental *in situ* measurements and postmortem microstructural observations. While the experimental techniques to determine the response of materials at strain rates in the range of $10^8\ s^{-1}$ and above do not exist in most of the research facilities, MD simulations can provide dynamic material properties (e.g. HEL and spall strength in various crystallographic orientations) and the intrinsic features of shock-induced mechanisms of deformation at high strain rates with atomic level resolution. These simulations rely on the use of a validated interatomic potential (against the experimental results on lattice constants, elastic moduli, cohesive energy, melting temperature, etc.) to extract the shock response of a domain of atoms in a material. Early studies focused on cubic or hexagonal crystal structures, which are amenable to simulating large rectangular systems with well-defined crystallographic planes along the two orthogonal edges of the study domain. For example, α-alumina, which has a hexagonal crystal structure, has been studied by Zhang et al. (2008). However, more complex crystal structures such as the rhombohedral structure of boron carbide do not have two well-defined perpendicular crystallographic planes and are difficult to simulate. Recently, DeVries et al. (2020) have investigated such structures and these results will be presented in Chapter 6.

Early work using the Tersoff potential to simulate the shock response of SiC were unsuccessful in properly describing the shock-induced SPT (Shimojo et al. 2000). This potential also overestimated the shock-induced stresses, wave velocities, and temperature in shocked SiC (Branicio et al. 2018a). In recent years, the availability of many-body potential developed by Vashishta (Vashishta et al. 2007)

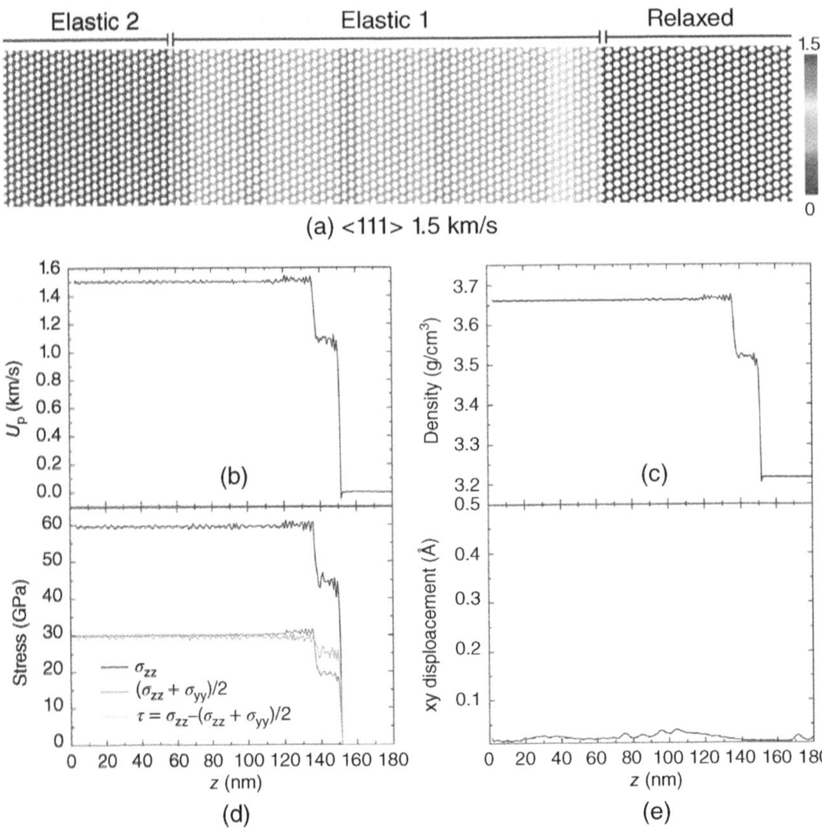

Figure 5.20 Illustration of elastic wave propagation in <111> SiC crystal at an impact velocity of 1.5 km/s. In (a), the undeformed structure (right) with coordination number of four and density of 3.2 g/cm^3 is elastically compressed (Elastic 1) first and then as the stress level raises, will transform to an intermediate five coordinated structure (Elastic 2). This structural transformation is seen as a split wave in the particle velocity plot shown in (b). The corresponding changes in density (c), stress state (d), and lateral displacements (e) are shown. Source: Branicio et al. (2018b). © 2018, AIP Publishing.

and ReaxFF potential by An and Goddard (An and Goddard III 2015) have motivated MD simulations of the shock response of structural ceramics. Using this potential by Vashishta et al. (2007), Branicio et al. (2018a, b) conducted shock simulations of SiC in three crystallographic orientations and in two nanocrystalline SiC ceramics with average grain size of 5 and 10 nm. Figure 5.20 illustrates the deformed structure in the <111> orientation in the elastic regime at a particle velocity of 1.5 km/s. The virgin material elastically compresses and then transforms to an intermediate five-coordinated structure as shown in Figure 5.20a. This structure formation is reflected as a split wave seen in the particle velocity, density, and stress signatures shown in Figure 5.20b–d, respectively. The wave corresponding to the intermediate structure formation travels at a slightly lower velocity than the elastic wave of the original crystal structure. The small lateral displacement of atoms (Figure 5.20e) and no loss in the shear stress (Figure 5.20d) indicate no energy dissipation during the formation of this this intermediate structure.

At higher particle velocities up to 2.5 km/s, a plastic wave follows the elastic wave and beyond that an SPT wave develops until 5 km/s as shown in Figure 5.21 (Branicio et al. 2018b). Twinning is the major plastic deformation mode (Figure 5.21a) in SiC, and in the <110> orientation, it occurs around 75 GPa axial stress (see Figure 5.21b). The lateral displacement of atoms during plastic deformation can be up to 6–8 Å as seen in Figure 5.21c. At 4 km/s impact velocity, the SPT wave evolves in the <111> orientation following the elastic precursor; see Figure 5.21d and e. The phase transformation from a four-coordinated zinc-blende structure to six-coordinated rock salt structure causes lateral displacement of atoms by as much as 10 Å as noted in Figure 5.21f. Unlike the elastic deformation in Figure 5.20, both the above inelastic deformation mechanisms (plasticity and phase transformation) cause the shear stress to be relieved rendering the axial and lateral stresses become the same, giving rise to the hydrostatic stress state for the rest of the shock duration. The material densifies from 3.21 to 4.21 g/cm^3 (a 24% volume reduction), and the high imposed shock further compresses the material to 4.91 g/cm^3 (an additional 14% reduction in volume).

MD simulations can also be effectively used to obtain the *Us–Up* relationships. For SiC with 5 and 10 nm grain size, this relationship (black curve) is shown in Figure 5.22a (Branicio et al. 2018b) along with the maximum shear stress (gray curve) at various impact velocities. At lower particle velocities, the shear stress reaches a high value due to elastic stress build up and drops once the plastic deformation initiates at about 1.5 km/s. In the smaller grain size ceramic, a lower threshold for plasticity is noted as it contains a higher density of grain boundaries (i.e. large number of atoms along the grain boundaries). Above 3.5 km/s, phase transformation evolves but the shear stress drop is not as significant and hence a small increase is seen at 4 km/s. At even higher particle velocities, an increase in temperature softens the material and the shear stress drops.

Unlike the monocrystalline samples where twinning was the primary deformation mode, the nanocrystalline SiC exhibits an additional deformation mechanism "grain boundary sliding." The associated lateral displacements are shown in Figure 5.22b and c, respectively. Li et al. (2017) investigated the shock-induced plastic deformation (twinning) in SiC single crystals along [001], [110], and [111] directions and depicted the anisotropy based on the lateral displacements of atoms in {111}, {110}, and {110} planes as shown in Figure 5.23.

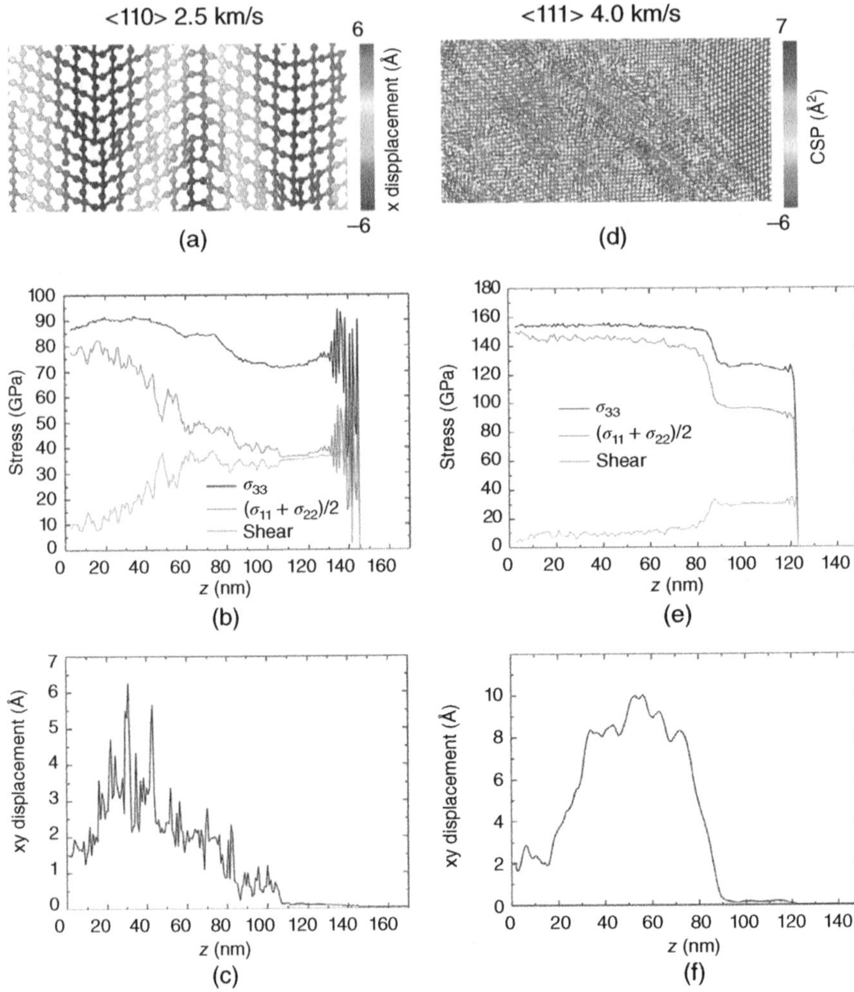

Figure 5.21 Structural changes in SiC at higher velocity. (a) Deformation twinning in the <110> orientation at 2.5 km/s and the corresponding (b) stress and (c) lateral displacement. (d) Phase transformation in the <111> orientation and the corresponding (e) stress and (f) lateral displacement. Source: Branicio et al. (2018a). © 2018, AIP Publishing.

Li et al. (2017) also extracted the Hugoniot of SiC along the three orientations as well as that of a nanocrystalline material and compared these results to those of experimental values. The relevant plots are shown in Figure 5.24a. The anisotropy during the early pressure–volume relationship is highlighted in Figure 5.24b, where the nanocrystalline SiC withstands the lowest shear stress and the single

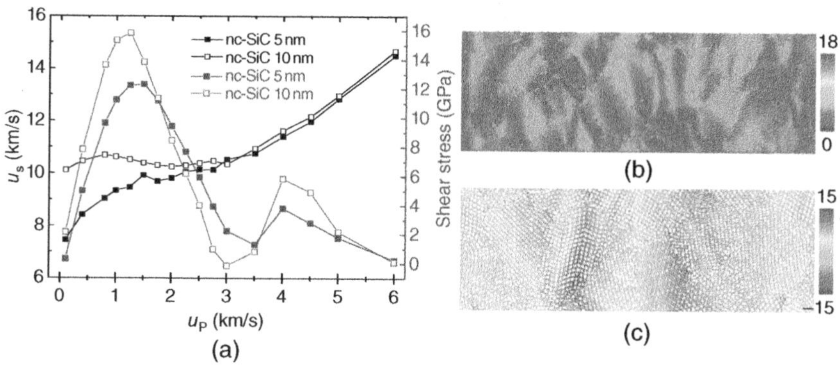

Figure 5.22 (a) The Us–Up relationship and shear stress values for nanocrystalline SiC and the corresponding lateral displacements for (b) grain boundary sliding and (c) deformation twinning. Source: Branicio et al. (2018a). © 2018, AIP Publishing.

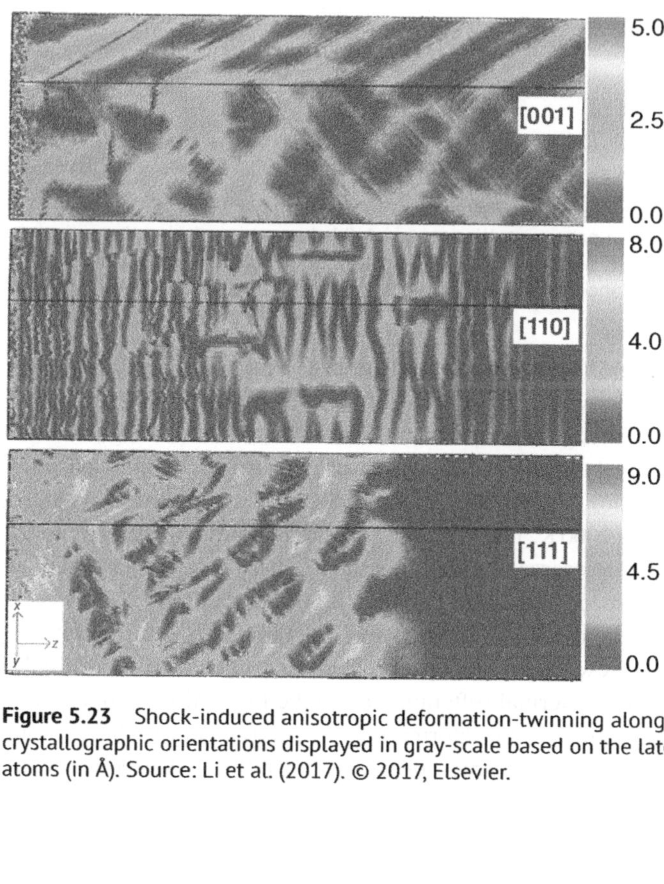

Figure 5.23 Shock-induced anisotropic deformation-twinning along the three crystallographic orientations displayed in gray-scale based on the lateral displacement of atoms (in Å). Source: Li et al. (2017). © 2017, Elsevier.

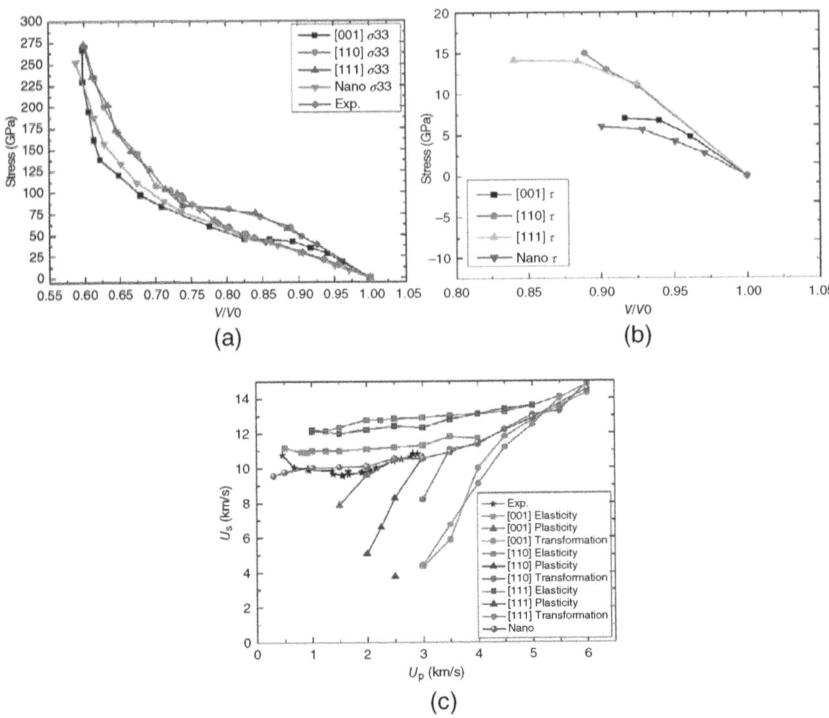

Figure 5.24 (a) Hugoniot stress versus specific volume for single crystal SiC along the three orientations and for a nanocrystalline material, (b) shear stress during the early deformation, and (c) elastic, plastic, and phase transformation *Us–Up* relationships for three orientations and a nanocrystalline material. Source: Li et al. (2017). © 2017, Elsevier.

crystal SiC exhibits significant anisotropy. The *Us-Up* relationships for various phases (elasticity, plasticity, and phase transformation) in three orientations and for nanocrystalline SiC along with experimental results are shown in Figure 5.24c.

Li et al. (2019) extended the above studies to investigate the shock-induced damage, phase transformation, and spall fracture in SiC in three orientations at an elevated temperature of 2000 K. The initial elevated temperature increases energy at the ground state and decreases the density of the material by about 6.12% which slows the elastic wave speed by almost 1 km/s in all three orientations. However, the velocities of plastic and transformation waves remained the same. Temperature rise as high as 8000 K was seen for impact velocities up to 6 km/s. Interestingly, despite the expectation of thermal softening due to the high initial temperature, the thresholds for generating plastic deformation remains the same as that at room temperature, but the threshold for structural phase transformation is reduced considerably. A significant anisotropy was also noted in phase transformation stress in different crystal orientations. Microstructurally, there was a significant reduction in twinning at the elevated temperature. The spall strength remained almost constant initially at about 24 GPa up to a velocity of 2.5 km/s and then decreased monotonically to less than 5 GPa, when the impact velocity increased to 6 km/s. The spall strength at 2000 K was 21% less compared to the strength at 300 K due to the softening effect in the bulk modulus.

In these atomistic simulations, the mechanism for spall fracture evolution is identified by observing the region with nearly zero density where complete material separation occurs. Figure 5.25 illustrates different spall fracture modes in SiC at low (1.6 km/s), intermediate (2.5 km/s), and high (6.0 km/s) particle velocities. At low velocities, the reflection of shock wave from the rear surface creates classical tensile brittle fracture on a single plane as seen in Figure 5.25a. Initially, many point defects are created along this plane, and with an increase in tensile stress, these point defects evolve into microcracks that eventually coalesce to form large cracks leading to complete brittle fracture. However, with an increase in particle velocity, the spall fracture is no longer brittle and is not caused by the tensile stress. The mechanism transitions to ductile fracture due to the rise in temperature during the shock process. At intermediate velocities, multiple spall planes evolve as seen in Figure 5.25b, and with further increases in velocity, the temperature rises significantly in the compressively stressed regions (colored in light-gray) where voids nucleate (micro spall) and grow rapidly at multiple locations to create a typical ductile fracture as shown in Figure 5.25c. It was found that this spall fracture process is not a strong function of initial temperature but of a rise in temperature during the shock process.

The characteristics of shock wave propagation at three velocities corresponding to spall or no spall are illustrated in Figure 5.26. At 1 km/s, no spall fracture occurs as the tensile stress generated due to the reflection of incoming shock wave is below the tensile strength of the material. The shock wave simply reflects back at the free end as a tensile wave with almost the same amplitude and duration; see Figure 5.26a. At a slightly higher particle velocity of 1.25 km/s seen in Figure 5.26b, the classical spall fracture along a single plane occurs exactly at the middle of the shock pulse width when it reaches the free end; see the x-t diagram in Figure 2.16a. This is reflected as the zero stress along the spall plane and waves traveling in two directions from the spall plane. Finally, at the over driven condition of 6 km/s in Figure 5.26c, the high temperature rise softens the material, and the reflected tensile stress is only a small fraction of the incident high amplitude compression as the spall fracture is due to the high temperature rise and not due to the reflected tension (Li et al. 2017). Thus, the MD simulations allow better understanding of the physical phenomenon occurring in situ at the length and time scales not possible to measure by experiments.

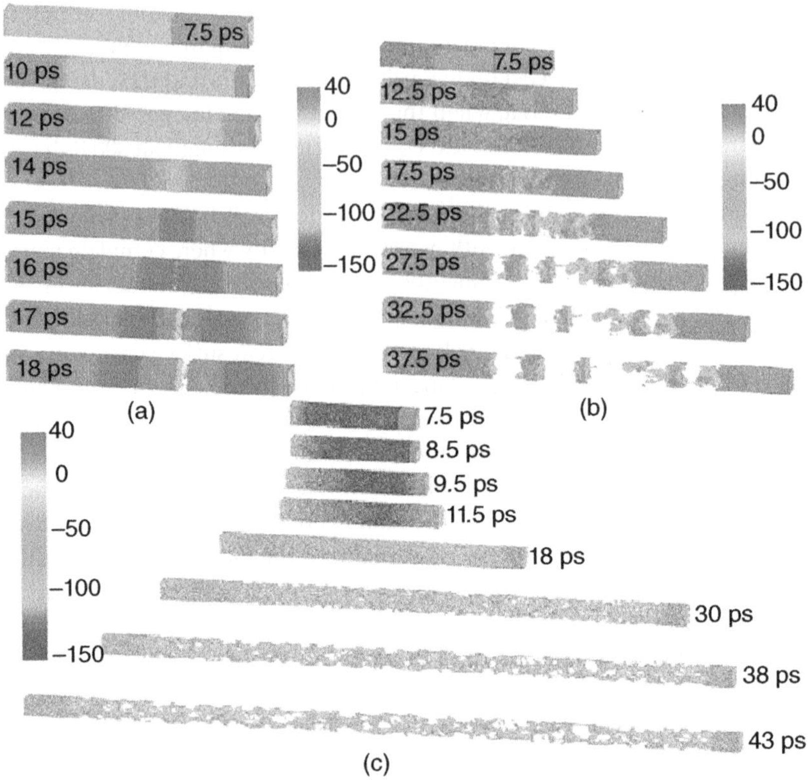

Figure 5.25 (a) Spall fracture simulation at 1.25 km/s (b) 3.5 km/s and (c) 6 km/s, illustrating, respectively, the classic tensile spall fracture on a single plane at low velocities, multi-plane fracture at intermediate velocities and temperature-induced void growth and fracture at higher velocities. Color range depicts stress in GPa. Source: Li et al. (2017). © 2017, Elsevier.

MD simulations can also be extended to gain insight at velocity ranges well beyond those possible by experiments. For example, hypervelocity impact at 18 km/s on alpha-alumina revealed an initial shock velocity at an impressive 30 km/s which soon slows to the elastic wave speed of 13.2 km/s (Zhang et al. 2008). The high temperature generated during this shock deformation amorphized the material and the pressure distribution around the impact location revealed that the shock front undergoes a transition from the initial threefold symmetry to sixfold symmetry. In addition to validating the occurrence of deformation slip and twinning along pyramidal, basal, and rhombohedral planes during penetration of a projectile into alumina ceramic, new deformation twin planes along $\left\{0\bar{1}11\right\}$ propagating roughly at a speed of 8 km/s were noted.

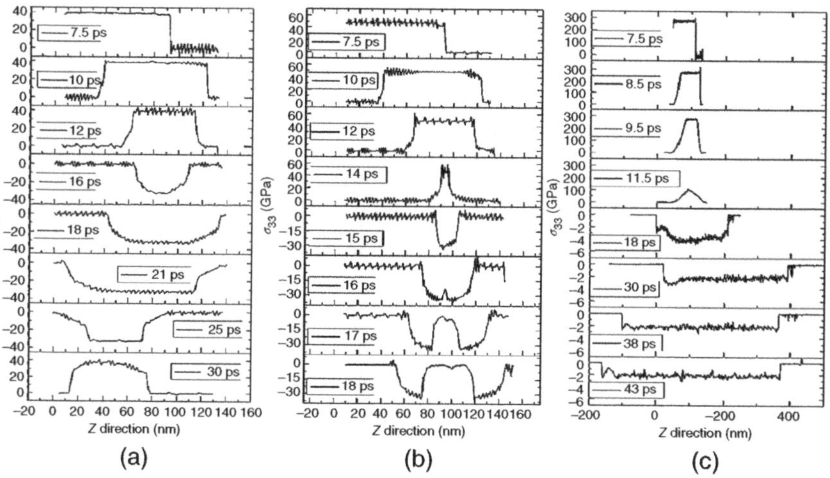

Figure 5.26 Shock profiles along the specimen length illustrating only wave reflection without spall in (a) at 1 km/s, (b) classical brittle spall due to reflected tensile wave on a single plane at 1.25 km/s, and (c) temperature-induced ductile micro-spall at 6 km/s along the [111] orientation of SiC. Source: Li et al. (2017). © 2017, Elsevier.

Localized defects nucleate in the vicinity of these twins and act as sources for crack nucleation during the tensile unloading.

Another ceramic that undergoes phase transformation is AlN, which has an initial hexagonal structure (wurtzite) and transforms into rock salt at a pressure between 20 and 25 GPa. Using the Vashishta potential, Branicio et al. (2006) simulated the response of AlN under shock loading in four crystallographic orientations at temperatures of 10 K and 300 K. Unlike SiC where the shock wave splits into elastic, plastic, and phase transformation fronts at sufficiently high particle velocities (> 2.5 km/s), the shock profile in AlN exhibited only a two-wave structure consisting of elastic and phase transformation fronts at an impact velocity as low as 0.8 km/s. In both cases, at velocities well above 4.0 km/s, a single overdriven phase transformation wave travels at higher velocities. A typical result for a particle velocity of 1.0 km/s is shown in Figure 5.27a where the elastic precursor is followed by the phase transformation front (Branicio et al. 2013). The particle velocity increases to 730 m/s in the elastic regime and maintains that value until the phase transformation wave arrives when the particle velocity jumps to 1 km/s. The associated density in the shock domain is shown in Figure 5.27b where the density of the elastically compressed region increases to 3.46 g/cm^3 from the unstressed density of 3.26 g/cm^3, which is then followed by an increased density in the phase transformation domain to 4.1 g/cm^3 of the rock salt structure (an increase of 26% from the relaxed wurtzite structure). The lateral displacement of atoms in the *xy*-plane reveals small displacement in the elastic wave and almost

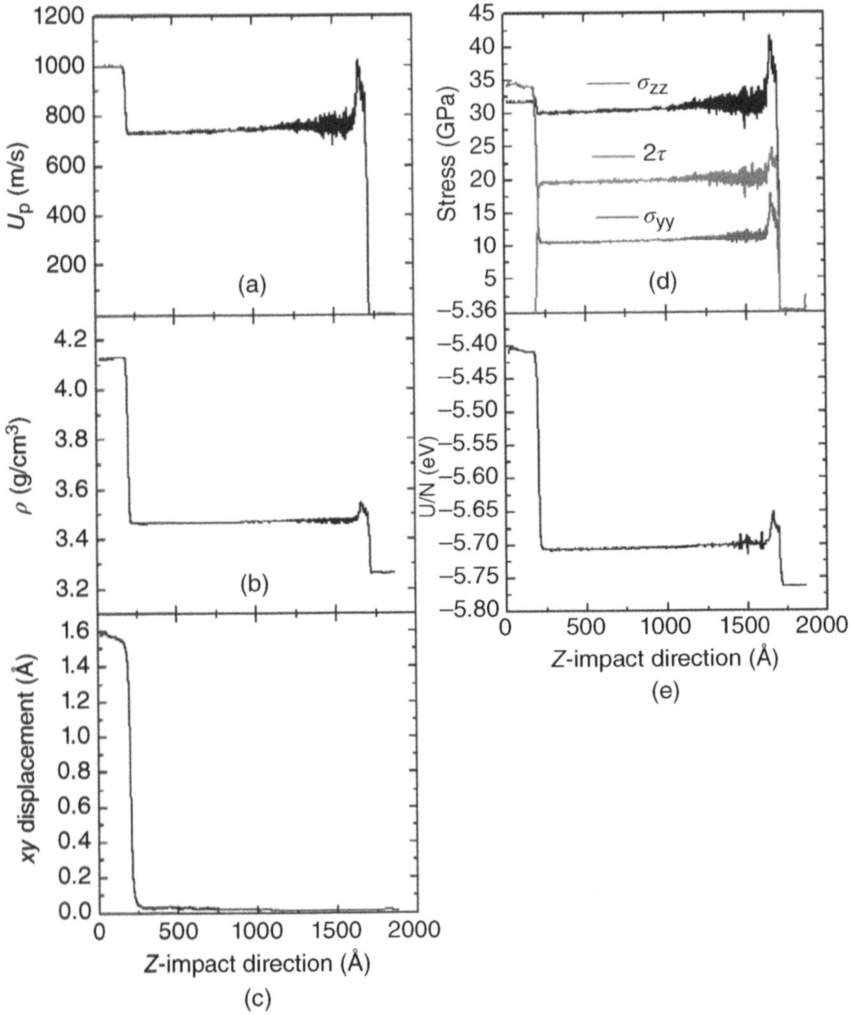

Figure 5.27 MD simulations of shock profiles in AlN at 1 km/s and at a temperature of 10 K. (a) Particle velocity as a function of distance revealing elastic and phase transformation waves and the corresponding (b) density variation, (c) lateral displacement, (d) stress state, and (e) energy. Source: Branicio et al. (2013). © 2013, Elsevier.

a steady displacement up to 1.6 Å in the transformation wave. Figure 5.27d illustrates the stress state. The applied shock loading induces a steady axial stress (σ_{zz}) of 30 GPa along the shock direction, a lateral stress of 10 GPa and a shear stress, 2τ of about 20 GPa. Upon phase transformation, the shear stress is quickly relieved, and the lateral stress increases to the same level as that of axial stress giving rise to almost hydrostatic condition. Finally, in Figure 5.27e the increase in the internal energy in the elastic and transformation waves is shown.

Figure 5.28 reveals the nature of microstructural changes for various particle velocities within the shocked region in AlN (Branicio et al. 2013). At 1 km/s, Figure 5.28a and b reveal the density changes as the shock propagates into the material. The undeformed material is shown in grey (wurtzite structure with density 3.26 g/cm^3 and coordination of four), and an elastically compressed material with a sharp front is shown in bright gray (density 3.46 g/cm^3 and coordination of five). It is followed by a dark region indicating the phase transformation wave at a higher density (rocksalt with density 4.1 g/cm^3 and coordination of six) and at a slower velocity. The transformation wave originates at the impact surface and grows into the material along the direction of the shock loading. This observation implies that the growth of the transformed phase occurs by nucleation of grains at the impact surface. Figure 5.28c reveals the grain structure in the transformed region (rocksalt), which seems to align along three crystallographic directions of the original hexagonal wurtzite crystal and is clearly a signature of heterogeneous nucleation. On the other hand, at 2 km/s in Figure 5.28d, the front becomes rough, and the elastic region is extremely short, which implies that

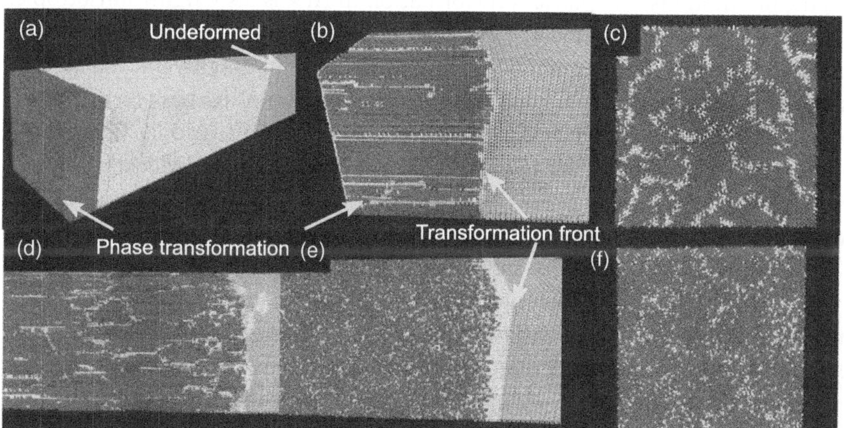

Figure 5.28 Simulated structure of shock wave in AlN at (a) 1 km/s revealing elastically deformed (bright gray) and phase transformed (dark) regions, (b) mostly planar phase transformation front nucleated from the impact surface, (c) a plane of phase transformation region revealing heterogeneous nucleation of rock salt structure, (d) rough phase transformation front at 2 km/s, (e) planar transformation front at 5 km/s as the temperature rises well above 1000 C. Nucleation of rock salt occurs within the front as opposed to at the impact plane, and (f) homogeneous nucleation of rock salt structure. Source: Reproduced with permission from Branicio et al. (2013).

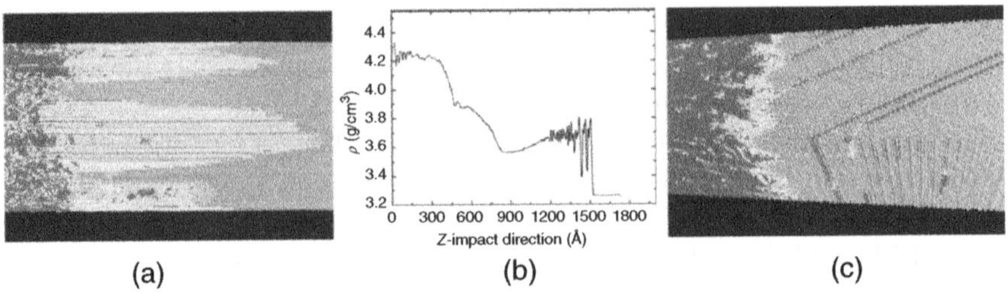

Figure 5.29 Shock front structure for impact at 2 km/s and $T = 10$ K in (a) $[11\bar{2}0]$ orientation revealing intermediate five-coordinated region growing ahead of the shock front, (b) corresponding density plot and (c) impact in $[10\bar{1}0]$ orientation revealing nucleation of partial dislocations ahead of the compressed region. Source: (a) Reproduced with permission from Branicio et al. (2013), (b) Branicio et al. (2013). © 2013, Elsevier.

the nucleation occurs at the shock front itself rather than at the impact surface as in Figure 5.28a. At even higher particle velocities of 5 km/s in Figure 5.28e, the transformation front once again becomes sharp with nucleation occurring immediately behind the shock front in a more homogeneous fashion; see Figure 5.28f.

Similar to the anisotropic plasticity in SiC (Figure 5.23) along three crystallographic orientations, the phase transformation in AlN is also highly anisotropic. For example., when impacted at 2 km/s along the $[11\bar{2}0]$ direction, the elastically stressed region grows ahead of the transformation front in a highly uneven fashion as shown in Figure 5.29a. Under these conditions, the roughness of the transformation front is reflected in the particle velocity and density looks like a ramp rather than a sharp jump; see Figure 5.29b. The nucleation of dislocations along preferred crystallographic planes as shown in Figure 5.29c may give rise to this uneven growth.

Branicio et al. (2013) also studied shock-induced phase transformation anisotropy in AlN. While the Hugoniot for various crystal orientations was the same in the velocity ranges studied, the sharp transformation wave front was seen only for the basal [0001] direction. For shocks along the $[10\bar{1}0]$ orientation, the transformation growth is preferred diagonally along the $[11\bar{2}0]$ direction, and the shock front again becomes rough. Partial dislocations nucleate ahead of the front and glide along the (0001) plane in the compressed region ahead of the transformation front.

Thus, MD simulations offer a unique insight into the mechanisms that operate at the atomic level. However, the biggest limitation in the widespread use of MD approach to dynamic deformation investigations is the lack of appropriate interatomic potentials. Rapid progress in this realm will greatly assist in unlocking the mysteries at the smallest length scales.

5.9 Summary and Conclusions

In this chapter, we have discussed the theory behind shock wave loading and described the material response for both ideal and real materials. In addition to plasticity, shock loading also induces phase transformation in some ceramics due to high pressure. While many ceramics exhibit post-yield hardening behavior under high shock pressure conditions, ceramics such as boron carbide exhibits post-yield softening behavior leading to amorphization. Materials such as calcite undergo multiple phase transformations under shock loading.

We conclude this chapter by emphasizing that the strength of a ceramic is a strong function of strain rate only in the intermediate strain rate regime, and at the two ends of this strain-rate spectrum, strength is almost rate independent. However, the mechanisms that control this behavior are vastly different. In the low and intermediate strain rate regimes only uniaxial stress conditions are imposed and brittle fracture dominates the deformation. In the shock regime, in addition to the extremely high strain rates, high confining pressures are also employed because of which deformation mechanisms that are not operative at low strain rates get activated. However, due to the lack of spatio-temporal resolution of the currently available experimental methods, the atomic level deformation mechanisms cannot be visualized. MD simulations offer a unique way to not only visualize dynamic deformation modes at the atomic scales but also help determine the relevant mechanistic measures such as HEL, stress for twin formation and phase transformation.

References

Achenbach, J.D. (1973). *Wave Propagation in Elastic Solids*. New York: North Holland Publishing Company.

An, Q. and Goddard, W.A. III, (2015). Atomistic origin of brittle failure of boron carbide from large-scale reactive dynamics simulations: suggestions toward improved ductility. *Physical Review Letters* 115: 105501.

Bourne, N., Rosenberg, Z., and Field, J. (1995). High-speed photography of compressive failure waves in glasses. *Journal of Applied Physics* 78: 3736–3739.

Bourne, N., Millett, J., Rosenberg, Z., and Murray, N. (1998). On the shock induced failure of brittle solids. *Journal of the Mechanics and Physics of Solids* 46: 1887–1908.

Branicio, P., Kalia, R., Nakano, A., and Vashishta, P. (2006). Shock-induced structural phase transition, plasticity, and brittle cracks in aluminum nitride ceramic. *Physical Review Letters* 96.

Branicio, P., Nakano, A., Kalia, R., and Vashishta, P. (2013). Shock loading on AlN ceramics: a large scale molecular dynamics study. *International Journal of Plasticity* 51: 122–131.

Branicio, P., Zhang, J., Rino, J. et al. (2018a). Plane shock loading on mono- and nano-crystalline silicon carbide. *Applied Physics Letters* 112: 111909.

Branicio, P., Zhang, J., Rino, J. et al. (2018b). Shock-induced microstructural response of mono- and nanocrystalline SiC ceramics. *Journal of Applied Physics* 123: 145902.

Cagnoux, J. and Longy, F. (1988). Spallation and shock-wave behavior of some ceramics. *Journal de Physique* 49: 3–10.

Carter, W. (1973). Hugoniot equation of state of some alkali halides. *High Temperatures - High Pressures* 5: 313–318.

Chen, M., McCauley, J., and Hemker, K. (2003). Shock-induced localized amorphization in boron carbide. *Science* 299: 1563–1566.

Dandekar, D.P. (2001). *Shock Response of Boron Carbide*, 1–34. Aberdeen Proving Grounds: US Army Research Laboratories.

DeVries, M., Subhash, G., and Awasthi, A. (2020). Shocked ceramics melt: an atomistic analysis of thermodynamic behavior of boron carbide. *Physical Review B* 101: 144107.

Ghosh, D., Subhash, G., Zheng, J., and Halls, V. (2012). Influence of stress state and strain rate on structural amorphization in boron carbide. *Journal of Applied Physics* 111: 063523.

Grady, D. (1994). Shock-wave strength properties of boron-carbide and silicon-carbide. *Journal de Physique IV Colloque* 4: 385–391.

Grady, D.E., 1995a. Dynamic properties of ceramic material. Sandia National Laboratories Report. SAND94-3266.

Grady, D. (1998). Shock-wave compression of brittle solids. *Mechanics of Materials* 29: 181–203.

Grady, D. (2017a). *Physics of Shock and Impact*, vol. 1. IOP Publishing.

Grady, D. (2017b). *Physics of Shock and Impact*, vol. 2. IOP Publishing.

Grady, D.E. (2017c). *The Shock Wave Equation of State, Physics of Shock and Impact*, vol. 2, 10–11–10–107. IOP Publishing.

Grady, D., Fenton, G., and Vogler, T. (2013). Equation of state and evidence of enhanced phase transformation for shock compression of distended compounds. *International Journal of Impact Engineering* 56: 19–26.

Koeppel, B. and Subhash, G. (1999). Characteristics of residual plastic zone under static and dynamic vickers indentations. *Wear* 224: 56–67.

Lankford, J. (2004). The role of dynamic material properties in the performance of ceramic armor. *International Journal of Applied Ceramic Technology* 1: 205–210.

Lankford, J., Predebon, W., Staehler, J. et al. (1998). The role of plasticity as a limiting factor in the compressive failure of high strength ceramics. *Mechanics of Materials* 29: 205–218.

Lawn, B. and Wilshaw, T.R. (1993). *Fracture of Brittle Solids*. Cambridge University Press.

Li, N., Wang, Y., Peng, R.L. et al. (2011). Localized amorphism after high-strain-rate deformation in TWIP steel. *Acta Materialia* 59: 6369–6377.

Li, W., Yao, X., Branicio, P. et al. (2017). Shock-induced spall in single and nanocrystalline SiC. *Acta Materialia* 140: 274–289.

Li, W., Hahn, E., Yao, X. et al. (2019). Shock induced damage and fracture in SiC at elevated temperature and high strain rate. *Acta Materialia* 167: 51–70.

Manghnani, M.H., Wang, Y., Li, F. et al. (2000). Elastic and vibrational properties of B4C to 21 GPa. *Science and Technology of High Pressure* 2: 25–30.

McQueen, R.G., Marsh, S.P., Taylor, J.W. et al. (1970). Chapter VII - The equation of state of solids from shock wave studies. In: *High Velocity Impact Phenomena* (ed. R. Kinslow), 293–417. NY: Academic Press.

Meyers, M.A. (1994). *Dynamic Behavior of Materials*. New York: Wiley.

Meyers, M., Xu, Y., Xue, Q. et al. (2003). Microstructural evolution in adiabatic shear localization in stainless steel. *Acta Materialia* 51: 1307–1325.

Munson, D. and Lawrence, R. (1979). Dynamic deformation of polycrystalline alumina. *Journal of Applied Physics* 50: 6272–6282.

Rajendran, A. and Grove, D. (1996). Modeling the shock response of silicon carbide, boron carbide and titanium diboride. *International Journal of Impact Engineering* 18: 611–631.

Ramesh, K., Hogan, J.D., Kimberley, J., and Stickle, A. (2015). A review of mechanisms and models for dynamic failure, strength, and fragmentation. *Planetary and Space Science* 107: 10–23.

Reinhart, W. and Chhabildas, L. (2003). Strength properties of Coors AD995 alumina in the shocked state. *International Journal of Impact Engineering* 29: 601–619.

Rice, M.H. and Walsh, J.M. (1957). Equation of state of water to 250 kilobars. *The Journal of Chemical Physics* 26: 824–830.

Rosenberg, Z. and Yeshurun, Y. (1985). Determination of the dynamic-response of AD-85 alumina with in-material manganin gauges. *Journal of Applied Physics* 58: 3077–3080.

Rosenberg, Z. and Yeshurun, Y. (1986). Release behavior of shock loaded commercial alumina. *Journal of Applied Physics* 60: 1844–1846.

Rosenberg, Z., Yeshurun, Y., and Brandon, D. (1985). Dynamic-response and microstructure of commercial alumina. *Journal de Physique* 46: 331–341.

Rosenberg, Z., Yaziv, D., Yeshurun, Y., and Bless, S. (1987). Shear-strength of shock-loaded alumina as determined with longitudinal and transverse manganin gauges. *Journal of Applied Physics* 62: 1120–1122.

Rosenberg, Z., Bourne, N., and Millett, J. (1996). Direct measurements of strain in shock-loaded glass specimens. *Journal of Applied Physics* 79: 3971–3974.

Sato, Y. and Akimoto, S. (1979). Hydrostatic compression of 4 corundum-type compounds- α-Al_2O_3, V_2O_3, CR_2O_3, and α-FE_2O_3. *Journal of Applied Physics* 50: 5285–5291.

Sato, T. and Akimoto, S. (1979). Hydrostatic compression of four corundum-type compounds. *Journal of Applied Physics* 50: 5285–5291.
Shimojo, F., Ebbsjo, I., Kalia, R. et al. (2000). Molecular dynamics simulation of structural transformation in silicon carbide under pressure. *Physical Review Letters* 84: 3338–3341.
Staehler, J.M., Predebon, W.W., Pletka, B.J., and Subhash, G. (2000). Micromechanisms of deformation in high-purity hot-pressed alumina. *Materials Science and Engineering A* 291: 37–45.
Subhash, G. (2000). *Dynamic Indentation Testing*, 519–529. ASM Handbook.
Tabor, D. (1948). A simple theory of static and dynamic hardness. *Proceedings of the Royal Society of London Series a-Mathematical and Physical Sciences* 192: 247–274.
Vashishta, P., Kalia, R., Nakano, A., and Rino, J. (2007). Interaction potential for silicon carbide: a molecular dynamics study of elastic constants and vibrational density of states for crystalline and amorphous silicon carbide. *Journal of Applied Physics* 101: 103515-12.
Xu, Y., Zhang, J., Bai, Y., and Meyers, M.A. (2008). Shear localization in dynamic deformation: microstructural evolution. *Metallurgical and Materials Transactions A* 39: 811.
Yan, X., Tang, Z., Zhang, L. et al. (2009). Depressurization amorphization of single-crystal boron carbide. *Physical Review Letters* 102: 075505.
Zhang, C., Kalia, R., Nakano, A. et al. (2008). Deformation mechanisms and damage in alpha-alumina under hypervelocity impact loading. *Journal of Applied Physics* 103: 083508.
Zhao, S., Kad, B., Remington, B. et al. (2016). Directional amorphization of boron carbide subjected to laser shock compression. *Proceedings of the National Academy of Sciences of the United States of America* 113: 12088–12093.

6

Dynamic Deformation of Icosahedral Boron-Based Ceramics

6.1 Introduction

Icosahedral boron-based ceramics are a class of materials whose crystal structure is predominantly comprised of 12-atom, boron-rich icosahedral units. These units organize themselves into 3D-space-filling lattices and are held together by strong covalent bonds, dictated by the chemistry of boron. Industrial compounds of interest in this category are boron carbide, boron suboxide (B_6O), and boron–aluminum–magnesium (BAM) materials. The two main characteristics that give an edge to these materials are their low density (~2.5 g/cm^3) and high hardness (>25 GPa). Boron carbide, perhaps the most studied icosahedral boron ceramic, has a Vickers hardness of about 30 GPa and ranks third in hardness behind diamond (about 100 GPa) and cubic boron nitride (50–70 GPa) (see Figure 1.1). Diamond becomes chemically and structurally unstable at high temperatures while icosahedral boron ceramics exhibit superior chemical and thermal stability at elevated temperatures. Moreover, diamond is expensive, and icosahedral boron ceramics, such as boron carbide, are already being produced on an industrial scale for armor applications. Their strength-to-weight ratio is the most attractive aspect that drives their use and development; however, their main demerit is a weakening phenomenon that occurs at high pressures, termed "amorphization." In this chapter, we focus on dynamic properties of boron carbide and some other strategic, boron-based icosahedral ceramics and provide insight into efforts underway for achieving amorphization resistance.

The high-strain-rate deformation of materials above a strain rate of $10^5\ s^{-1}$ is intimately connected with deformation under high-pressure loading because of the lateral confinement that is inherent in these experimental methods. As such, the magnitude of pressure plays a dominant role in shock experiments. This is well reflected in the shock Hugoniot equations which do not contain any terms that represent strain rate or time. As early as 1969, Wilkins et al. (1969), Grady (1998), and Johnson and Holmquist (1999) recognized that the

Dynamic Response of Advanced Ceramics, First Edition.
Ghatu Subhash, Amnaya Awasthi, Dipankar Ghosh.

influence of strain rate is not significant (beyond a strain rate above $10^5\,s^{-1}$, see Figure 4.3) and that pressure alone contributes to the high-strain-rate constitutive response of materials. This statement was further supported by depth of penetration (DOP) experiments on confined alumina ceramic tiles (Anderson and Royal-Timmons 1997) whose ballistic performance was less dependent on the impact velocity but more on the confinement pressure. Thus, pressure plays a much significant role in high-strain-rate deformation than strain rate. Therefore, in this chapter, we also discuss high pressure response of icosahedral ceramics, although these experiments were conducted in the low-strain-rate regime. Examples of such experiments include indentation by a sharp indenter and diamond anvil experiments where pressures in excess of 20 GPa can be achieved (Ghosh et al. 2012; Parsard and Subhash 2017). It is also interesting to note in studies by Parsard and Subhash (2017) that the amorphization zone depth was independent of strain rate and was only dependent on applied pressure.

6.2 Atomic Structure of Icosahedral Boron-Rich Ceramics

The structure of icosahedral compounds is a key contributor to their unique combination of light weight and high hardness. Composed mostly of boron, which predominantly forms strong covalent bonds, these compounds exhibit high thermal and mechanical stability. At the crystal structure level, they are comprised of the common basic unit: the boron-rich 12-atom icosahedron. The all boron-containing, 12-atom and 20-faced icosahedrons are the basic repeat unit in α-boron referred to as $\alpha\text{-}B_{12}$ (see Figure 6.1a). The structure of boron carbide, shown in Figure 6.1b consists of boron-rich icosahedra (here 11 boron atoms and one carbon atom) interconnected by three atom chains with carbon atoms on terminal positions, each connecting to different icosahedra. A similar structure is that of boron suboxide (Figure 6.1c), which does not contain chain interconnects as those in boron carbide. Instead, one oxygen atom connects to three different icosahedra and there is no bond between the nearest-neighbor oxygen atoms. Structural complexities increase in "BAM" materials, whose structure is shown in Figure 6.2, where the boron-rich icosahedra are interconnected by metal atoms, Mg and Al.

Differences between icosahedral boron-rich compounds is either the composition of the icosahedra or covalent interconnects that join the neighboring icosahedra. Since the icosahedra are extremely stable (Emin 1987), the overall structure comprising of interconnected icosahedra is not space-filling. Because of this open structure, icosahedral boron compounds not only possess low density but also large volumes of interstitial "empty space" in their crystal make up. Figure 6.3

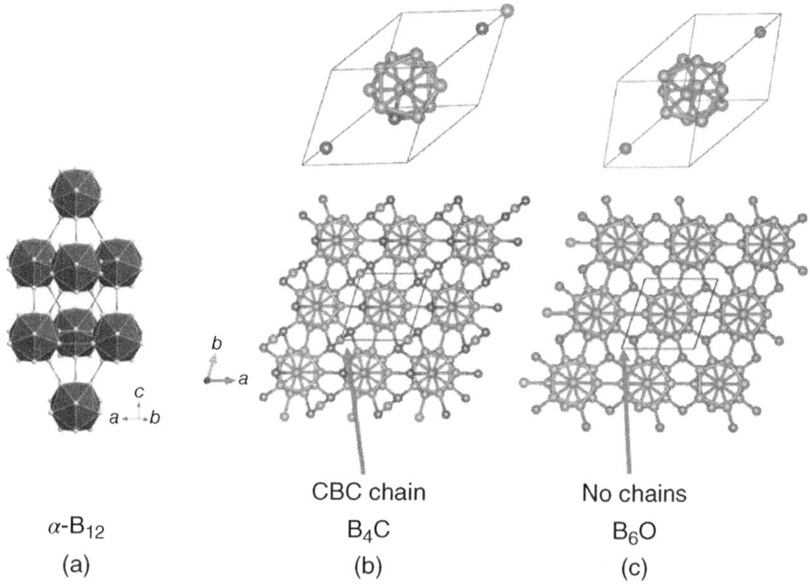

Figure 6.1 Crystal structures of (a) α-B_{12} Source: Albert and Hillebrecht (2009). © 2009, John Wiley & Sons., (b) boron carbide, (with stoichiometry ($B_{11}C_p$)CBC, hence B_4C), and (c) boron suboxide (B_6O).

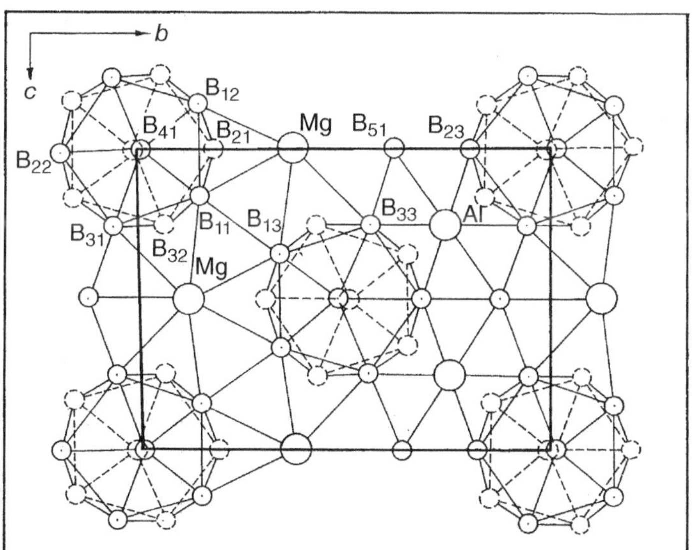

Figure 6.2 Structure of BAM ($MgAlB_{14}$) showing icosahedral arrangement of boron atoms. Source: Reproduced with permission from Matkovich and Economy (1970). © 1970, International Union of Crystallography.

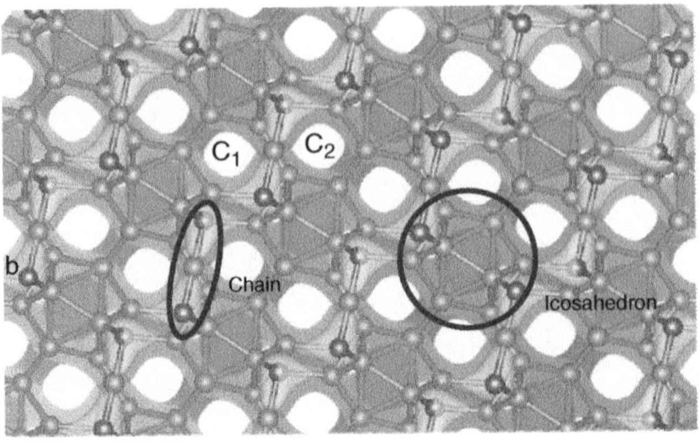

Figure 6.3 Boron carbide with $(B_{12})CBC$ stoichiometry. Two electron density surfaces evaluated by DFT depicted by dark and light grayscale. [001] view of the lattice showing icosahedra, chains, and two distinct cage spaces marked "C1" and "C2."

shows two electronic density isosurfaces (evaluated by the authors using density functional theory (DFT) calculations), where light gray indicates high density and dark gray indicates low density. Through such analysis of electronic density, we can infer that the crystal structure contains high amounts of free spaces called "cage spaces" (Subhash et al. 2016). Incidentally for the periodic structure of boron carbide shown in Figure 6.3, these cage spaces are highly directional and tend to be oriented along the [001] direction.

The interstitial space formed as a result of the icosahedral packing can accommodate additional atoms that serve as inter-connections between adjacent icosahedra. Studies show that p-block elements of the Periodic Table, mostly from Groups IV, V, and VI, could likely occupy such positions. Depending on the element and its properties, up to four atoms can be accommodated: one interstitial atom per icosahedron such as $B_{12}S$ (Lundstrom and Andreev 1996), two interstitial atoms per icosahedron including $B_{12}OO$ (or B_6O) and $B_{12}PP$ (or B_6P) (Herrmann et al. 2013; Lundstrom and Andreev 1996), three interstitial atoms per icosahedron such as $B_{12}CBC$ and $B_{12}NBN$ (Solozhenko et al. 2006; Will and Kossobutzki 1976), and four interstitial atoms per icosahedron such as $(B_{12})BBBB$ (Shirai et al. 2014). These interstitial atoms may be connected by covalent bonds (as in the CBC chains in boron carbide) or may not be bonded at all (for example, the two oxygen atoms in the B_6O are not bonded to each other and do not form a chain, see Kunka et al. (2018a and 2018b)). In these structures, the atoms of one of the above p-block elements replaces one or two atoms of boron in the icosahedron, and hence these compounds are called boron-rich icosahedral compounds.

Since the number of atoms that constitute a unit cell in a typical icosahedral compound is large (15 for boron carbide), density functional theory (DFT) calculations (Aydin and Simsek 2011; Fanchini et al. 2006) have revealed that atomic permutations at crystallographic sites is permissible, giving rise to closely resembling structures that have common stoichiometry but different atomic arrangements. This feature is known as polymorphism and is widely exhibited by boron carbide. Additionally, these compounds also have stoichiometric variations, where the proportion of boron and carbon is not fixed. The maximum carbon percentage has been found to be 33% corresponding to $B_{10}C_5$ (Konovalikhin and Ponomarev 2010).

6.3 Boron Carbide

6.3.1 Properties

The family of icosahedral boron-rich ceramics is large and boron carbide, also called "black diamond" due to its high hardness (Thevenot 1990), is perhaps the most investigated material in the collection. Combined with its potential to be mass-produced at industrial scale along with its low mass density (about 2.52 g/cm^3 (Thevenot 1990)), high hardness, high strength (about 440 MPa flexural strength, 500 MPa tensile strength (Thevenot 1990)) well above 3 GPa in compression (Hogan et al. 2016; Swab et al. 2019; Vargas-Gonzalez et al. 2010; Wereszczak et al. 2010), boron carbide has been attractive for armor applications. Boron carbide has high resistance to abrasion and that property finds application as abrasive powders and coatings, cutting and grinding tools (Lipp 1970). It has impressive wear resistance, and finds application as structural material for blasting nozzles (Jianxin 2005) and wire-drawing dies. Ceramic bearings are also made of boron carbide. The melting point of boron carbide is quite high (about 2450°C (Bauccio 1994; Pierson 1996)) and it exhibits high thermal stability rendering it suitable for refractory applications (Matkovich et al. 1977) and as strengthening agent in aluminum-based composites (Onoro et al. 2009). It has applications in the nuclear industry as well due to its neutron absorption cross section (600 barns (Lipp 1970; Matkovich et al. 1977)), neutron detection (Emin and Aselage 2005; Höglund et al. 2012; Subramanian et al. 2010), and radioactive shielding applications (Kipcak et al. 2013). Due to its p-type semiconducting behavior, high-temperature stability, and high Seeback coefficient (300 μVK^{-1}), boron carbide is also used in thermoelectric applications and high temperature electronic devices (Lee et al. 1992; Sasaki et al. 2005). This includes radiothermal generation of electricity in space applications (Kanatzidis et al. 2012). Finally, T cell-guided boron neutron capture therapy relies exclusively on boron carbide nanoparticles (Mortensen et al. 2006). For the interested reader, an excellent one-stop resource for state-of-the art in boron carbide structure, processing, properties, and applications is given by Reddy (2020).

6.3.2 Nomenclature

The complex structures of icosahedral boron-rich ceramics warrant special nomenclature to effectively differentiate them. It is unlikely for a fabricated sample of boron carbide to have the exact stoichiometry of B_4C or $B_{12}C_3$. This ceramic possesses a unique propensity to exist in multiple stoichiometries, wherein the proportion of carbon and boron may statistically vary from 8% carbon (Bouchacourt and Thevenot 1981) to 20% carbon (Niihara et al. 1984) depending on the local variations in the synthesis conditions (Shirai 2014). The high variability in atomic makeup can be challenging, especially for the atomistic modeling community. Thus, a new nomenclature with the notation (B_mC_{fz})ijk (Kunka et al. 2016), includes the required site-specificity of constituent atoms in a crystallographic unit. The notation is better described in Figure 6.4. Here, the subscript "m" refers to the total number of boron atoms in the icosahedra, "f" indicates whether the carbon occupies an equatorial ("e") or polar ("p") site, while "z,", a number, denotes the location of the icosahedral carbon atom(s) based on a site ordering scheme. Additionally, "i," "j," and "k" pertain to the elemental composition of chain atoms whose order adheres to same ordering scheme. Based on this notation, $(B_{11}C_{p5})$BCC and $(B_{11}C_{p6})$CCB have distinct representations even though stoichiometrically, each represents B_4C. Unfortunately, symmetry means that representations having distinct nomenclature could typify the same atomistic structure: e.g. $(B_{11}C_{p4})$BCC, $(B_{11}C_{p5})$BCC, $(B_{11}C_{p6})$BCC, and $(B_{11}C_{p3})$CCB all

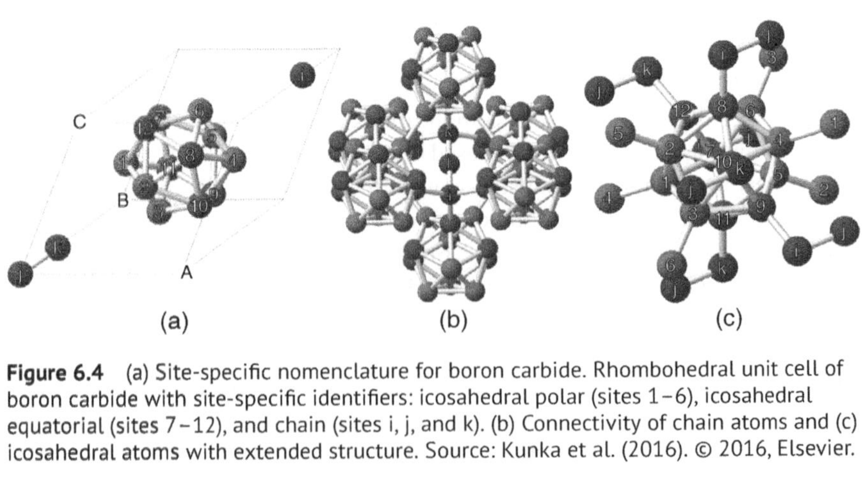

Figure 6.4 (a) Site-specific nomenclature for boron carbide. Rhombohedral unit cell of boron carbide with site-specific identifiers: icosahedral polar (sites 1–6), icosahedral equatorial (sites 7–12), and chain (sites i, j, and k). (b) Connectivity of chain atoms and (c) icosahedral atoms with extended structure. Source: Kunka et al. (2016). © 2016, Elsevier.

Table 6.1 Grouping of polymorphs for boron carbide variants.

Group	Equivalent B_4C polymorph(s)
–	zijk of $(B_mC_{fz})ijk$
i	1CBC | 2CBC | 3CBC | 4CBC | 5CBC | 6CBC
ii	7CBC | 8CBC | 9CBC | 10CBC | 11CBC | 12CBC
iii	CCC
iv	7BCC | 10CCB | 8BCC | 11CCB | 9BCC | 12CCB
v	4BCC | 1CCB | 5BCC | 2CCB | 6BCC | 3CCB
vi	1BCC | 4CCB | 2BCC | 5CCB | 3BCC | 6CCB
vii	10BCC | 7CCB | 11BCC | 8CCB | 12BCC | 9CCB
viii	3, 5BCB | 2, 6BCB | 3, 4BCB | 1, 6BCB | 1, 5BCB | 2, 4BCB
ix	5, 6BCB | 2, 3BCB | 4, 6BCB | 1, 3BCB | 4, 5BCB | 1, 2BCB
x	1, 4BCB | 2, 5BCB | 3, 6BCB

Groups of identical structures codified by different symbolic representations as per naming scheme developed by Kunka et al. (2016).
Source: Kunka et al. (2016). © 2016, Elsevier.

describe the same structure if we include full periodicity of the lattice. Kunka et al. (2016) grouped a few of such representations depicting identical atomic structures as shown in Table 6.1. This is the most effective naming scheme till date in distinguishing boron carbide polymorphs.

6.4 Amorphization in Boron Carbide

6.4.1 History of Amorphization Studies

At compressive pressures close to the Hugoniot elastic limit (HEL), boron carbide undergoes "pressure-induced amorphization" which manifests as atomic-scale catastrophic collapse of the structural order. Due to this behavior, boron carbide exhibits a loss of strength at high pressures. The phenomenon was first observed by Grady (1994) in shock experiments as an anomalous "post-HEL softening" behavior as opposed to "post-HEL hardening" in other structural ceramics (e.g. SiC and Al_2O_3) (see Figure 5.13). Later, Dandekar (2001) also noted similar behavior in shock experiments and concluded that boron carbide likely undergoes a phase transformation of some kind. Domnich et al. (2002) first conducted Raman spectroscopy on indented specimens of boron carbide and found some unique spectral characteristics of amorphized material. New spectral bands were found to appear at 1340, 1520, and 1810 cm^{-1}, and the peak at 1340 cm^{-1} was particularly broad and relatively high in intensity than the remaining spectrum. That overall broadening of the most intense peak and perhaps based on its proximity to spectral positions of amorphous carbon, Domnich et al. (2002) termed this behavior "amorphization." Chen et al. (2003) conducted transmission electron microscopy (TEM) analysis of recovered fine particles of boron carbide after ballistic impact and noted the presence of amorphous zones in boron carbide. These zones were slender bands measuring a few nanometers in thickness whose diffraction pattern contained a diffused halo, typical of amorphized phases. These zones were surrounded by crystalline material whose diffraction pattern revealed a rhombohedral structure. Surprisingly, no separation of material (cracking) between the crystalline and amorphous phases was observed, indicating a smooth transition between the two phases. Chen et al. (2003) also noted that low velocity impacts below 900 m/s on boron carbide created large fragments (same as other structural ceramics), while high-velocity impact above 900 m/s caused a higher fraction of comminuted material near the impact zone which was associated with amorphization-induced excessive cracking.

In 2004, Ge et al. (2004) expanded upon the work of Domnich et al. (2002) and Chen et al. (2003) and published both TEM imagery as well as more elaborate Raman spectra (Figure 6.5) of five different test samples of boron carbide: (i) pristine B_4C, (ii) indented, single-crystal B_4C, (iii) indented polycrystalline B_4C,

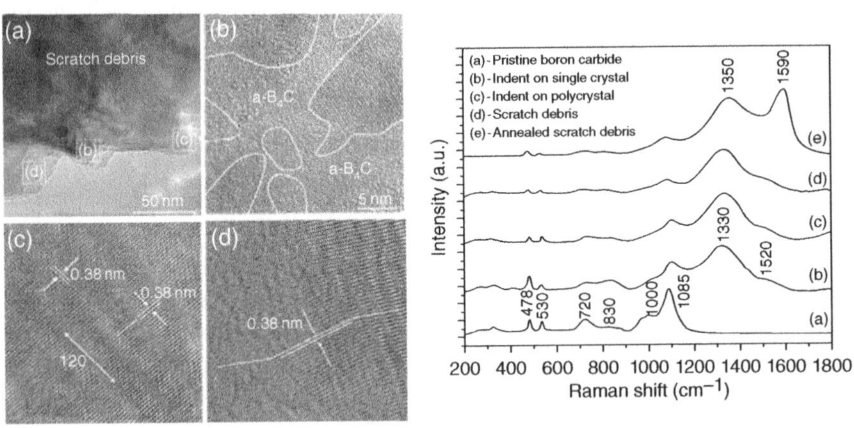

Figure 6.5 (Left) TEM images of amorphized regions of scratch debris of boron carbide. (b)–(d) are magnifications of boxed regions depicted in (a). (Right) Raman spectra for five different boron carbide material systems: (a) pristine single crystal boron carbide, (b) indented single crystal, (c) indented hot-pressed polycrystal boron carbide, (d) scratch debris of single crystal, and (e) annealed scratch debris. Source: (a) Reproduced with permission from Ge et al. (2004), (e) Ge et al. (2004). © 2004, Elsevier.

(iv) scratch debris of B_4C, and (v) annealed scratch debris of B_4C. It was found from TEM images of both indented and scratch debris samples that the amorphization zones were similar to those in shock-recovered samples of Chen et al. (2003). They were slender in shape and measured a few nanometers across. Raman spectral data obtained by a low-energy laser for crystalline as well as indented or scratch-recovered samples demonstrated similar results as those of Domnich et al. (2002). When a high-energy laser was deployed, it resulted in a new peak situated around $1590\,cm^{-1}$ with a high intensity, suggestive of creation of graphite. When scratch debris was viewed with TEM, it was inferred that scratching of single crystal boron carbide produces different results: (i) formation of nanocrystals, (ii) lattice shearing, and (iii) lattice rotation. It was thought that at least one of two mechanisms could be under play: (i) either bonds break, followed by some local rotation to accommodate shear strain, and/or (ii) a local phase transformation occurred under high pressure, and upon unloading a new equilibrated structure with randomly oriented nanocrystalline grains appeared. Based on additional TEM studies on nanoindented single crystal on the (001) surface, it was speculated that the single crystal was likely broken under pressure, perhaps into nanocrystals. When amorphized regions in a polycrystalline sample were analyzed by electron energy loss spectroscopy (EELS), the presence of an enhanced π^* peak in the carbon K-edge indicated the occurrence of sp^2 bonded carbon. Perhaps this feature added more impetus to the speculation about the presence of graphite in the amorphous domain.

Vogler et al. (2004) performed a series of planar impact experiments to assess the shock and release behavior of boron carbide. The HEL was in the range of 15.4–17.3 GPa. The analysis of experimental results argued for the existence of a phase change beyond the HEL at about 40 GPa, yet no conclusive evidence validated that hypothesis. It was noted that the material exhibits elastic/plastic behavior in both loading as well as unloading phases from the shocked state and that shear stress in the shocked state was low (up to about 2 GPa). Moreover, shear strength in the shocked state was high (10–15 GPa), maintained about the same as that at HEL.

Though Domnich et al. (2002) hesitated to label new spectral peaks of amorphized B_4C as D- and G-modes emanating from graphite and free carbon respectively, yet subsequent studies used this terminology rather abundantly. The temperature dependence of Raman spectra by Yan et al. (2006) demonstrated temperature sensitivity of amorphized Raman spectrum. In this work, low- and high-temperature Raman spectroscopy of amorphous boron carbide produced by nanoindentation were studied to assess nanoscale structural changes due to temperature. Indented boron carbide was heated from 83 to 873 K and *in-situ* Raman spectroscopy was used for characterization of evolved structural changes. From 83 K to room temperature (Figure 6.6), the peak shifts associated with

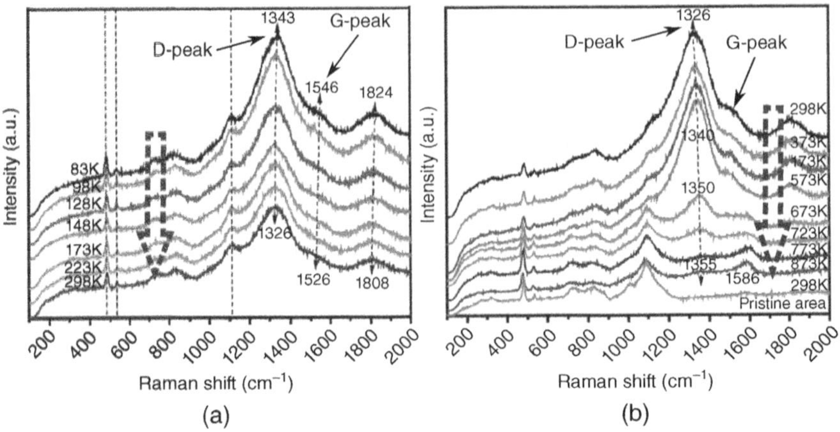

Figure 6.6 Effect of heating amorphized boron carbide from (a) 83 to 298 K and (b) from 298 to 873 K. Dashed arrows depict increasing temperature. Experiments conducted by Yan et al. (2006). Source: Awasthi and Subhash (2020). © 2020, Elsevier.

amorphized material, namely the D- and G-modes, were found to be more sensitive to temperature change, undergoing significant shift than crystalline peaks, which barely demonstrated any response to temperature change. Interestingly, all these peak shifts were left shifts. Upon heating the sample from room temperature to 673 K, the intensity of the D-band was found to diminish significantly while a clearly discernible G-band appeared. Further heating to 873 K destroyed the D-band as the G-band became more prominent. Heating from room temperature up to 873 K also made the $1080\,cm^{-1}$ crystalline peaks prominent, and the $1800\,cm^{-1}$ band gradually disappeared. The authors stressed that the appearance of the G- and D-bands in amorphized boron carbide is due to the separation of carbon atoms from chains. This produces two products, sp^2-hybridized carbon that exhibits the so-called G-band, and clustered carbon, which results in the appearance of the so-called D-band. They also added that boron atoms belonging to chains may reside inside the sp^2 hybridized carbon cluster due to the potential impact of temperature rise. It is important to note that there is no strong experimental scientific evidence that proves the emergence of D- and G-bands in amorphized boron carbide. There are no reports of compositional characterization of the amorphous material to validate existence of segregated phases whether graphitic or carbon clusters. Some speculations were given to the appearance of a Raman-band at $1810\,cm^{-1}$ in this work, suggesting it could arise from stretching of the carbon–boron double bond between carbon aromatic rings. This claim has never been validated till date and the appearance of a Raman peak near $1810\,cm^{-1}$ in amorphized boron carbide has been a mystery and has never been resolved.

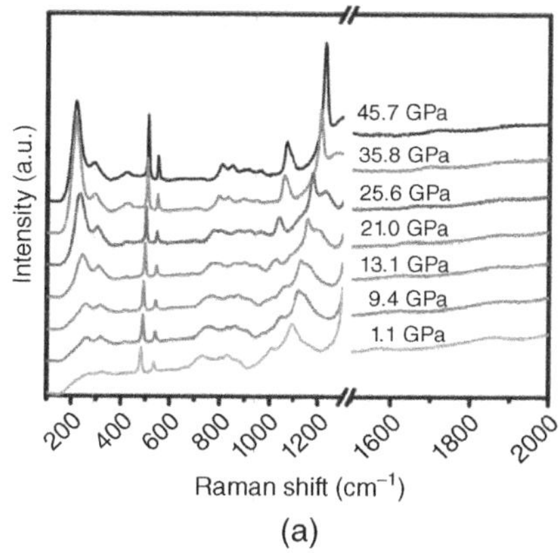

Figure 6.7 Depressurization amorphization experiment of Yan et al. (2009) involving loading and unloading boron carbide sample in a diamond anvil. Raman spectra for different levels of pressure during depressurization cycle for (a) hydrostatic loading (b) non-hydrostatic loading. Amorphization occurs in the non-hydrostatic case as evident from lack of reproduction of crystalline peaks and emergence of new peaks when unloaded beyond 20.2 GPa. Source: Yan et al. (2009). © 2009, American Physical Society.

Chronologically, the next evidence of amorphization in boron carbide was found in high-pressure diamond anvil experiments conducted by Dandekar et al. (2008) and Yan et al. (2009). In these experiments, large pressures up to several tens of GPa can be applied and the Raman spectral shifts (intensities of peaks and their positions) as a function of applied pressure is simultaneously monitored. When depressurized from high, non-hydrostatic loads up to 50 GPa, boron carbide exhibited a drastic change in its Raman signature (see Figure 6.7) at about 20 GPa, which closely matched that of the Raman spectrum of indented boron carbide (Domnich et al. 2002), as well as TEM observations of Chen et al. (2003) and Ge et al. (2004). Hydrostatic analog of this experiment showed no evidence of amorphization for loads up to 45 GPa Yan et al. (2009). It was also found that the Raman bands associated with amorphization do not show if the maximum pressure was less than about 25 GPa. At a pressure beyond 45 GPa, the distribution of amorphous regions was heterogeneous. The authors also observed that conventional deformation-induced defects such as twins, stacking faults, and dislocations were absent. *Ab initio* molecular dynamics (AIMD) was used to further investigate the atomistic origins of amorphization. It was found that hydrostatic stresses do not cause discontinuities in the pressure–volume relationship, but rather the uniaxial loading response consisted of several steep volume drops as the pressure increased. This was attributed to the bending of C–B–C chains and to their bonding with atoms of neighboring icosahedra. Note that it was formation of these new bonds that prevented recovery of the original structure and caused permanent deformation, which was argued to be amorphization in the bulk material. The structures were also analyzed by X-Ray diffraction (XRD) to understand deformation of the lattice (Dandekar et al. 2008). It was found that XRD did not show any evidence of phase transition up to a pressure of 47 GPa. It was inferred that the loss in crystallinity happened at submicron levels and was not detected by XRD. In fact, it was concluded that amorphization in boron carbide is not substantial or widespread, but, rather, it occurs in small domains.

Once identification of amorphization behavior of B_4C through TEM and Raman spectroscopy was accomplished, newer works focused more on its characterization and evaluating some key metrics associated with this phenomenon. Ghosh et al. (2012) and Parsard and Subhash (2017) studied the influence of the loading rate on amorphization characteristics by utilizing static and dynamic indentations on boron carbide and found that amorphization is actually a pressure-driven phenomenon, independent of strain rate. Static and dynamic indentation tests produced the same amorphous Raman signature, regardless of the strain-rate deployed. By incrementally removing material over submicron depths from the indented surface and obtaining Raman signature of the exposed surface at each depth, Subhash et al. (2013) noted that the amorphization (D-peak) intensity initially increases and then decreases, as seen in Figure 6.8a. By plotting the area beneath the D-peak at successive depths from the indented surface and stacking these planes, they generated a 3-D map of the amorphized zone beneath the Vickers indentation (see Figure 6.8b), which indicates size, shape as well as distribution of amorphization intensity. It was found that the amorphized zone in boron carbide extended to a depth of almost seven times the indentation depth.

Later studies by Parsard and Subhash (2017) revealed that the amorphized zone size beneath the indentation is independent of loading rate and increased linearly

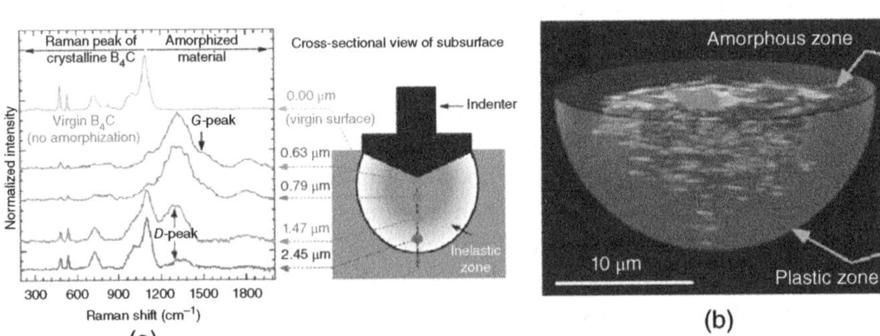

Figure 6.8 (a) Raman spectra evolution as a function of depth beneath the indentation and (b) 3D map of area under the D-peak of Raman spectrum beneath the indentation. Source: Subhash et al. (2013). © 2013, Elsevier.

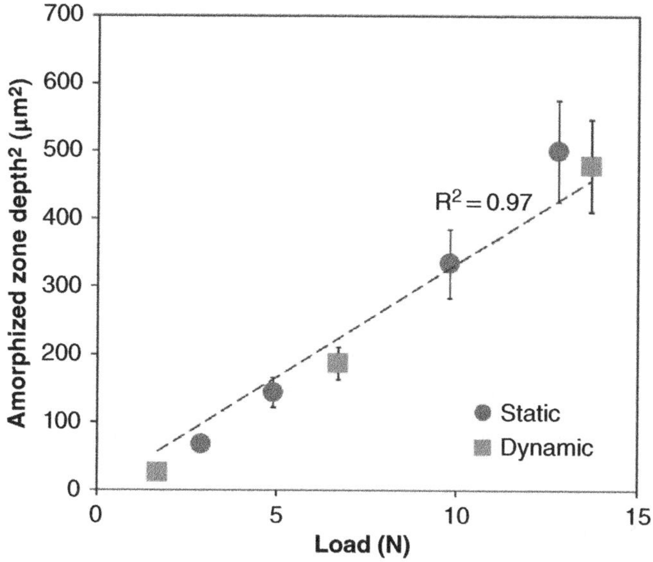

Figure 6.9 Amorphization zone depth as a function of load for static and dynamic indentations for boron carbide. Source: Parsard and Subhash (2017). © 2017, Elsevier.

with load (Figure 6.9). Further analysis of Raman spectra from virgin and indented surfaces also revealed that the 1088 cm^{-1} peak shifts a few wavenumbers to the right, revealing compressive stress build up due to amorphization. Since a completely amorphized region (i.e. lack of atomic order) does not show well-defined peaks in the Raman spectrum, it is further inferred that amorphized state causes a change in volume, and this volume expansion introduces compressive stresses in the surrounding crystalline matrix.

Zhao et al. (2016) performed laser shock on boron carbide, and its TEM imagery showed the same amorphization features as in Figure 6.5a. The conclusion from all these experiments points to the inference that the three tests, indentation, shock, and diamond anvil, all produce similar TEM imagery while indentation and diamond anvil tests produce identical Raman spectra for boron carbide. It is now well known that the effect referred to as "post-yield stress softening" (Grady 1994) or possible "phase transformation" (Dandekar 2001; Grady 1994) was in fact the occurrence of amorphization, which has been confirmed extensively through TEM and Raman spectra. It is important to note that this behavior occurs in isolated nanoscale islands or long streaks and is not uniformly or homogeneously distributed in the specimen (Figure 6.10) (Subhash et al. 2016). Also note from this figure that dislocations can be seen within the amorphization bands indicating the dominance of shear for their occurrence.

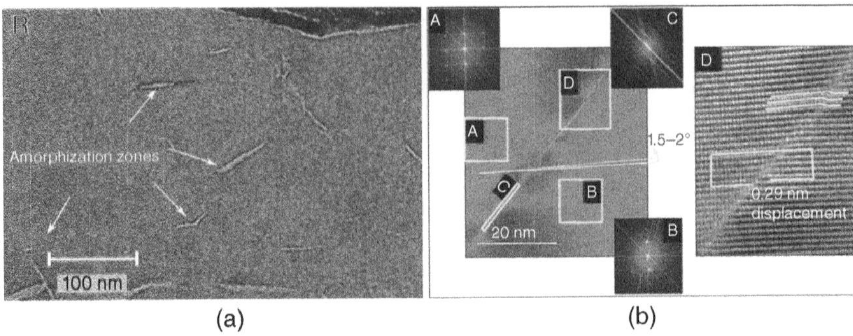

Figure 6.10 (a) Typical TEM image of amorphized islands and long streaks beneath an indentation. Source: Reproduced with permission from Subhash et al. (2016). (b) Close up of TEM and Fast Fourier Transform of TEM image depicting evidence of amorphization. Also note dislocations along the amorphization bands in the image on the right end.

6.4.2 Theoretical/Computational Works

The need for understanding amorphization fueled several computational works focused on obtaining atomic-scale knowledge of the amorphization process. The first work was by Fanchini et al. (2006) who performed DFT-based, Gibbs free-energy calculations to assess possible origins for the failure of boron carbide just above the HEL. Calculations showed that the energetic barrier for shock amorphization was weakest for the polymorph $B_{12}CCC$. It was suggested that a meta-phase may exist energetically during high-pressure conditions, consisting of layers of graphite (Figure 6.11) situated along inter-icosahedral voids. This work focused on determining the transition pathway and associated energetics involved in the structural collapse of $(B_{12})CCC$ to finally produce a combination of pure α-boron (α-B_{12}) and graphite. Using Gibbs free-energy results, it was further demonstrated that boron carbide polytypes (e.g. $(B_{11}C)CBC$) went through distinct steps of energetically favored site exchanges of specific carbon and boron atoms, to achieve configurations of a segregated graphite phase. The process of "$(B_{12})CCC => B_{12}$ + graphite" was termed amorphization, which different polytypes and polymorphs took, after first transforming into $(B_{12})CCC$. It was further concluded that the failure threshold in an experimental sample might largely depend on the amount of $(B_{12})CCC$ polytype in the mix. This theory of formation of meta-phases was built using evidences of Raman spectral features of amorphized boron carbide including appearance of the so-called G-band to be linked with formation of segregated graphite from Domnich et al. (2002).

Aryal et al. (2011) conducted DFT simulations of B_4C under hydrostatic and uniaxial loading conditions, demonstrating that boron carbide undergoes a phase transformation at high stress levels, especially above pressures with a

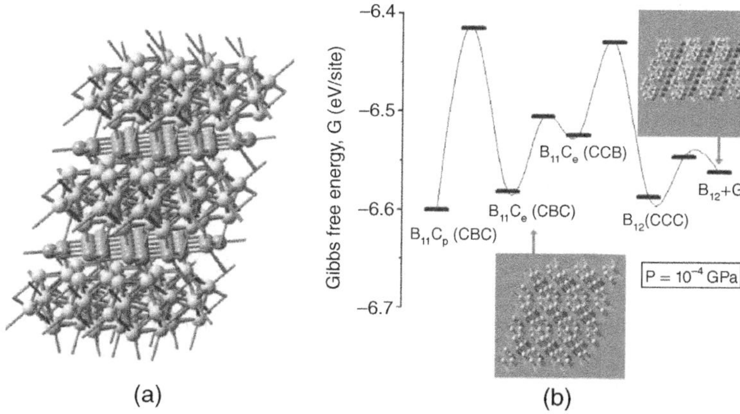

Figure 6.11 (a) Proposed layered structure of amorphized boron carbide consisting of alternating layers of graphene and alpha-boron. (b) DFT-computed energy pathway for obtaining the alternating layered structure starting with pristine ($B_{11}C_p$)CBC. Source: Fanchini et al. (2006). © 2006, American Physical Society.

corresponding strain of 0.2. In this work, amorphization was identified as the phenomenon causing sudden jumps in energy and volume which were accompanied by severe distortion of icosahedra and formation of new bonds between chain and icosahedra around the threshold strain of 0.2.

More recently, the DFT work of An et al. (2014) utilized shear loading along the weakest slip direction $\left(01\overline{1}\overline{1}\right) < \overline{1}101 >$ of boron carbide to arrive at the shear strength of boron carbide. In this work, the definition of amorphization was that as the external load increases, there is a predominance of local failure characterized by breaking of inter-icosahedral B—C bonds and causing formation of "shear transformation zones." Additional application of load causes distortion within this zone, further leading larger planar bands and ultimately, "amorphous bands," which are much larger. Atomistic structures residing in these shear transformation zones undergo bond rupture both within the icosahedra and the chains. Thus, according to this definition, amorphization involves complete destruction of original crystal structure. It is important to note that while this theory admits a complete breakage of bond, it is not compliant with other DFT-based analyses, where amorphization did not include bond-breaking events (Aryal et al. 2011; Yan et al. 2009) or where bond breaking was followed by a unique recombination (Fanchini et al. 2006).

Some studies were performed on boron carbide polymorphs using the AIMD method by Taylor (2015), making use of the Hugoniotstat ensemble. The primary definition of amorphization here was a sudden energy jump in the Hugoniot, caused due to irreversible bending of chains. These studies hypothesized that icosahedral deformations occurred at a larger time scale and hence are not captured in the AIMD framework. Those suggestions have not been validated. Additionally, the use of a Hugoniotstat ensemble for softening materials like boron carbide has not been verified. In the realm of AIMD simulations on boron carbide, Shamp et al. (2017) evaluated liquid Hugoniot for boron carbide for pressures up to 1.5 TPa. These investigations assumed that high pressure causes discontinuities in Hugoniot, first around 50 GPa and then in the range 90–100 GPa, which segregate solid boron carbide into elemental components, boron and carbon, beyond 100 GPa pressure. These studies did not comment much about qualitative aspects of structural transformation or amorphization that occurs between solid and liquid phases. Amorphization was rather assumed to lead to a complete segregation between solid and liquid phases for convenience of simulations.

The literature on boron carbide also contains large-scale system (over 200 000 atoms) simulated by reactive molecular dynamics (RMD) using the ReaxFF potential (An and Goddard III 2015), capable of modeling bond breaking and formation events, to examine shear deformation (Figure 6.12). These studies examined large shear deformations, (up to 175%) for two specific slip-systems, $(0001) / < 10\bar{1}0 >$ and $\left(01\bar{1}\bar{1}\right) / < \bar{1}101 >$, known to be key players in deformation twinning and amorphous band formation in boron carbide. The mechanism for amorphization was suggested as: under large deformations, the $(0001) / < 10\bar{1}0 >$slip system causes twin formation, leading to amorphization bands, cavitation, and ultimately crack opening. On the other hand, the $\left(01\bar{1}\bar{1}\right) / < \bar{1}101 >$ slip-system formed amorphous bands directly without producing any intermediate twins. It was further theorized that once the amorphous bands form, cavitation occurs, leading to crack opening. These studies indicated that inclusion of bond breaking and formation can capture vital information about the amorphization phenomenon. On similar lines, it is also known that the development of potentials for α-boron for large-scale computational molecular dynamics (MD) is in progress (Pokatashkin et al. 2015).

The twinning feature in B_4C has been investigated by Xie et al. (2015) using DFT simulations and TEM, and it was noted that shear deformation can lead to symmetric as well as asymmetric twins. Two polymorphs, namely $(B_{11}C)CBC$ and $(B_{12})CBC$, were fabricated as part of this study with respective stoichiometries of B_4C and $B_{13}C_2$. Of these polymorphs, B_4C possessed both symmetric and asymmetric twinning features while $B_{13}C_2$ was found to exhibit only symmetric twinning. Perhaps because the atomistic structure of $(B_{11}C)CBC$ contains asymmetric icosahedra (due to the presence of a polar C-atom), it is possible to obtain both symmetric as well as asymmetric twinning driven by shear. Since there are no polar carbon sites in $B_{13}C_2$, it supports only symmetric twins.

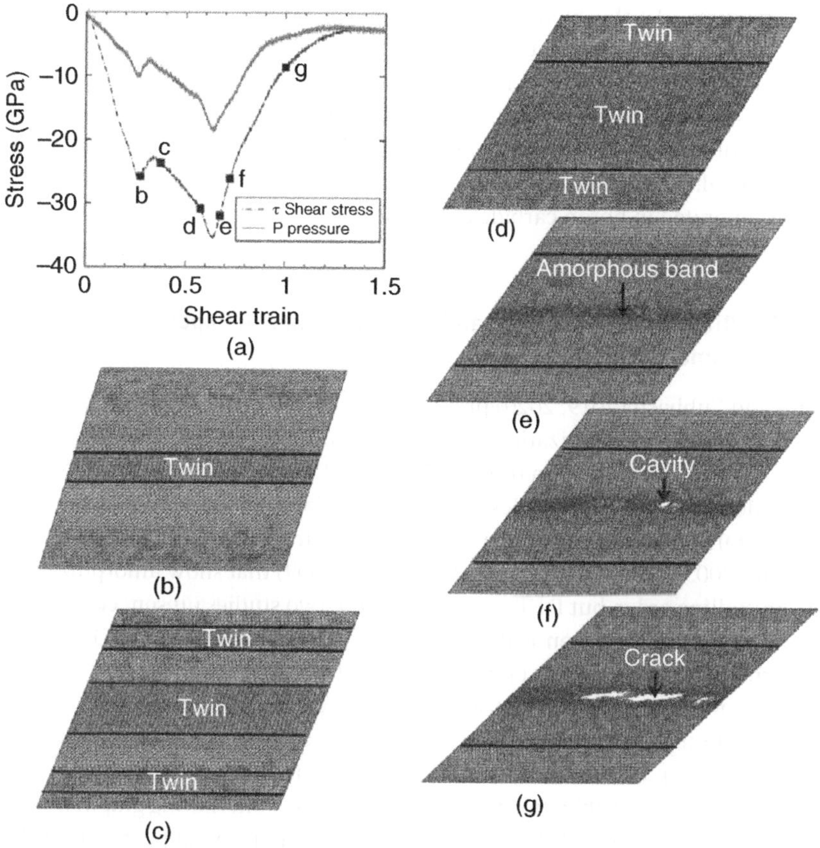

Figure 6.12 Simulation of boron carbide dynamic response to shear loading using molecular dynamics with the ReaxFF potential. (a) Stress–strain response, (b)–(g) Evolution of damage in the simulated domain with increasing shear load, including formation of twins, amorphous band, cavitation, and cracking. Source: An and Goddard III (2015). © 2015, American Physical Society.

Zhao et al. (2016), developed an analytical representation of temperature versus pressure of boron carbide during shock. They hypothesized two drivers of temperature increase: homogeneous shock temperature rise and temperature rise due to localized shear. At sufficiently low-shock pressures (below the HEL), local temperature increases arise from homogeneous compression of the material, and it is insufficient to cause melting. However, as pressures increase above the HEL of boron carbide, shear localization and amorphization occur, and some amount of the accompanying lattice friction from the shear band contributes to the overall temperature rise. This shear-driven temperature rise increases for higher pressures. As the melting temperature of boron carbide reduces fairly substantially with increasing pressure, Zhao et al. (2016) argued that at sufficiently high shock pressures, the shear localization heating might surpass the melting point at least in highly localized "hotspots," causing melting and amorphization of boron carbide. Recent investigations through molecular dynamics simulations by DeVries et al. (2020) indeed revealed that temperature well in excess of melting point are generated in boron carbide as will be shown later.

6.4.3 Conflicts Among Theories and Commonalities in Experimental Evidence of Amorphization

Awasthi and Subhash (2019, 2020) promulgated that there are many gaps among attempts to explain amorphization in boron carbide. On one hand, there is good evidence to spot amorphization through TEM and Raman spectroscopy; however, if we lay down all attempts of describing amorphization and mechanisms proposed, several conflicts are noticed. For instance, the framework of TEM studies by Chen et al. (2003), Ge et al. (2004), and Yan et al. (2009) that show amorphization has no crystalline order, but DFT simulations from two studies present conflicting models where amorphization is due to chain bending (Aryal et al. 2011) or due to the formation of new bonds (Taylor 2015). Further, both of these DFT studies directly disagree with An et al. (2014), who proposed amorphization to be a complete destruction of icosahedra. These later studies also contradict the definition of amorphization put forward by one of the early works in this area by Fanchini et al. (2006) where amorphization was coined as layers of alternating graphene and α-B_{12}. Awasthi and Subhash (2020) also highlighted that the appearance of the so-called D- and G-bands in the Raman spectral signature of amorphized boron carbide is likely speculative too, since existence of carbon in amorphized regions, whether in the form of graphite or free carbon is a mere hypothesis; it has not been confirmed or proved by any deterministic experimental method.

Despite the differences in approaches toward amorphization, there are three unique commonalities in the pressure-induced amorphization in boron carbide across several experiments: (i) amorphous islands are of nanometer size (see Figure 6.13a), (ii) the Raman spectrum of amorphized boron carbide is same across different experiments (Figure 6.13b), and (iii) the diffraction pattern of the amorphized regions shows a distinct "halo," indicative of the loss of local order (Figure 6.13c). Note that Figure 6.13b also depicts a lesser discussed feature of the spectrum: the dominant peak at 1090 cm^{-1} drifts right irreversibly upon amorphization. In light of these observations, a new theory has been proposed recently from Awasthi and Subhash (2019) which is based on thermodynamics of the amorphization phenomena and makes use of MD and DFT simulations.

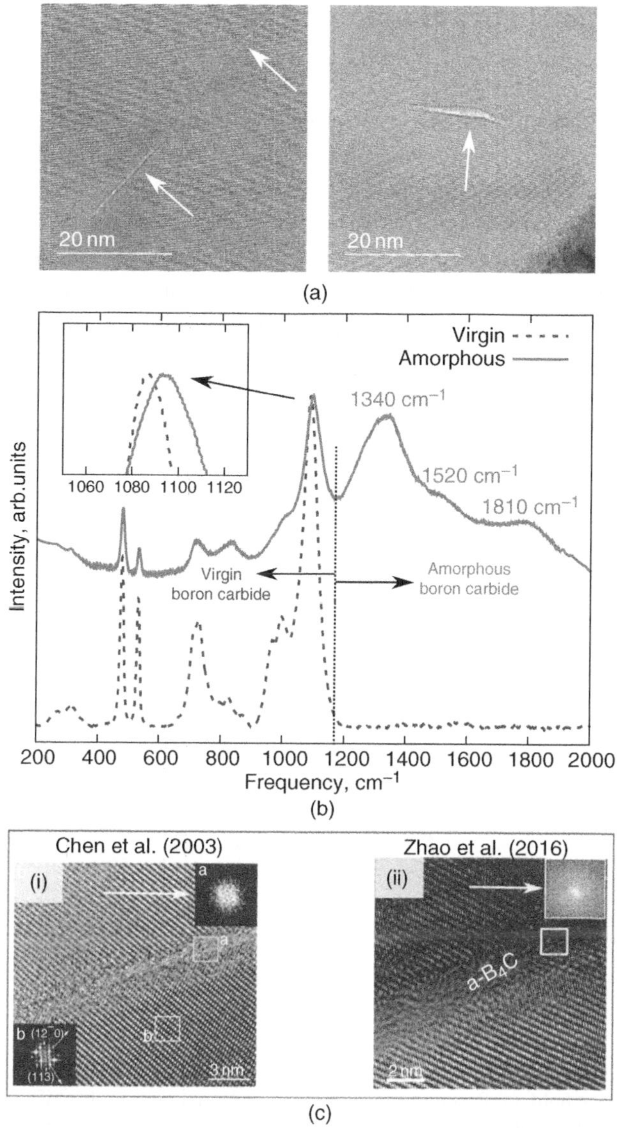

Figure 6.13 (a) Experimental evidences of amorphization in boron carbide. (a) TEM images of boron carbide depicting nanosized amorphous domains. (b) Raman spectra of crystalline and amorphous boron carbide. Amorphous spectrum shows three new spectral features, appearing at 1340, 1520, and 1810 cm^{-1}. Inset shows a less-commonly-discussed feature wherein the crystalline peak (centered around 1090 cm^{-1}) right-shifts in the amorphous material by a few wave numbers. (c) Identical Fourier analysis of amorphous domains created due to (i) mechanical shock and (ii) laser shock showing significant loss of structural order. Source: (a) Reproduced with permission from Awasthi and Subhash (2019), (b) Awasthi and Subhash (2019). © 2020, Elsevier.

6.5 Thermodynamic Theory of Amorphization and Raman Spectra

Awasthi and Subhash (2019) rationalized amorphization in boron carbide to be a complex phenomenon with thermodynamics as the driving agent (as also proposed by Zhao et al. (2016)). They used MD simulations to hydrostatically compress a domain of boron carbide to demonstrate that a high pressure of the order of 40–50 GPa is sufficient to raise the temperature to nearly the melting point and observed loss of order (amorphization) due to bond breaking. These results laid the framework for studying the interplay between temperature and pressure, wherein it was identified that high pressure causes a rise in temperature, which leads to a loss of integrity in local structure through several bond-breakage events (amorphization). It was argued that high pressure brings atoms together, forcing them against their interatomic repulsion, thereby increasing kinetic energy and hence the rise in temperature. For example, in Figure 6.14 it is shown that for a volume compression of 30%, a pressure of 70 GPa and a temperature rise of 2100 K was observed (Awasthi and Subhash 2020). Thus, amorphization

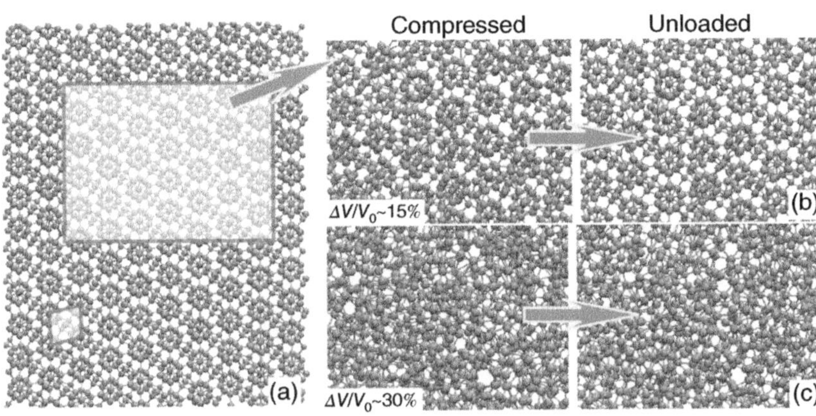

Figure 6.14 Levels of amorphization in volumetric compression tests performed using MD utilizing the ReaxFF force field. (a) Periodic cuboidal computational domain for MD analysis showing the smallest rhombohedral unit cell. The marked rectangular window of observation is same in (b) and (c) that depict the atomic structure for volumetric compressive strain, $\left(\frac{\Delta V}{V_0}\right)$, up to 15% and 30%, respectively and then after unloading. For $\left(\frac{\Delta V}{V_0}\right) = 15\%$, the temperature and pressure went up to about 725 K and 30 GPa, respectively, and residual structural damage is low. For $\left(\frac{\Delta V}{V_0}\right) = 30\%$, temperature and pressure levels were about 2100 K and 70 GPa and the unloaded structure has no residual order. Source: Awasthi and Subhash (2019). © 2019, AIP Publishing.

of the kind seen in high pressure shock experiments in boron carbide, which is also termed pressure-induced amorphization, is the result of portions of the material having sufficient temperatures to favor bond-breaking (Awasthi and Subhash 2019, 2020). Note that such behavior is different from amorphization caused in a shear test (An and Goddard III 2015), where the test-structure fails in tension and never undergoes high-pressure conditions such as those in a shock test. Recently, DeVries et al. (2020) substantiated the proposed thermodynamic rationale through simulated shock analysis of boron carbide using MD, where it was observed that temperature-mediated structural changes due to shock produced highly comparable radial pair distribution function (RPDF) plots to those from MD-based volumetric compression studies of boron carbide (Awasthi and Subhash 2019). These results were earlier discussed briefly at the end of Chapter 5.

After the external pressure is removed, the temperature ceases to rise at the same rate, while atoms with dangling bonds cannot re-bond to their original ordered configuration. They rather remain free and come under the influence of each other's non-bonded (van der Waal's type) interactions. Using force terms from van der Waals interactions, it has been shown that free carbon and boron atoms exert internal pressures up to 150 and 450 GPa, respectively, if spatially constrained in a volume to produce the same density as boron carbide under ambient conditions, $3.52\,\mathrm{g\,cm^3}$ (see Supplementary Information in Awasthi and Subhash (2019)). This occurs because equilibrium separations are larger for unbonded atoms than covalently bonded ones. It is argued (Awasthi and Subhash 2019) that such conditions prevail in nanoscale pockets, especially near defects with a higher stress-concentration. Given the high levels of internal pressure in nanoscale pockets created upon unloading, it can be easily rationalized that the nanoscale amorphous domains exert that high pressure on the surrounding crystalline matrix. This process causes the nanoscale amorphous domain to expand, while creating compressive stresses in the surrounding crystalline domain. Such a physical explanation of the amorphization process rationalizes the experimental observation of "depressurization amorphization" of Yan et al. (2009) and "volume expansion" hypothesis for amorphous domains (Parsard et al. 2018). For additional discussion about pressures caused by amorphous atoms, the reader is referred to section 5.4 of Awasthi and Subhash (2020).

Exhibition of unique Raman spectral features in amorphous boron carbide is as integral to the phenomenon as observation of loss of structural order in TEM. As discussed before, this topic too has been shrouded with speculations. Following their model of amorphization of boron carbide, Awasthi and Subhash (2019) gave a new perspective of these unique spectral features in amorphized material. According to this explanation, there exists a locked-in state of stress near the amorphous zone, which includes the amorphized material in nanoscale islands and its immediate neighborhood of crystalline material. The amorphized zone, likely consisting of free atoms, exerts pressure on the boundary between the phases, causing compressive stresses in the crystalline domain. When this tiny, isolated, nanoscale-amorphized material surrounded by a large volume of compressed crystalline phase is studied under a Raman spectrometer, the majority of the data collected from a scan is representative of the compressed crystalline material. Awasthi and Subhash (2019) argued that given the diameter of laser used in current Raman instruments, which is about 1 μm, information from the nanoscale amorphous regimes is merely noise and the majority of the response is reflective of the surrounding crystalline material, which, based on the arguments above, is under a state of compression, including very high levels near the interface, of up to 150–450 GPa. Such a locked-in state of material is not new; in fact analytical solutions have been proposed for similar cases in the mechanics community through the work of Eshelby (1957), followed by the micromechanics field of research (Budiansky 1965; Christensen 1990; Hashin 1960; Mori and Tanaka 1973) that focuses on evaluation of stress distribution in composite materials. Awasthi and Subhash (2019) also evaluated Raman spectra of boron carbide for different levels of hydrostatic pressures using DFT, which correlated well with the experimental literature (Yan et al. 2009). Notable features of this comparison were the nature of shifts of some chosen peaks (see Figure 6.15). Overall Raman spectrum was hypothesized to be a superposition of spectra based on a radially decaying stress from an amorphization island. It was assumed that all the spectral response emanates from crystalline regions surrounding and the amorphized island contributes negligible to the overall spectrum. An example of such a superposition is shown in Figure 6.16 (Awasthi and Subhash 2019). This is the closest any theory of amorphization of boron carbide has ever compared with experimental data. Note that the locations and relative intensities of peaks in the overall predicted spectrum lie in good proximity of experimental results. Additionally, the unexplained peak at 1810 cm^{-1} was predicted too. For more details of DFT modeling of Raman spectra of boron carbide and locked-in state of stresses following amorphization, refer to Awasthi and Subhash (2020).

This theory answers many ambiguities of nanoscale attributes of material structure and their contribution to Raman spectra that have given a new insight into understanding of amorphization in boron carbide. More evidence of the validity of the model of Awasthi and Subhash (2019) reflects from the experimental work of Yan et al. (2006) who showed that the transition from crystalline to amorphous phases in boron carbide reverses by heating to over 800 K and then cooling. According to the model of Awasthi and Subhash (2019), amorphized atoms, which had separated from the parent structure, cannot return to their original locations, but upon heating they absorb energy to enhance that probability. If the cooling cycle were slower, likely the same response would be achieved as well. The model of Awasthi and Subhash (2019) also explains the less-discussed small right shift

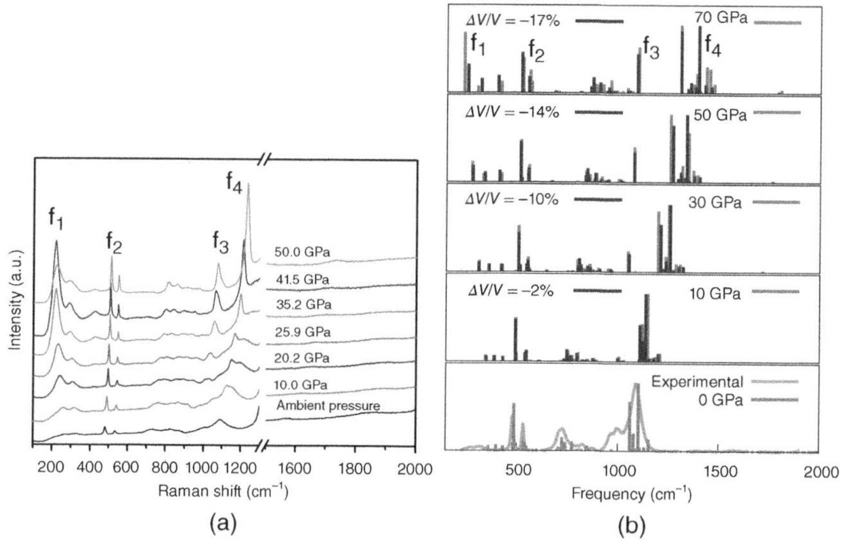

Figure 6.15 Comparison between (a) experimental (Yan et al. 2009) Raman spectra of boron carbide at different levels of hydrostatic pressures using a diamond anvil test set up and (b) DFT-computed impact of volumetric compression (dark gray) (Awasthi and Subhash 2020) and hydrostatic pressure (light gray) (Awasthi and Subhash 2019) on the Raman spectral signature of boron carbide. Four similarities are identified between experimental behavior and computational results and are depicted by f_1–f_4. f_1, left shift of low frequency peaks; f_2, mid-spectrum peaks situated near 480 cm^{-1} exhibit slight right shift; f_3, a new peak emerges near 1050 cm^{-1} and drifts right; and f_4, most dominant peak of ~1090 cm^{-1} grows in intensity and shifts right. Source: Awasthi and Subhash (2020). © 2020, Elsevier.

of the dominant crystalline peak of 1080 cm^{-1} in the experimental Raman spectrum of amorphized specimen (Figure 6.13b). This shift arises due to relatively lower levels of compressive stresses (about 1–5 GPa) in the matrix material adjacent to the amorphized zone, which can also be understood from the perspective of the superposition process. This rationalization is also consistent with TEM images showing only tiny nanoscale islands of amorphized material (Chen et al. 2003; Zhao et al. 2016), along with XRD studies of Dandekar et al. (2008) which proved that amorphization is not widespread.

6.6 Simulating Shock in Boron Carbide with MD Simulations: Amorphization and Thermodynamics

Most recently, MD simulations of shock response of amorphization in boron carbide have emerged. Grady (1994) noted many unique features in the shock

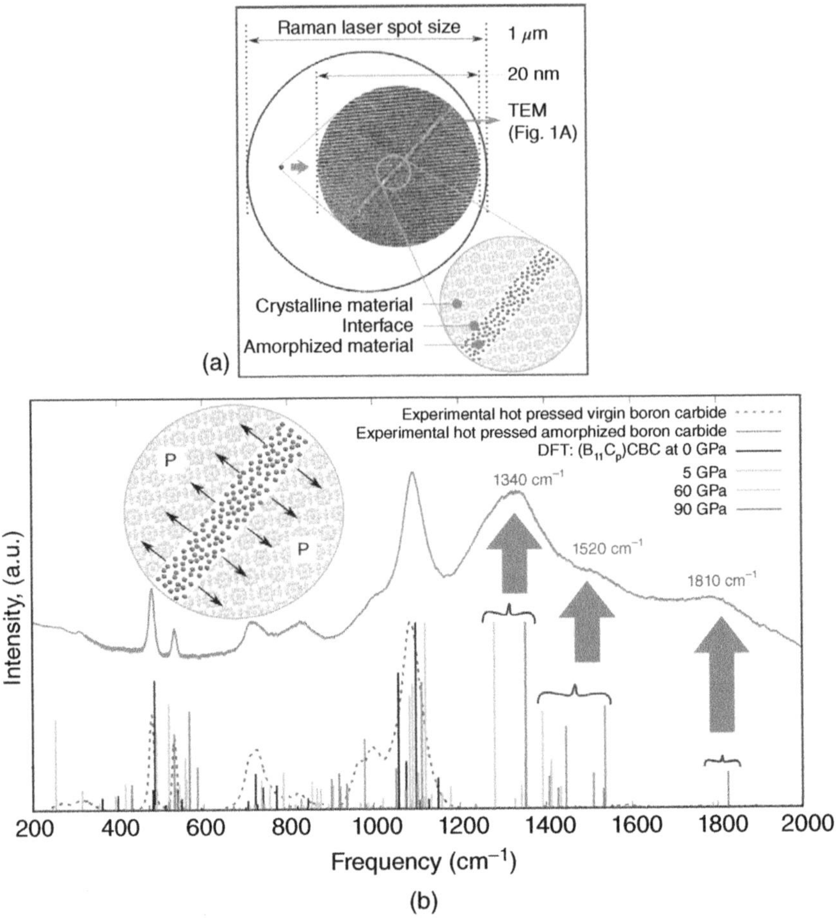

Figure 6.16 (a) Mechanistic layout of amorphization in boron carbide showing atomistic detail and the associated length scales of the amorphous zone, interface, and the surrounding crystalline material. Notice the difference in length scales between the amorphous material (nanometers) and laser spot size (a micron). (b) Through the proposed rationale of the mechanistic layout of amorphization, all three salient features of amorphized spectrum, occurring at 1340, 1520, and 1810 cm^{-1} are captured. Additionally, the less-discussed right shift of the 1090 cm^{-1} peak is also rationalized (compare the dominant peak of 0 GPa and 5 GPa DFT results). Source: Reproduced with permission from Awasthi and Subhash (2019). © 2020, Elsevier.

response of boron carbide: loss of shear strength, near fluid-like behavior, heterogeneous deformation, and anomalous volume compression. These issues remained unresolved for almost a quarter century. Through molecular dynamics simulations of the shock process of a domain of boron carbide atoms impacting

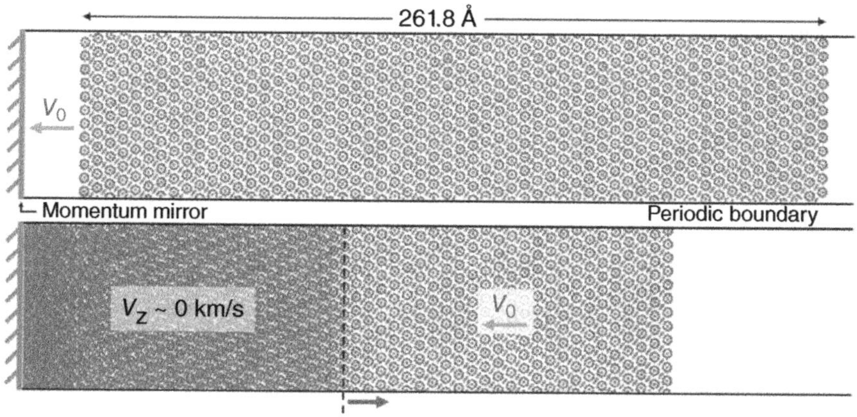

Figure 6.17 Domain of boron carbide used for simulation of shock response before impact (top) against a momentum mirror and after impact (bottom), which shows a compressive front traveling to the right. Source: DeVries et al. (2020). © 2020, American Physical Society.

against a rigid stationary surface (momentum mirror), DeVries et al. (2020) resolved these long-standing issues utilizing the ReaxFF potential developed by An and Goddard (2015). Although the atomic arrangement of rhombohedral unit cell of boron carbide does not have two mutually orthogonal planes, they advanced a procedure to build a repeatable supercell containing 9000 atoms in an appropriate orientation and then created copies of this supercell in the Cartesian system to build a simulation domain consisting 108 000 atoms with dimensions of 49.4 Å × 56.0 Å × 269.2 Å, as shown in Figure 6.17.

Impact velocities ranging from 0.5 to 6 m/s were chosen, and the evolved shock response was analyzed to extract pressure, temperature, and density changes within the domain using the Lagrangian and Eulerian binning methods with different bin sizes. For the first time, they were able to simulate the Rayleigh lines using MD, and from the end-states of these lines at different velocities they constructed the Hugoniot of boron carbide as shown in Figure 6.18a. By comparing this Hugoniot to the hydrostat (pressure–volume relationship in the elastic regime), the post-HEL softening behavior can be clearly visualized in Figure 6.18b. The deviation of the Hugoniot from the hydrostat occurs just after the impact velocity of 1 km/s. This behavior has been attributed to the loss of shear strength during the shock deformation.

The maximum shear stress ($\tau = \frac{1}{2}\left[\sigma_{zz} - \frac{1}{2}\left(\sigma_{xx} + \sigma_{yy}\right)\right]$) during the deformation, shown in Figure 6.18c, reveals that the material sustains this level of shear stress for the entire loading duration up to an impact velocity of 1.5 km/s, after which a

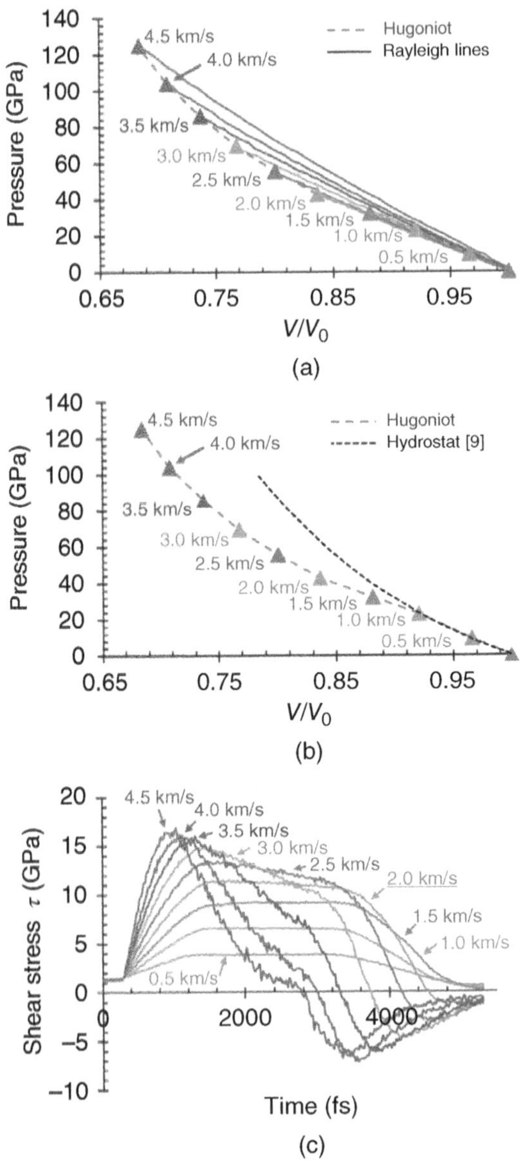

Figure 6.18 (a) MD-simulated Rayleigh lines for various velocities and the corresponding Hugoniot, (b) comparison of the Hugoniot with Hydrostat reveals softening and (c) loss of shear strength at higher velocity during the shock deformation. Source: DeVries et al. (2020). © 2020, American Physical Society.

loss in the shear stress is noticeable. With an increase in shock velocity, a dramatic loss in material's ability to sustain the maximum shear stress for the entire duration of shock is clearly seen.

To probe the origins of this loss in shear strength of the material, DeVries et al. (2020) plotted the rise in temperature during the shock deformation, as shown in Figure 6.19. Superposed on this plot are also the lines for the melting point (T_{melt}) and homogeneous rise in temperature (T_{homo}) of the bulk solid with pressure, and rise in temperature of the material due to the conversion of plastic work to heat for various conversion factors (fraction of work converted to heat (β) = 0 to 1) as determined by Zhao et al. (2016). Temperature data from these MD simulations clearly shows that at low impact velocities up to 1 km/s, the rise in temperature is consistent with T_{homo}, but it rises rapidly reaching the melting point of the material at 2.5 km/s. Interestingly, the temperature rise follows the predicted trend of increasing temperature for $\beta = 0.9$ all the way up to an impact velocity of 6 km/s where the predictions of temperature rise made by Dandekar (2001) based on the work of Pavlovsky (1971) are also compared. Also note that the impact velocity corresponding to the observed deviation of Hugoniot from the hydrostat in Figure 6.18b occurs at about 1.5 km/s, which coincides with the deviation in temperature from the bulk material rise in Figure 6.19. Similar comparison between these two figures shows that the loss in shear strength seems to occur when the temperature rise reaches close to that of melting point. Thus, we can now attribute the observed post-yield

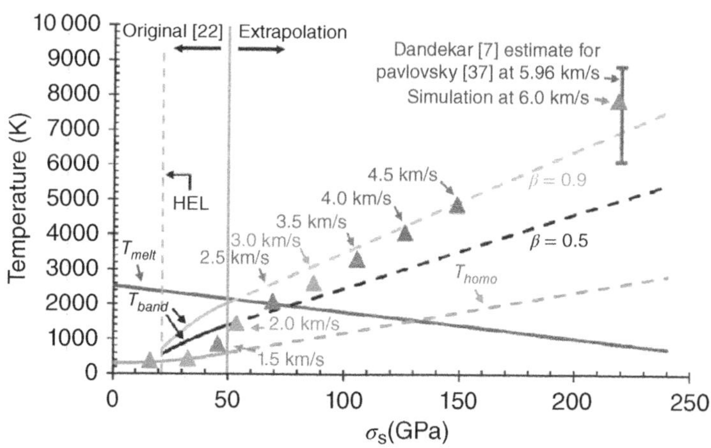

Figure 6.19 Comparison of MD simulated temperature rise for various impact velocities with that extrapolated trends of Zhao et al. (2016) for different values of plastic work conversion rates (β) to heat. Source: DeVries et al. (2020). © 2020, American Physical Society.

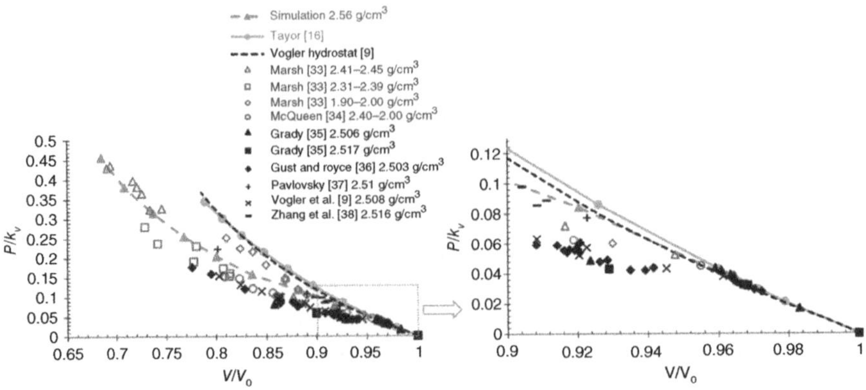

Figure 6.20 Comparison of MD-simulated Hugoniot with experimental data from various sources collected by DeVries et al. (2020). The values shown in the legend next to the reference correspond to the density of boron carbide used in each investigation. The enlarged portion of the plot on the right shows that the MD simulated Hugoniot deviates from the hydrostat much later than the experimental data because a defect-free domain is considered in the simulations. Source: DeVries et al. (2020). © 2020, American Physical Society.

softening by Grady (1994) to the temperature rise in the material, a phenomenon that has not been addressed in the literature for this ceramic.

Finally, DeVries et al. (2020) compared the MD-simulated Hugoniot with experimental data available in the literature (normalized to align the elastic portions of all the data from various sources) and noted a good comparison as shown in Figure 6.20. The enlarged section of the plot on the right shows that the defect-free simulation system reveals higher HEL value than the experimental data at various densities.

The MD simulations also shed light on the atomic level deformation mechanisms. Figure 6.21a shows the radial distribution function of atoms versus their pair separation distance for various impact velocities. The small peaks at a separation distance of approximately 1.5 Å correspond to the chain atoms, and the tallest peak next to them at 1.8 Å corresponds to atoms in the icosahedra of boron carbide. With increasing impact velocity, first the peak corresponding to chain atoms vanishes followed by a reduction in the amplitude of the icosahedral peak. Simultaneously, other peaks at greater separation distances vanish, and the distribution becomes almost a straight line. This indicates the disorder in atomic arrangement, which is interpreted as amorphization.

Figure 6.21b illustrates that with increasing impact velocity, the lateral displacements between the atoms also increase, which is reflective of large atomic vibrations associated with increased temperature and melting of the material. Finally, Figure 6.22a shows the initial atomic rearrangement before the shock wave arrival,

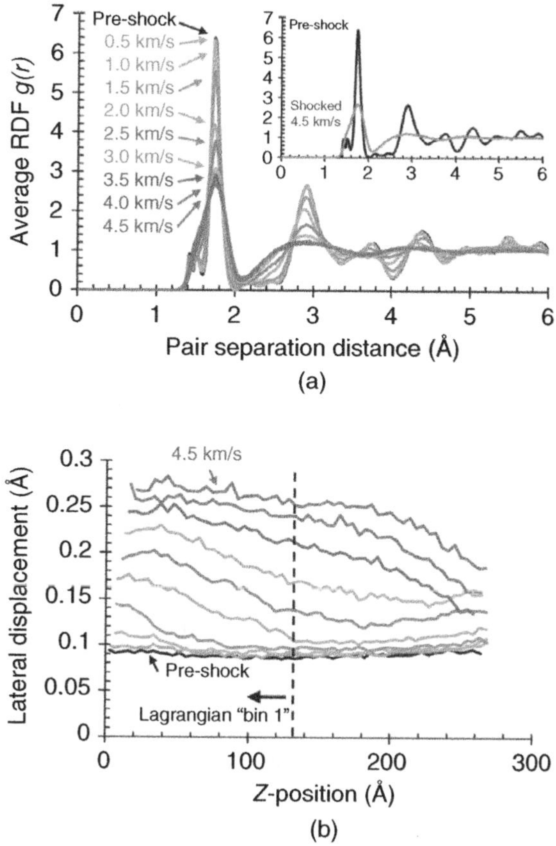

Figure 6.21 (a) Radial pair distribution function depicting the atom separation distances for various impact velocities and (b) and lateral displacements of atoms indicating melting behavior at higher impact velocities. Source: DeVries et al. (2020). © 2020, American Physical Society.

and Figure 6.22b through f shows the material after the passage of the shock wave. The atomic arrangement is completely destroyed, leading to amorphization.

6.7 Amorphization-Resistant Ultrahard Icosahedral Boron-Rich Ceramics

Recognizing that amorphization is a weakening phenomenon, Subhash et al. (2016) proposed several potential avenues for making the boron carbide crystal structure amorphization resistant. The crystal structure of boron carbide consists of cage spaces (see Figure 6.3), and in the presence of high compressive stress, these spaces accommodate the bending displacement of the CBC chain atoms during the initial amorphization process. Subhash et al. (2016) postulated that if the cage spaces can be filled with foreign atoms, the propensity for amorphization

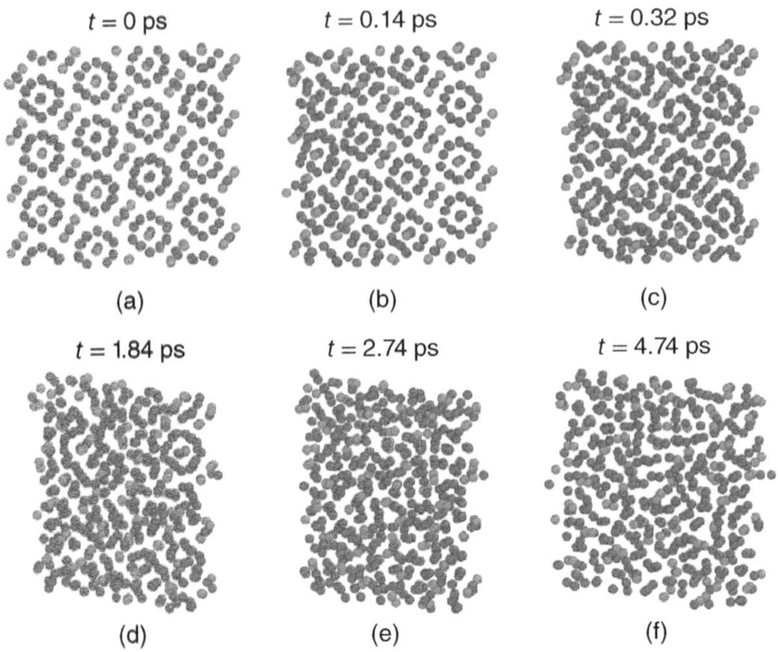

Figure 6.22 Snap shots of atomic positions (a) before the arrival of the shock wave, (b) at the time when shock wave front is at the center of the domain, and (c–f) atomic rearrangement during the passage of the shock wave. Source: DeVries et al. (2020). © 2020, American Physical Society.

can be reduced and hence the strength and hardness of amorphization can be further improved. The presence of cage spaces is responsible for the low density of boron carbide, so occupation of those are likely to raise the density. It was predicted that atoms as large as nickel (Ni) can be accommodated in the cage space. Currently, there are no known experimental methods to precisely introduce atoms into the cage spaces. Preliminary investigations into the stress–strain behavior of such structures has shown improvement in stiffness for the doped system using DFT (Subhash et al. 2016) (see Figure 6.23).

Using spatial Raman mapping, DeVries et al. (2016) demonstrated that reducing the grain size to 300 nm produced significantly less amorphization compared to grain size of 10 μm in boron carbide (Figure 6.24). Thus, nanograining is yet another means of achieving amorphization resistance in boron carbide (Subhash et al. 2016). Another method proposed by Subhash et al. (2016) and Awasthi and Subhash (2020) aims at producing combinatorial synthesis of icosahedral ceramics by mixing polymorphic variants with high amorphization resistance. This procedure requires substantial experimental as well as theoretical research to come up

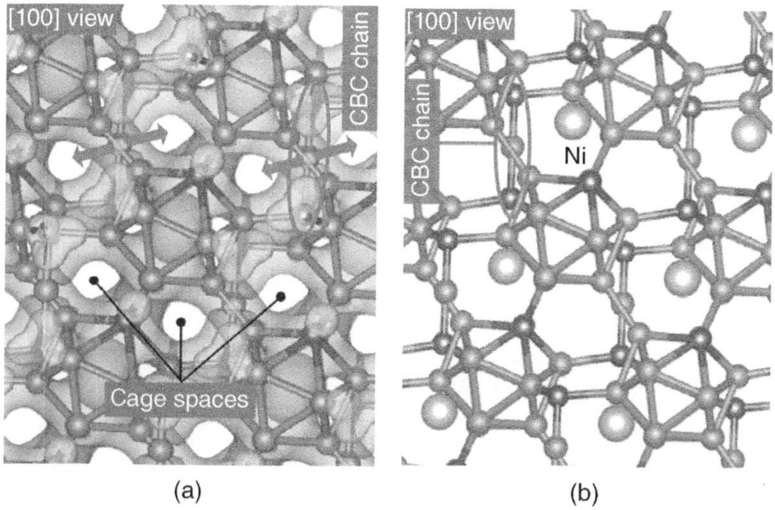

Figure 6.23 Proposed strategy of doping "cage spaces" in boron carbide with foreign atoms. (a) 3D contour maps of two density isosurfaces and depiction of cage spaces. (b) Doping one of the cage spaces by Ni atoms. Source: Subhash et al. (2016). © 2016, Elsevier.

Figure 6.24 Impact of reducing grain size to nanometer scale on propensity of amorphization in boron carbide assessed using indentation experiments. Propensity of amorphization is significantly reduced by nanograining. Source: Reproduced with permission from Subhash et al. (2016).

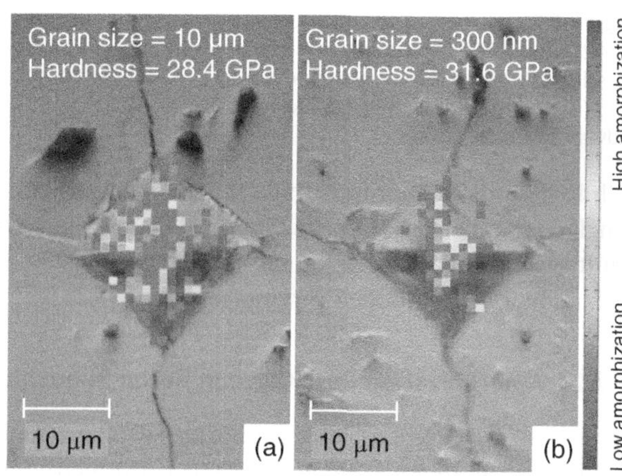

with adequate combinations of material mixes, and the method is discussed in more detail in Chapter 8.

Recently, Xiang et al. (2019) demonstrated that an addition of about 1 at.% silicon to boron carbide could arrest pressure-induced amorphization to a noticeable degree. In this work, a diffusion couple was established using large-grained boron carbide (millimeter grain-size) and silicon hexaboride SiB_6 to create a Si gradient.

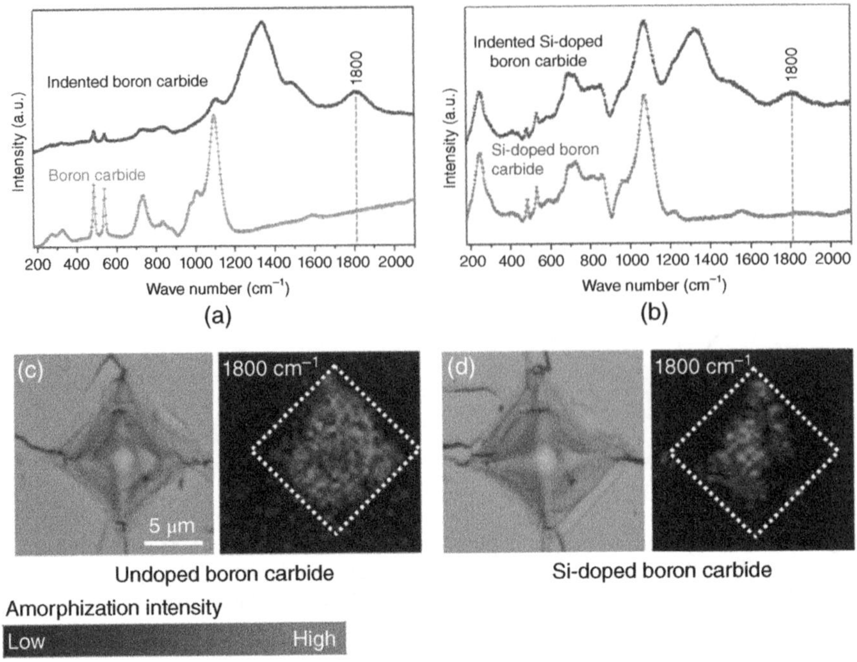

Figure 6.25 Arrest of amorphization by doping boron carbide with silicon. (a) Raman spectrum of pristine and amorphized un-doped boron carbide (b) Raman spectrum of pristine and amorphized Si-doped boron carbide (c) Raman mapping in un-doped sample and (d) doped sample. Source: Xiang et al. (2019). © 2019, American Association for the Advancement of Science.

While the pristine boron carbide sample showed long, distinct, and dominantly single-variant shear faults, the Si-doped counterpart exhibited short, diffuse, and multivariant shear faults. Raman mapping showed a distinct reduction in amorphization for the Si-doped variant (see Figure 6.25).

6.7.1 Amorphization Resistance in Boron Suboxide

In the past two decades, boron suboxide (B_6O) has emerged as a potential superhard material. In 2002, He et al. (2002) performed Vickers hardness tests on this material and found the average hardness to be about 45 GPa. Given that B_6O could be synthesized at ambient pressure or at a relatively low pressure compared to diamond or cubic boron nitride, there is strong interest in this compound (Chen et al. 2007). Interest in B_6O dates to as early as 1973, when Petrak et al. (1973) developed experimental data on B_6O for distention (porosity)-dependence on Young's modulus and shear modulus. Perhaps the strongest case was made in 1988 by Badzian (1988), who demonstrated that B_6O could easily scratch wear {001} tips of

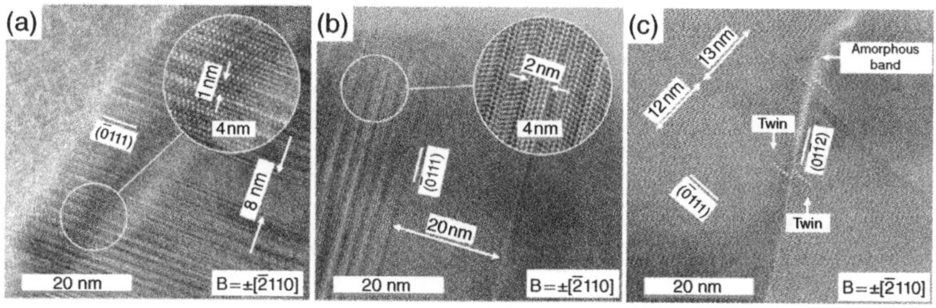

Figure 6.26 HRTEM imagery showing nanotwinning in boron suboxide by Kunka et al. (2018a). Distinct twin-spacings of (a) $\lambda = 1$ nm, (b) $\lambda = 2$ nm, and (c) $\lambda = 13$ nm were observed. Amorphization evident in (c), where an amorphization band shears a few sparsely spaced nanotwins. Source: Reproduced with permission from Awasthi and Subhash (2020).

diamond octahedra and scratch the {111} face of diamond. This discovery was in direct conflict with the contemporary belief that "diamond is the hardest of all substances and can only be scratched by another diamond" (Wilks and Wilks 1976). See Figure 6.1c for a description of the atomic-level crystal structure of boron suboxide.

Unlike boron carbide, boron suboxide does not have the chain structure, and so it was initially speculated that boron suboxide does not form amorphization bands. However, amorphization bands were observed beneath indentations by Kunka et al. (2018a). Compared to boron carbide, boron suboxide exhibits two unique features: (i) the Raman spectrum of B_6O does not show any new features upon amorphization and (ii) B_6O is extremely susceptible to the formation of nanotwins (twins separated at a nanometer spacing); see Figure 6.26a and b. Therefore, only TEM observations can provide direct evidence of amorphization. Kunka et al. (2018a), noted that although B_6O forms amorphization bands as seen in Figure 6.26c, such bands were less prevalent in zones containing a high density of nanotwins. Therefore, investigations are focused on the possibility of utilizing nanotwins to arrest the propensity of amorphization at the nanoscale in B_6O (An et al. 2016a,b; Kunka et al. 2018a). This approach is scientifically sound because nanotwin formation is energetically favorable compared to the nontwinned structure in B_6O (Kunka et al. 2018b). However, the difference in formation energy is less than 3 meV, and a slight fluctuation in the processing environment can yield a nontwinned structure. Therefore, strict control of the processing environment is necessary to achieve a fully nanotwinned structure of boron carbide.

An and Goddard (2015) performed electronic structure calculations for shear deformation to investigate three crystals, B_6O ($B_{12}O_2$), B_6P ($B_{12}P_2$), and B_4C ($B_{12}C_3$). The propensity for deformation was studied along the most plausible

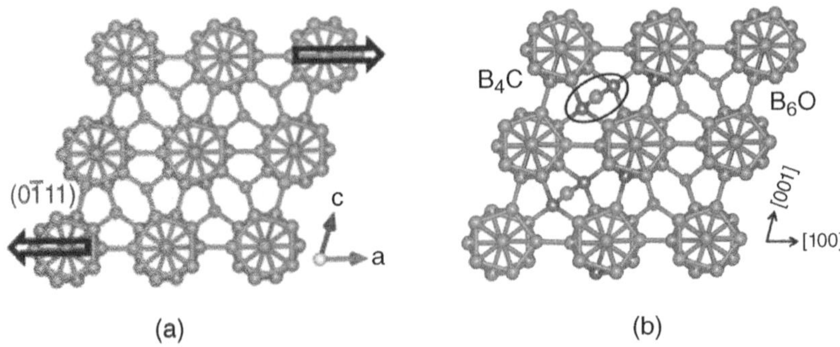

Figure 6.27 Atomistic models for (a) shear deformation of B_6O (An and Goddard 2015) Source: An and Goddard (2015). © 2015, American Chemical Society. and (b) a composite of B_4C and B_6O (Tang et al. 2015). Source: Tang et al. 2015. © 2015, American Chemical Society.

slip system: (001)/⟨100⟩. Quite interestingly, icosahedra in B_6O and B_6P did not fail up to strain levels of 120%. B_4C was susceptible to chain bending, resulting in structural collapse at high pressures. Since chains are absent in B_6O and B_6P, these materials likely have less propensity to failure, and more studies are needed to fully establish this feature. The structural similarity between B_4C and B_6O-type materials led Tang et al. (2015) to investigate composite structures with alternating B_4C and B_6O components using DFT (see Figure 6.27). The composite systems were deformed by shear in six slip systems: (001)/[010], (001)/[100], (100)/[001], (100)/[010], (010)/[001], and (010)/[100]. The (001)/[100] slip system was found to be the most susceptible. However, the composite system exhibited 40% higher strength in that slip direction compared to the B_4C system. Therefore, such multicomponent systems may have the potential to mitigate amorphization in these high hardness icosahedral compounds.

6.7.2 Other Potential Icosahedral Structures

Boron phosphide ($B_{12}P_2$ or B_6P) and boron arsenide ($B_{12}As_2$ or B_6As) are additional icosahedral boron-rich ceramics with potential to be superhard materials (Kunka et al. 2018b). Compression of boron phosphide in a diamond anvil setup revealed that it does not amorphize up to 43.2 GPa (Gao et al. 2017). Studies for $B_{12}As_2$ (Wu et al. 2011) showed that it does not demonstrate any evidence of phase transformation or amorphization up to a threshold pressure of 25.5 GPa. These materials should be further explored for their propensity for amorphization and applications in dynamic loading. Similarly, BAM materials are also promising structures for ultrahigh hardness. However, processing of these materials to full density continues to be a challenge.

6.8 Summary and Conclusions

In this chapter, we discussed recent developments pertaining to research on icosahedral boron ceramics, a class of ceramics which are both lightweight as well as good contenders for superhardness. Their crystal structures are composed of boron-rich icosahedral arrangements, and the presence of strong covalent bonds gives them their unique characteristics. We discussed how the spatial arrangement of icosahedral units allows vacancies in the crystal structure, also termed as "cage spaces." This crystal structure, owing to the presence of icosahedra and chains along with cage spaces, is susceptible to catastrophic damage at high pressures through a phenomenon called "amorphization." The topic of amorphization has remained a mystery for a long time, and this chapter discussed recent developments demystifying amorphization in boron carbide using multiscale atomistic modeling, quantum simulations of Raman spectra, and molecular dynamics with the ReaxFF potential. The role of temperature in amorphization, especially during shock loading is highlighted. Recent atomistic simulations demonstrate how high pressure leads to elevation of temperature, which could even reach near-melting, hence drawing the conclusion that amorphization is actually a thermodynamic phenomenon. Dynamic behavior leading to amorphization has been studied for additional icosahedral boron ceramics including boron suboxide, boron arsenide, and boron phosphide. Finally, new ideas on improving the mechanical performance of these materials to an ultrahard domain by mitigating amorphization were discussed as well. Additional visions on the future of icosahedral boron ceramics is presented in Chapter 8.

References

Albert, B. and Hillebrecht, H. (2009). Boron: elementary challenge for experimenters and theoreticians. *Angewandte Chemie International Edition* 48: 8640–8668.

An, Q. and Goddard, W. (2015). Boron suboxide and boron subphosphide crystals: hard ceramics that shear without brittle failure. *Chemistry of Materials* 27: 2855–2860.

An, Q. and Goddard, W.A. III, (2015). Atomistic origin of brittle failure of boron carbide from large-scale reactive dynamics simulations: suggestions toward improved ductility. *Physical Review Letters* 115: 105501.

An, Q., Goddard, W., and Cheng, T. (2014). Atomistic explanation of shear-induced amorphous band formation in boron carbide. *Physical Review Letters* 113: 095501.

An, Q., Reddy, K., Dong, H. et al. (2016a). Nanotwinned boron suboxide (B_6O): new ground state of B_6O. *Nano Letters* 16: 4236–4242.

An, Q., Reddy, K., Qian, J. et al. (2016b). Nucleation of amorphous shear bands at nanotwins in boron suboxide. *Nature Communications* 7: 11001.

Anderson, C. and Royal-Timmons, S. (1997). Ballistic performance of confined 99.5%-Al_2O_3 ceramic tiles. *International Journal of Impact Engineering* 19: 703–713.

Aryal, S., Rulis, P., and Ching, W. (2011). Mechanism for amorphization of boron carbide B_4C under uniaxial compression. *Physical Review B* 84: 184112.

Awasthi, A. and Subhash, G. (2019). High-pressure deformation and amorphization in boron carbide. *Journal of Applied Physics* 125: 215901.

Awasthi, A. and Subhash, G. (2020). Deformation behavior and amorphization in icosahedral boron-rich ceramics. *Progress in Materials Science* 112: 100664.

Aydin, S. and Simsek, M. (2011). First-principles calculations of elemental crystalline boron phases under high pressure: orthorhombic B-28 and tetragonal B-48. *Journal of Alloys and Compounds* 509: 5219–5229.

Badzian, A.R. (1988). Superhard material comparable in hardness to diamond. *Applied Physics Letters* 53: 2495–2497.

Bauccio, M. (1994). *ASM Engineered Materials Reference Book*. CRC.

Bouchacourt, M. and Thevenot, F. (1981). Analytical investigations in the B-C system. *Journal of the Less-Common Metals* 82: 219–226.

Budiansky, B. (1965). On the elastic moduli of some heterogeneous materials. *Journal of the Mechanics and Physics of Solids* 13: 223–227.

Chen, M., McCauley, J., and Hemker, K. (2003). Shock-induced localized amorphization in boron carbide. *Science* 299: 1563–1566.

Chen, C., He, D., Kou, Z. et al. (2007). B_6O-based composite to rival polycrystalline cubic boron nitride. *Advanced Materials* 19: 4288.

Christensen, R.M. (1990). A critical evaluation for a class of micro-mechanics models. *Journal of the Mechanics and Physics of Solids* 38: 379–404.

Dandekar, D.P. (2001). *Shock Response of Boron Carbide*, 1–34. Aberdeen Proving Grounds, MD: US Army Research Laboratories.

Dandekar, D., Ciezak, J., and Somayazulu, M. (2008). *Compression and Associated Properties of Boron Carbide*. Aberdeen Proving Ground, MD: Army Research Lab.

DeVries, M., Pittari, J., Subhash, G. et al. (2016). Rate-dependent mechanical behavior and amorphization of ultrafine-grained boron carbide. *Journal of the American Ceramic Society* 99: 3398–3405.

DeVries, M., Subhash, G., and Awasthi, A. (2020). Shocked ceramics melt: An atomistic analysis of thermodynamic behavior of boron carbide. *Physical Review B* 101: 144107.

Domnich, V., Gogotsi, Y., Trenary, M., and Tanaka, T. (2002). Nanoindentation and Raman spectroscopy studies of boron carbide single crystals. *Applied Physics Letters* 81: 3783–3785.

Emin, D. (1987). Icosahedral boron-rich solids. *Physics Today* 40: 55–62.

Emin, D. and Aselage, T. (2005). A proposed boron-carbide-based solid-state neutron detector. *Journal of Applied Physics* 97: 013529.

Eshelby, J.D. (1957). The determination of the elastic field of an ellipsoidal inclusion, and related problems. *Proceedings of the Royal Society of London. Series A: Mathematical and Physical Sciences* 241: 376–396.

Fanchini, G., McCauley, J., and Chhowalla, M. (2006). Behavior of disordered boron carbide under stress. *Physical Review Letters* 97: 035502.

Gao, Y., Zhou, M., Wang, H. et al. (2017). The high-pressure compressibility of B12P2. *Journal of Physics and Chemistry of Solids* 102: 21–26.

Ge, D., Domnich, V., Juliano, T. et al. (2004). Structural damage in boron carbide under contact loading. *Acta Materialia* 52: 3921–3927.

Ghosh, D., Subhash, G., Zheng, J., and Halls, V. (2012). Influence of stress state and strain rate on structural amorphization in boron carbide. *Journal of Applied Physics*: 111, 063523.

Grady, D. (1994). Shock-wave strength properties of boron-carbide and silicon-carbide. *Journal de Physique IV* 4: 385–391.

Grady, D. (1998). Shock-wave compression of brittle solids. *Mechanics of Materials* 29: 181–203.

Hashin, Z. (1960). *The Elastic Moduli of Heterogeneous Materials*. Cambridge, MA: Harvard University.

He, D., Zhao, Y., Daemen, L. et al. (2002). Boron suboxide: as hard as cubic boron nitride. *Applied Physics Letters* 81: 643–645.

Herrmann, M., Sigalas, I., Thiele, M. et al. (2013). Boron suboxide ultrahard materials. *International Journal of Refractory Metals & Hard Materials* 39: 53–60.

Hogan, J.D., Farbaniec, L., Sano, T. et al. (2016). The effects of defects on the uniaxial compressive strength and failure of an advanced ceramic. *Acta Materialia* 102: 263–272.

Höglund, C., Birch, J., Andersen, K. et al. (2012). B_4C thin films for neutron detection. *Journal of Applied Physics* 111: 104908.

Jianxin, D. (2005). Erosion wear of boron carbide ceramic nozzles by abrasive air-jets. *Materials Science and Engineering A* 408: 227–233.

Johnson, G. and Holmquist, T. (1999). Response of boron carbide subjected to large strains, high strain rates, and high pressures. *Journal of Applied Physics* 85: 8060–8073.

Kanatzidis, M.G., Hogan, T., and Mahanti, S. (2012). *Chemistry, Physics, and Materials Science of Thermoelectric Materials: Beyond Bismuth Telluride*. Springer Science & Business Media.

Kipcak, A.S., Gurses, P., Derun, E.M. et al. (2013). Characterization of boron carbide particles and its shielding behavior against neutron radiation. *Energy Conversion and Management* 72: 39–44.

Konovalikhin, S. and Ponomarev, V. (2010). Estimation of the upper limit of carbon concentration in boron carbide crystals. *Russian Journal of Physical Chemistry A* 84: 1445–1448.

Kunka, C., Awasthi, A., and Subhash, G. (2016). Crystallographic and spectral equivalence of boron-carbide polymorphs. *Scripta Materialia* 122: 82–85.

Kunka, C., An, Q., Rudawski, N. et al. (2018a). Nanotwinning and amorphization of boron suboxide. *Acta Materialia* 147: 195–202.

Kunka, C., Yang, X., An, Q., and Subhash, G. (2018b). Icosahedral superstrength at the nanoscale. *Physical Review Materials* 2: 063606.

Lee, S., Mazurowski, J., Ramseyer, G., and Dowben, P.A. (1992). Characterization of boron carbide thin films fabricated by plasma enhanced chemical vapor deposition from boranes. *Journal of Applied Physics* 72: 4925–4933.

Lipp, A. (1970). *Boron Carbide: Production, Properties, and Applications*. Battelle Northwest Laboratories.

Lundstrom, T. and Andreev, Y. (1996). Superhard boron-rich borides and studies of the B-C-N system. *Materials Science & Engineering, A: Structural Materials: Properties, Microstructure and Processing* 209: 16–22.

Matkovich, V. and Economy, J. (1970). Structure of $MgAlB_{14}$ and a brief critique of structural relationships in higher borides. *Acta Crystallographica, Section B: Structural Crystallography and Crystal Chemistry* 26: 616–621.

Matkovich, V.I., Samsonov, G.V., and Hagenmuller, P. (1977). *Boron and Refractory Borides*. Springer.

Mori, T. and Tanaka, K. (1973). Average stress in matrix and average elastic energy of materials with misfitting inclusions. *Acta Metallurgica* 21: 571–574.

Mortensen, M.W., Sørensen, P., Björkdahl, O. et al. (2006). Preparation and characterization of boron carbide nanoparticles for use as a novel agent in T cell-guided boron neutron capture therapy. *Applied Radiation and Isotopes* 64: 315–324.

Niihara, K., Nakahira, A., and Hirai, T. (1984). The effect of stoichiometry on mechanical properties of boron carbide. *Journal of the American Ceramic Society* 67: C-13-C-14.

Onoro, J., Salvador, M., and Cambronero, L. (2009). High-temperature mechanical properties of aluminium alloys reinforced with boron carbide particles. *Materials Science & Engineering, A: Structural Materials: Properties, Microstructure and Processing* 499: 421–426.

Parsard, G. and Subhash, G. (2017). Raman spectroscopy mapping of amorphized zones beneath static and dynamic Vickers indentations on boron carbide. *Journal of the European Ceramic Society* 37: 1945–1953.

Parsard, G., Subhash, G., and Jannotti, P. (2018). Amorphization-induced volume change and residual stresses in boron carbide. *Journal of the American Ceramic Society* 101: 2606–2615.

Pavlovsky, M. (1971). Shock compressibility of six very hard substances. *Soviet Physics - Solid State* 12: 1736–1737.

Petrak, D.R., Ruh, R., and Shook, W.B. (1973). *Elastic Properties of Boron Suboxide and The Diametrical Compression of Brittle Spheres*. Wright-Patterson Air Force Base, OH: Air Force Materials Lab.

Pierson, H.O. (1996). *Handbook of Refractory Carbides and Nitrides: Properties, Characteristics, Processing and Apps*. William Andrew.

Pokatashkin, P., Kuksin, A., and Yanilkin, A. (2015). Angular dependent potential for alpha-boron and large-scale molecular dynamics simulations. *Modelling and Simulation in Materials Science and Engineering* 23: 045014.

Reddy, M.K. (2020). *Boron Carbide: Structure, Processing, Properties and Applications*. Nova Science Publishers Inc.

Sasaki, S., Takeda, M., Yokoyama, K. et al. (2005). Thermoelectric properties of boron-carbide thin film and thin film based thermoelectric device fabricated by intense-pulsed ion beam evaporation. *Science and Technology of Advanced Materials* 6: 181.

Shamp, A., Zurek, E., Ogitsu, T. et al. (2017). Properties of B_4C in the shocked state for pressures up to 1.5 TPa. *Physical Review B* 95: 184111.

Shirai, K. (2014). Prediction of phase diagrams for hard materials: application to boron crystals. *International Journal of Thermophysics* 35: 1888–1899.

Shirai, K., Sakuma, K., and Uemura, N. (2014). Theoretical study of the structure of boron carbide $B_{13}C_2$. *Physical Review B* 90: 064109.

Solozhenko, V.L., Le Godec, Y., and Kurakevych, O.O. (2006). Solid-state synthesis of boron subnitride, B6N: myth or reality? *Comptes Rendus Chimie* 9: 1472–1475.

Subhash, G., Ghosh, D., Blaber, J. et al. (2013). Characterization of the 3-D amorphized zone beneath a Vickers indentation in boron carbide using Raman spectroscopy. *Acta Materialia* 61: 3888–3896.

Subhash, G., Awasthi, A., Kunka, C. et al. (2016). In search of amorphization-resistant boron carbide. *Scripta Materialia* 123: 158–162.

Subramanian, C., Suri, A., and Murthy, T. (2010). Development of Boron-based materials for nuclear applications. *Barc Newsletter* 313: 14.

Swab, J.J., Pittari, J.J. III, and Gamble, W.R. (2019). Uniaxial tensile strength and fracture analysis of a hot-pressed boron carbide. *Journal of the European Ceramic Society* 39: 1965–1973.

Tang, B., An, Q., and Goddard, W.A. III, (2015). Improved ductility of boron carbide by microalloying with boron suboxide. *The Journal of Physical Chemistry C* 119: 24649–24656.

Taylor, D.E. (2015). Shock compression of boron carbide: a quantum mechanical analysis. *Journal of the American Ceramic Society* 98: 3308–3318.

Thevenot, F. (1990). Boron carbide - a comprehensive review. *Journal of the European Ceramic Society* 6: 205–225.

Vargas-Gonzalez, L., Speyer, R.F., and Campbell, J. (2010). Flexural strength, fracture toughness, and hardness of silicon carbide and boron carbide armor ceramics. *International Journal of Applied Ceramic Technology* 7: 643–651.

Vogler, T., Reinhart, W., and Chhabildas, L. (2004). Dynamic behavior of boron carbide. *Journal of Applied Physics* 95: 4173–4183.

Wereszczak, A.A., Kirkland, T.P., Strong, K.T. et al. (2010). Size scaling of tensile failure stress in boron carbide. *Advances in Applied Ceramics* 109: 487–492.

Wilkins, M., Cline, C., and Honodel, C. (1969). *Fourth Progress Report Of Light Armor Program*. Livermore: California University Lawrence Radiation Lab.

Wilks, E.M. and Wilks, J. (1976). *The Properties of Diamond*. New York, NY: Academic Publishers.

Will, G. and Kossobutzki, K. (1976). An X-ray structure analysis of boron carbide, $B_{13}C_2$. *Journal of the Less-Common Metals* 44: 87–97.

Wu, J., Zhu, H., Hou, D. et al. (2011). High pressure X-ray diffraction study on icosahedral boron arsenide (B12As2). *Journal of Physics and Chemistry of Solids* 72: 144–146.

Xiang, S., Ma, L., Yang, B. et al. (2019). Tuning the deformation mechanisms of boron carbide via silicon doping. *Science Advances* 5: eaay0352.

Xie, K.Y., An, Q., Toksoy, M.F. et al. (2015). Atomic-level understanding of "asymmetric twins" in boron carbide. *Physical Review Letters* 115: 175501.

Yan, X., Li, W., Goto, T., and Chen, M. (2006). Raman spectroscopy of pressure-induced amorphous boron carbide. *Applied Physics Letters*: 131905, 88.

Yan, X., Tang, Z., Zhang, L. et al. (2009). Depressurization amorphization of single-crystal boron carbide. *Physical Review Letters* 102: 075505.

Zhao, S., Kad, B., Remington, B. et al. (2016). Directional amorphization of boron carbide subjected to laser shock compression. *Proceedings of the National Academy of Sciences of the United States of America* 113: 12088–12093.

7

Dynamic Behavior of Brittle Transparent Materials

7.1 Introduction

Glass is one of the oldest human-made materials dating from the time of Mesopotamia and ancient Egypt (Cooper and Swiler 2019). Early forms of glass included jewelry (e.g. beads), decorative vessels, ornaments for worship and royalty, architectural windows, and mosaic tiles in churches. In the modern era, transparent materials have emerged as structural materials for many civilian, aerospace, and military applications including windows for automobiles, residential and commercial buildings, armored vehicles, airplanes, and communication and electronic devices. Glass is also a major component of solar panels, electronic and computer display cases, visors, impact- and blast-resistant windows for buildings and space structures, riot protection gear for law-enforcement personnel, etc. Applications requiring transparency in infrared (IR) and other laser wavelengths also use transparent windows. For successful deployment in these applications, the material must be resistant to impact and yet be transparent to a range of wavelengths of light. In recent years, glass technology has advanced to meet many of these demands. Glasses are now relatively low-cost, lightweight, possess good thermal and chemical stability, and can be produced with no residual porosity. Glass can be fabricated in large sizes and into complex curved geometries. Current melting and heat-treatment technologies offer good reproducibility as compared to the powder processing and sintering methods used for opaque ceramics, which are prone to microstructural inhomogeneities. Glass-forming technology also allows compositional modifications to be made to customize properties through a variety of techniques including chemical strengthening, tempering, and controlled crystallization processes. Another advantage of glass is that surface coatings can be applied to provide desirable features such as polarization, scratch resistance, and protection against harmful rays (including lasers, ultra violet (UV), and IR light) without affecting geometry, weight, and transparency. In addition,

Dynamic Response of Advanced Ceramics, First Edition.
Ghatu Subhash, Amnaya Awasthi, Dipankar Ghosh.

glass plates and sheets can be used in tandem with other clear laminates of various thicknesses and properties to provide greater impact resistance.

In recent years, interest in the mechanical behavior and failure response of transparent materials under dynamic loads has increased substantially not only due to their widespread use in the military and transportation industries but also due to their applicability in structures (e.g. buildings and bridges) that require impact and blast resistance. This interest has sparked a surge in glass research that endeavors to address several fundamental issues. Of prime importance is identification of deformation modes under dynamic loads, including the behavior beyond the Hugoniot elastic limit (HEL), damage or failure front (FF) formation, fragmentation characteristics, and spall strength. Also, the density of glasses can vary over a broad range (2.5–6 g/cm^3) due to compositional variations and can influence their dynamic deformation behavior. Mechanical loading on glasses can induce densification, phase transformation, damage, and failure, all of which must be considered in the design of blast- and impact-resistant glass structures. In this chapter, a detailed discussion on a range of relevant topics including pressure- and strain-rate dependence of strength and fracture in transparent materials, interface effects of layered glasses, and impact response and damage propagation in monolithic and laminated glasses, shock response, and laser-pulse-induced damage will be presented. In addition to the behavior of low- and high-density glasses, chemically strengthened glasses, spinels and polycrystalline ceramics, and single crystal sapphire will also be discussed.

7.2 Types of Transparent Materials

There are a variety of transparent materials such as monolithic glasses (e.g. soda-lime [SL] silica, borosilicate [BS], lead glasses, and fused silica), single-crystal sapphire, and polycrystalline glass ceramics (e.g. aluminum oxynitride or AlON, aluminum oxynitride spinel ($Al_{23}O_{27}N_5$), magnesium aluminate spinel ($MgAl_2O_4$), transparent alumina (Al_2O_3)), and polymers (e.g. polycarbonate, PMMA, transparent nylon). A brief discussion on their structure and composition are provided with an intent to establish a basis for understanding their mechanical behavior.

7.2.1 Glass

Glass is one of the most widely used transparent material. Silicon dioxide (SiO_2) or silica is the primary constituent of many glasses used for bottles and windows. A form of crystalline silica is quartz, a familiar mineral found in sand. Quartz has a hexagonal crystal structure and exhibits long-range order as shown in Figure 7.1a (Alexander et al. 2008). In α-quartz (a polymorph of quartz), the silicon (Si) and

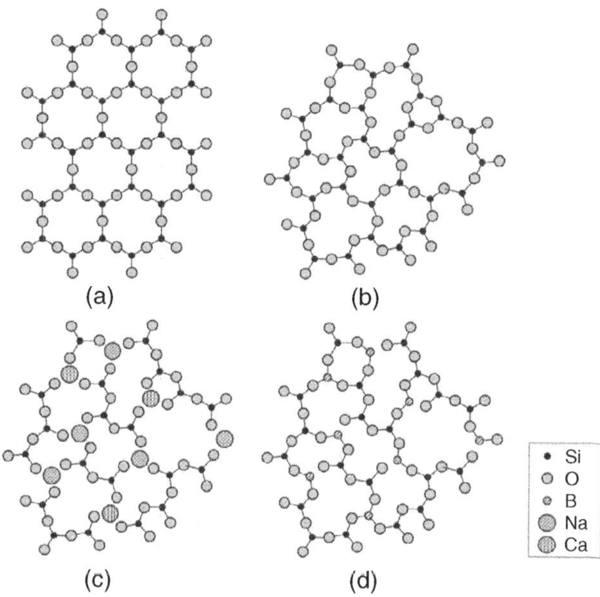

Figure 7.1 Atomic arrangement of (a) single crystal quartz, (b) fused silica, (c) fused silica modified with sodium and calcium, and (d) fused silica with boron replacing silicon. Source: Alexander et al. (2008). © 2008, Elsevier.

oxygen (O) atoms are covalently bonded and arranged in a tetrahedral formation with strict Si–O–Si bond angles of 144°. Depending on temperature and pressure, crystalline silicon can exist in a range of structural configurations with varying bond lengths and angles. For instance, in β-tridymite (a polymorph of silica), the Si–O–Si bond angle is 180°. This is significant because the majority of glasses do not have a consistent bond angle in their tetrahedra, and this disorder renders them amorphous without a long-range order (Kingery et al. 1976; Alexander et al. 2008). This variability in structure and bond angle also impacts density; amorphous silica has lower density than the most common SL glass (2.2 versus 2.52 g/cm^3) because of the open spaces between the interlocking SiO_2 tetrahedra. Early studies using X-ray diffraction (XRD) clearly revealed this amorphous structure (Warren 1934). Although similar Si and O bonds are present in fused quartz and amorphous silica, the tetrahedral formation is partially disrupted in amorphous silica, where the Si–O–Si bond angles are no longer uniformly 180°; see Figure 7.1b. Instead, the bond angles vary over a broad range and the structure lacks long-range order, i.e. exhibits a loss of connectivity. The network exhibits ring-like structures over a length scale of only a few angstroms. These variations in bond length, bond angle, and connectivity result in amorphous nature of the structure with large differences in local density.

The term "glass" refers to several different materials. Basic glass consists of simply silica, but most forms of glass production incorporate other elements. These can be broadly classified into two large groups: normal glasses (e.g. SL silicate glass) and anomalous glasses (BS or Pyrex® or fused silica glass). To produce normal or SL glass, alkaline oxides (e.g. Na_2O and CaO) are added. These alkali ions act as network modifiers, and they reduce the number of Si–O bonds, making the entire structure more open (loss of connectivity). In this modified structure, there are more oxygen ions that are bonded to a single Si atom, as shown in Figure 7.1c. Typical soda lime glass has a density greater than that of regular glass, approximately 2.41 g/cm^3, due to the increased mass of the larger alkaline ions. The reduction of Si ions prevents network formation (or formation of larger network rings), and, therefore, Si is considered to be the network former in SL glass.

Incorporation of lighter elements such as boron into silica framework can improve the strength of glasses without increasing its density. Boron forms stronger covalent bonds with oxygen than Si and acts as a network former (see Figure 7.1d). When 8–15% of boron trioxide (B_2O_3) is added to silica, BS glass (trade name Pyrex®) with a density of about 2.23 g/cm^3 is formed. However, the coordination number of boron is three (while that of Si is four), and thus BO_3 triangles are formed instead of tetrahedrons. Hence, unfilled oxygen bonds remain in the structure. This weakness is compensated by the increased strength of the B–O bonds as compared to the Si–O bonds. BS glasses have low coefficient of thermal expansion (CTE, ~3.3×10^{-6} K^{-1} at 20 °C, about a third that of common SL glass) and are used for applications requiring high thermal shock resistance, similar to the Pyrex® bakeware used in household or laboratory instruments. Addition of 5–10% alumina (Al_2O_3) to silica glass improves its thermal resistance and durability. This aluminosilicate glass is extensively used for stove-top cookware, fiberglass and in glass-reinforced plastics for boats and fishing rods. This glass is the base material for chemical strengthening which is then used for cell phone covers, cartridges, and many other applications.

Finally, when the heavy metal lead is incorporated into the silica glass, it increases the refractive index and this property gives the leaded glass the sparkle seen in beautiful cut crystal pitchers, wine glasses, chandeliers, and figurines. They also exhibit a distinctive ring-one when struck. Flint glass contains anywhere from 8% to 70% of lead and can have a density greater than 5 g/cm^3, over double that of regular glass.

7.2.2 Chemically Strengthened Glass

Glasses are sensitive to surface flaws, and under tensile loads cracks initiate from these flaws, leading to premature fracture of the material. Thermal tempering or surface chemical strengthening by ion exchange is often used to induce in-plane surface compressive stresses and reduce this sensitivity to surface defects. Chemically strengthened glass (e.g. Gorilla Glass®) has become increasingly popular in recent years due to its ability to resist impact-induced fracture originating from these surface defects. This glass is used for a variety of civilian applications including hurricane- and earthquake-resistant architectural windows, electronic touch panels, covers for display cases and digital electronic gadgets (e.g. cell phone, TV, laptop, etc.), pharmaceutical autoinjector cartridges, airplane and helicopter vehicle windshields, bullet- and blast-resistant laminate windows for motor vehicles and bank tellers.

The process of surface strengthening begins with a commercially available, common glass such as lithium aluminosilicate glass used for window panels and in kitchen cooking tops. The glass is successively placed in molten salt baths consisting of sodium nitrate ($NaNO_3$) and potassium nitrate (KNO_3), respectively, for several hours. First, in the molten sodium salt bath, the lithium ions (Li^+) from the original glass are exchanged with larger Na^+ ions. Next, in the molten potassium salt bath, the Na^+ are exchanged with even larger K^+ ions. The "wedging" of the larger potassium and sodium ions into the spaces previously filled by smaller lithium ions generates large surface (in-plane) compressive stress. The smaller Na^+ ions infuse to a depth of about 1 to 2 mm below the surface of the glass and create compressive stresses up to several hundred megapascals (MPa), whereas the larger K^+ ions can migrate only to a depth of few microns into the surface but create even greater surface compressive stresses up to a gigapascal (GPa) close to the surface. More details of this chemical strengthening process can be found in Varshneya and Spinelli (2009).

A comparison of the properties of a traditional glass and a commercially available chemically strengthened glass called IonArmor® is provided in Table 7.1 (Jannotti et al. 2012). Note that the strengthened glass has a slightly greater density and a noticeable increase in elastic modulus and hardness values. The biggest improvement comes from the threshold load for crack initiation during static indentation. The strengthened glass has almost 15 times the load capacity of the regular glass before indentation cracks initiate. This improvement can be translated in its enhanced ability to withstand tensile load through reduced sensitivity to surface flaws. The large surface compressive stress suppresses the propensity for crack initiation from the preexisting surface defects, thus allowing for a greater threshold for failure load.

Many transparent materials (e.g. quartz, sapphire, strengthened glass) exhibit optical anisotropy or birefringence (double refraction), i.e. dependence of refractive index on polarization direction of light. By taking advantage of this property along with photoelastic techniques, one can determine the stress state under load or the induced residual stresses in chemically strengthened glasses. Details of the theory on how to determine stress using photoelastic technique can be found in

Table 7.1 Properties of unstrengthened and chemically strengthened glass.

Property	Unstrengthened	Strengthened	% Change
Density (g/cm^3)	2.41	2.45	+1.4
Longitudinal wave velocity (km/s)	6.13	6.23	+1.7
Shear wave velocity (km/s)	3.69	3.79	+2.6
Elastic modulus (GPa)	80.9	84.6	+4.6
Shear modulus (GPa)	33.3	35.0	+5.3
Poisson's ratio	0.22	0.21	−4.0
Avg. static hardness (GPa)[a)]	5.7	6.3	+6.8
Avg. dynamic hardness (GPa)[a)]	5.4	6.0	+5.5
Critical load threshold (kgf)[a)]	1.0	15.0	+1400

a) Vickers indentation.
Source: Jannotti et al. (2015). © 2015, Elsevier.

Dally and Riley (1991). A typical compressive stress profile of IonArmor® determined using photoelasticity is given in Figure 7.2a (Jannotti et al. 2011). A compressive stress of about 1 GPa is induced close to the edge on the surface and the stress magnitude quickly drops to 250 MPa at a depth of 30 μm before gradually tapers to zero within 0.8 mm from the surface. To maintain force equilibrium with in the specimen, tensile stress is generated in the interior of the specimen. The magnitude of this tensile stress depends on the cross-sectional area supporting this stress. Since a large magnitude of compressive stress is limited to only a small region closest to the surface, the generated tensile stress magnitude in the interior of the specimen is significantly smaller, about 39 MPa. The monochromatic photoelastic fringe patterns corresponding to the induced residual stress profile due to chemical strengthening are shown in Figure 7.2b, c. A thicker sample is used to obtain the global stress profile in Figure 7.2b and a thin sample is used to obtain stress profile close to the edge. The in-plane principal stress difference (twice the magnitude of maximum shear stress) is calculated from the number of fringes (N), wavelength of light (λ), specimen thickness (h), and the stress-optic coefficient (c) using (Jannotti et al. 2011):

$$\sigma_1 - \sigma_2 = 2\tau_{max} = \frac{N\lambda}{hc} \tag{7.1}$$

From the above equation, it is seen that by reducing the thickness of the specimen, for the same stress difference, one can observe higher number of fringes close to the edge of the sample as seen in Figure 7.2c, and hence obtain a more accurate estimate of stress.

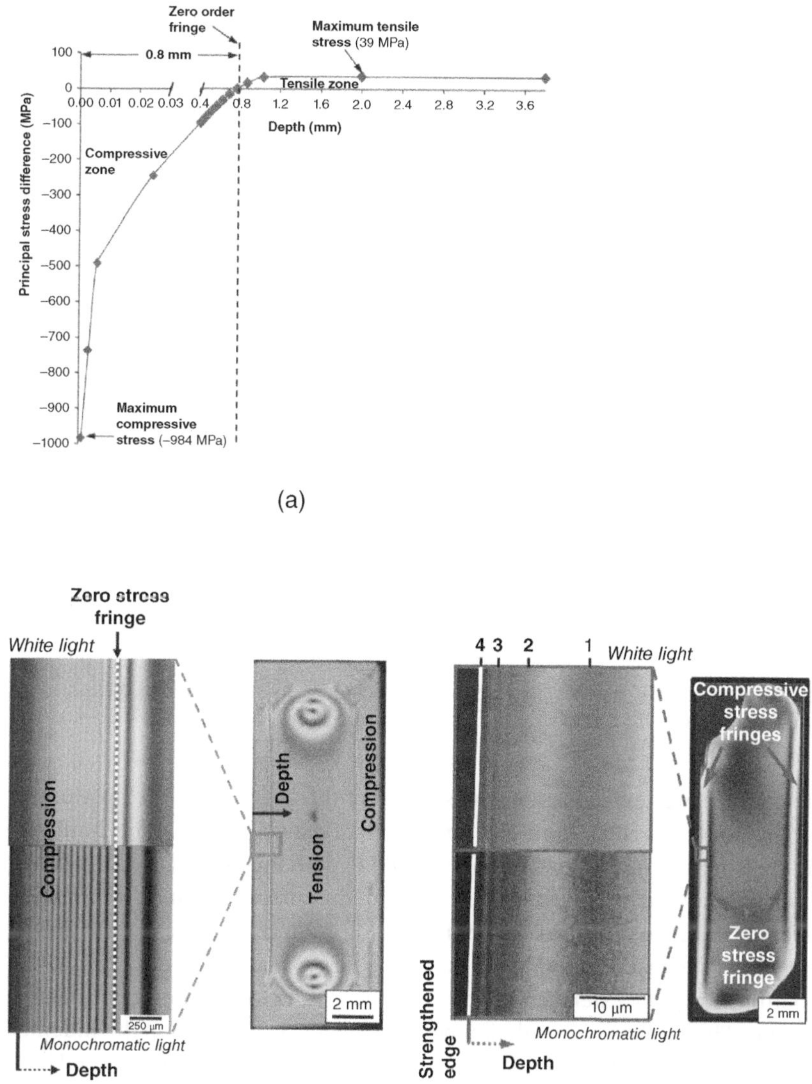

Figure 7.2 (a) Residual stress profile in a chemically strengthened glass. (b) Photoelastic fringe pattern of the entire top surface of a thick specimen showing outer compressive stress region and an interior tensile stress region. (c) High resolution photoelastic fringe pattern on a thin specimen to obtain the magnitude of stress close to the edge. Source: Jannotti et al. (2011). © 2011, John Wiley & Sons.

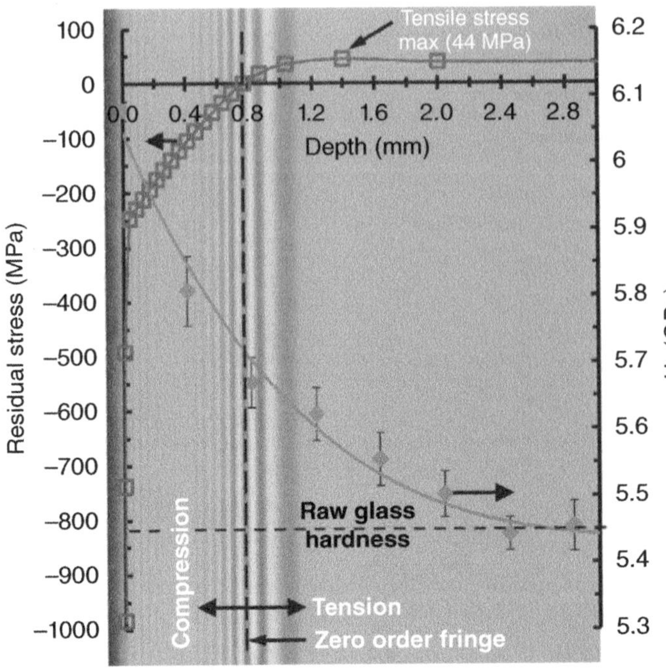

Figure 7.3 Photoelastic fringe-pattern of stress in chemically strengthened glass and the associated Vickers hardness values. Source: Jannotti et al. (2012). © 2012, Elsevier.

Indentation hardness values on the surface of the glass provide a measure of the strengthening arising from the chemical exchange process. The hardness profile measured using Vickers indentation along with the monochromatic fringe pattern is superposed in Figure 7.3. As expected, the surface hardness is improved by almost 11% due to the chemical strengthening compared to the central tensile region, which shows the same hardness as the virgin material. Interestingly, the high-hardness region extends well beyond the region containing compressive stress state. This surprising result can be rationalized as follows: As the indenter penetrates the surface, it displaces atoms on either side of the indentation. However, on the highly compressive side of the indentation, heavier atoms are closely packed (compressed) due to the surface strengthening than on the other side (tensile side) of the indentation. Although the indenter is in a tensile region (beyond 0.8 mm from the edge), this differential is still felt (up to 1.2 mm) beyond the null-stress region (the dark dashed line representing the zero-order fringe). Thus, chemically strengthened glasses provide not only reduced sensitivity to surface flaws but also higher hardness to the surface. However, as will be seen later, these stresses result in large accumulated strain energy which has profound influence on the impact resistance and fracture characteristics of strengthened glasses.

7.2.3 Transparent Crystalline Ceramics

Unlike the amorphous structure of glass, many transparent ceramics (e.g. spinels and AlON) have a crystalline structure and are optically isotropic (not birefringent). In addition to having high hardness and erosion resistance, crystalline ceramics allow transmission of electromagnetic wavelengths from 0.25 to 6 μm and do not exhibit polymorphic transformations (no thermally induced phase changes) (McCauley et al. 2012). Transparent crystalline ceramics are also much stronger than glasses and have better impact resistance. In this chapter, we will deal with three such transparent crystalline ceramics: aluminum oxynitride (AlON or $Al_{23}O_{27}N_5$), spinel (magnesium aluminate spinel ($MgAl_2O_4$)), and sapphire (single crystal alumina, Al_2O_3). An example of transparent AlON is shown in Figure 7.4 (McCauley et al. 2009).

7.2.3.1 Aluminum Oxynitride (AlON or $Al_{23}O_{27}N_5$)

Aluminum oxynitride (γ-AlON) is a unique compound in the aluminum oxide and aluminum nitride (Al_2O_3-AlN) system with a cubic spinel structure and an approximate composition of $Al_{23}O_{27}N_5(9Al_2O_3{\cdot}5AlN)$. It is a polycrystalline high-performance structural ceramic material that can be produced in a transparent form. It has a density of $3.69\,g/cm^3$, a hardness close to that of sapphire (~15 GPa), and is optically transparent from near-ultraviolet to IR wave lengths. It can be manufactured into complex shapes using traditional powder sintering techniques. An excellent review of the AlON history, processing and characterization, and impact response are provided in McCauley et al. (2009).

Figure 7.4 Example of transparent crystalline ceramic AlON. Source: Reproduced with permission from McCauley et al. (2009).

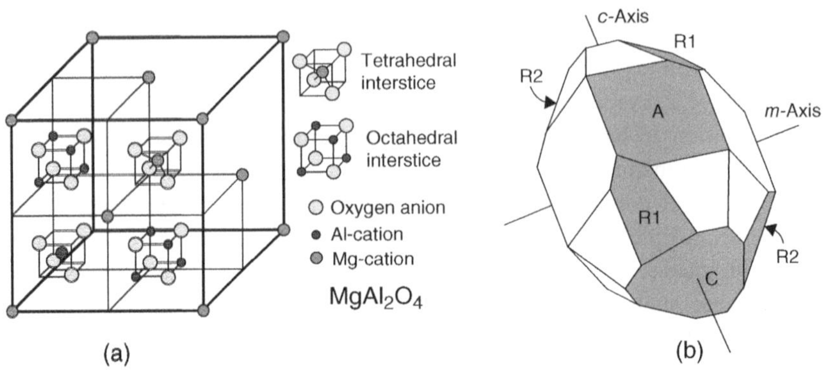

Figure 7.5 Structure of (a) spinel. Reproduced with permission from Haney and Subhash (2012) and (b) sapphire. Source: (a) Haney and Subhash (2012). © 2012, Elsevier, (b) Haney and Subhash (2011a). © 2011, Elsevier.

7.2.3.2 Spinel ($MgAl_2O_4$)

Spinel or magnesium aluminate ($MgAl_2O_4$) is composed of a stoichiometric mixture of MgO and Al_2O_3 arranged in an isotropic face-centered-cubic (FCC) structure (see Figure 7.5a). The magnesium and aluminum oxides occupy the octahedral and tetrahedral interstitial locations in the FCC structure, respectively. These materials are produced through hot pressing methods and are polycrystalline. Spinels can be produced with a range of grain sizes (from a few tens of microns to hundreds of microns) depending on the processing conditions.

7.2.3.3 Sapphire

Sapphire is single-crystal aluminum oxide with a hexagonal crystal structure as shown in Figure 7.5b. It is manufactured using single-crystal growth methods such as the heat exchanger method (HEM), edge-defined film-fed growth (EFG), and the Czochralski processes. It has high hardness (9 on the Mohs scale or Knoop hardness of 22 GPa), moderate fracture toughness (2 MPa-$m^{1/2}$), high scratch resistance, excellent thermal conductivity, and good chemical stability (Rioux et al. 2007). Sapphire can also be modified to block laser threats with dielectric coatings. It can transmit light effectively and, unlike glass shields that can lose visibility, sapphire is unaffected by exposure to chemicals and solvents. Typical properties of Mg-spinel, sapphire, and AlON are summarized in Table 7.2.

7.2.4 Laminate Composites

These composites are intended for impact resistance and are typically comprised of high fracture toughness polymer inner layers with a hard glass or ceramic outer layer. The outer hard layer is intended to damage or erode an incoming projectile

Table 7.2 Properties of spinel and sapphire from Haney and Subhash (2013) and Strassburger (2009). Properties of AlON from Strassburger et al. (2013).

Property	Mg-spinel	Sapphire	AlON
Density (kg/m^3)	3580	3970	3670
Young's modulus (GPa)	273	435	315
Shear modulus (MPa)	110	145	135
Poisson's ratio	0.26	0.27	0.24
Biaxial flex. strength (MPa)	172	758	
Compressive strength (GPa)	2.69	2	
Vickers hardness @ 1 kgf (GPa)	14.1	17.4	18.50 Knoop @ 200 g
Fracture toughness (MPa-m$^{1/2}$)	1.95	2.14	2.4
Fracture surface energy (J/m^2)	6.5	7.3	
Brittleness (μm^{-1})	1027	1647	
Impedance (kg/m^2-s)	4.51×10^7	6.13×10^7	
Spall strength (GPa)	0.63	7	
Hugoniot elastic limit (GPa)	11.3	23	10.4–11.6
Longitudinal wave speed (m/s)	9730	11 500	10 250
Shear wave speed (m/s)	5540	7100	6040

Source: Haney and Subhash (2013). © 2013, Springer Nature.

tip, and the high-toughness inner layer prevents fragments from flying off and causing injury or damage on the other side. The glasses are joined by polymer interlayers such as polyurethane (PU) or polyvinylbutral (PVB). The material for the interlayer is carefully chosen to minimize loss of transparency and possible residual stress build up due to the differences in the CTE as well. The polymer interlayer may also help prevent crack propagation from one layer to the next during projectile penetration as will be discussed later.

7.3 Deformation Mechanisms of Glass

7.3.1 Densification Under Compression

One of the earliest investigations on glass was conducted by Bridgman and Simon (1953) and Bridgman (1948) on its paradoxical behavior compared to other brittle materials. The investigations included its unique characteristics such as high compressibility, decreasing bulk modulus over a specified pressure region, and unrecoverable volumetric strain due to permanent densification. The effects of pressure on the densification (or compressibility) of glass were later investigated by Mackenzie (1963), Sakka and Mackenzie (1969), Uhlmann (1973), and Rouxel et al. (2008). These studies indicated that glasses can undergo a permanent densification up to 20% under shear loads and high temperatures, and their Poisson's ratio can change significantly (Rouxel et al. 2008). Such deformation can be related to the open structure of glass (see Figure 7.1), which consists of significant interstitial space. In the following discussion, we will present the effects of structure, defects, and residual stress state on the densification, fracture, and fragmentation behavior of glasses.

7.3.2 Sensitivity to Defects

One of the prime weaknesses of glasses is their extreme sensitivity to surface and volume (internal) defects, which are inadvertently introduced during manufacturing process or subsequent use in service. The surface defects often include surface roughness, microscratches, and surface voids. Volume defects include impurity atoms (chemical defects), microvoids, and microcracks. The nature of defects (size, shape, location and their distribution) influences the wave propagation and damage evolution during the impact process. Sands et al. (2009) intentionally introduced various defects in glass ceramics and conducted impact tests. Three major conclusions were made: (i) mass loss due to defects (introduced prior to testing) is more deleterious to the impact performance than mass loss due to the thickness reduction of the target, (ii) surface defects are more detrimental than bulk defects, and (iii) defect geometry plays a major role on impact response. For instance, elliptical defects caused more degradation of impact performance than circular and rectangular defects. The influence of surface defect geometry on flexural strength was investigated by Nie et al. (2009) who prepared BS glass specimens with different surface roughnesses by grinding with 200 and 1500 grit sandpaper and also by polishing and etching the surface with hydrofluoric acid. The surface profiles are shown in Figure 7.6. For ground samples, the flexural strength decreased with an increase in surface roughness as shown in Figure 7.7. Grinding-induced defects were more deleterious than etching-induced surface pits, which have lower stress concentration. Also, cracks initiated from the sharper grinding-induced surface flaws at lower tensile stress levels than from blunt surface defects on a chemically etched surface despite having higher surface roughness. Thus, one must consider factors beyond surface roughness to rationalize the tensile strength of glass materials. It was also noted that with increased loading rates, the surface tensile strength increased for all the surfaces, but the strength of the etched surfaces was higher by almost an order of magnitude as compared to that of specimens with rough grinding surfaces, as noted in Figure 7.7.

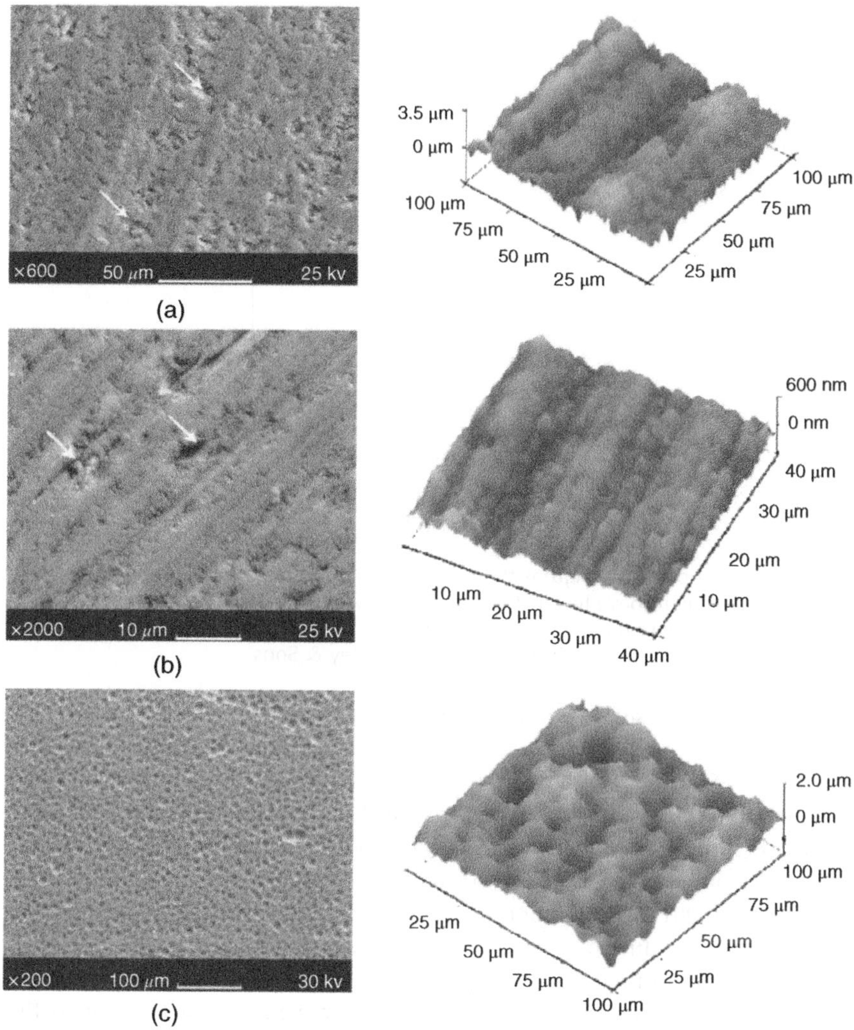

Figure 7.6 Surfaces of glass after sanding and etching. SEM images of the altered surface are presented in the left column, and corresponding surface roughness maps are shown in the right column for glass (a) sanded with 220-grit paper, (b) sanded with 1500-grit paper, and (c) chemically etched. Source: Reproduced with permission from Nie et al. (2009). © 2009, John Wiley & Sons.

7.3.3 Deformation of Chemically Strengthened Glass

Surface chemical strengthening of a traditional glass significantly improves its properties and performance. Upon chemical strengthening, Vickers hardness

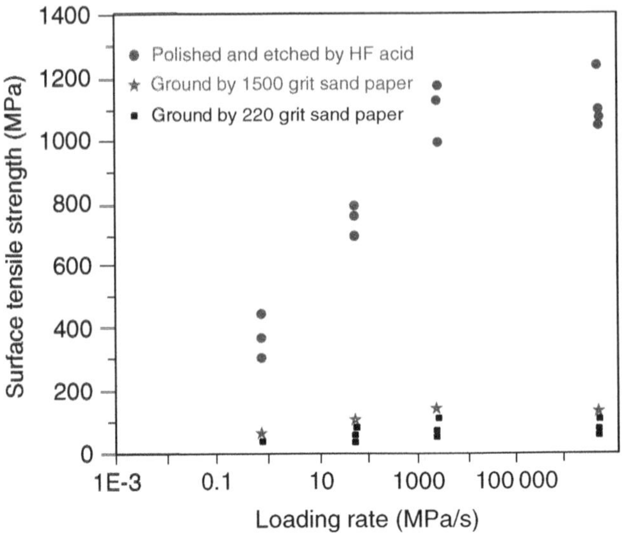

Figure 7.7 Variations in strength with loading rate and surface treatment of glass. Note that ground specimens have lower strength as compared with etched specimens, and the strength increases more rapidly with loading rate for etched specimens than ground specimens. Source: Nie et al. (2009). © 2009, John Wiley & Sons.

measurements on a surface perpendicular to the strengthened surface revealed that the hardness increased from 5.4 GPa (for the unstrengthened glass) to 6.1 GPa (13% increase). Images of representative indents on strengthened and virgin (raw) glass are shown in Figure 7.8, which reveal the influence of strengthening on the deformation response; there is significantly lower damage on the strengthened glass compared to the regular (raw) glass for equivalent indentation loads (Jannotti et al. 2012). In addition to reduced radial and lateral cracking in the strengthened glass due to surface compression, shear faulting is also observed, which is indicative of reduced brittleness. The lack of surface compressive stress causes undue cracking in the regular glass. However, as will be seen in the ensuing discussions on impact response of chemically strengthened glass, that the interior tension can have unintended consequences.

7.4 Impact Testing of Plates and Rods

Damage induced by pressing a sharp tool (indenter) against a transparent material has been extensively investigated in the literature (Fischer-Cripps 1997; Lawn et al. 1976; Lawn and Swain 1975; Lawn et al. 1975; Lawn and Wilshaw 1975; Swain and Hagan 1976). Such studies under quasi-static loads have established fundamental

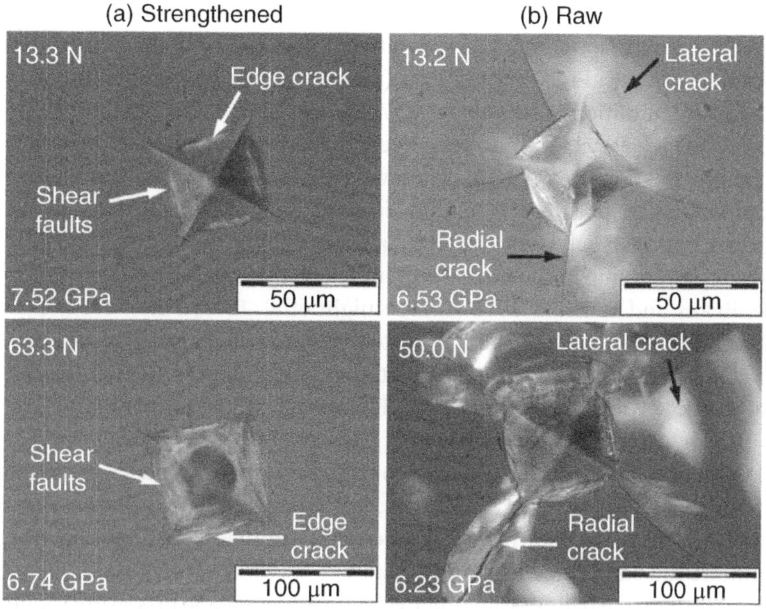

Figure 7.8 Comparison of indentation induced cracking between strengthened glass and regular glass. Source: Reproduced with permission from Jannotti et al. (2012).

theories for damage initiation, fracture and material removal during machining, indentation, and low-velocity impact (Lawn and Wilshaw 1993). We now extend these studies to the dynamic-strain-rate regime, where the time to initiate and propagate damage is severely limited and the applied pressures are high.

7.4.1 Ball-on-Rod Impact of Glass

A ball impact test consists of a pressurized gas gun and a target (see Figure 2.19). The pressurized gas gun launches a foam sabot that holds the spherical metal projectile. A rigid block situated at the end of the gun barrel impedes the high velocity sabot, allowing the ball to launch toward the target. The target can be a plate, a rod, or a composite laminate. A high-speed camera captures the impact process and the incipient damage in the specimen. Both reflected (white) light and transmitted (shadow) light imaging can be used to capture the evolving damage patterns and failure front (FF) propagation characteristics. When slender rods are used as targets, the damage propagation is primarily along the length of the rod, which can be considered as 1-D damage propagation, and hence, the relationship between the projectile velocity and damage front velocity can be evaluated by capturing the damage mode with high-speed videography.

Damage induced in glass by low-velocity (<50 m/s) sphere impact shares several common features with static-indentation-induced damage by a sharp indenter tip. For instance, a critical indentation load (velocity, in the case of ball impact) is required for the initiation of a cone crack (Andrews 1931; Longchambon 1934; Tillett 1956). This load was found to be related to the ball (indenter) radius, and the relationship is known as Auerbach's law (Auerbach 1891). At sufficiently high-impact velocities, plastic deformation initiates beneath the surface of the target material, and the analysis by the Hertzian contact mechanics approach fails to establish the stress state in the plastic region. Early investigations (Cherepanov et al. 1979; Evans and Wilshaw 1976, 1977; Knight et al. 1977; Tillett 1956; Tsai and Kolsky 1967) regarding the impact of steel spheres on Pyrex® (BS) and SL glasses revealed that, similar to static indentation, cone cracks, median cracks, and lateral cracks evolve at low velocities. Persson et al. (1993) compared the damage caused by an indentation with that of low-velocity impact (below 45 m/s), and noted that Hertzian theory can be used for the analysis of the elastic response when cone cracks emanate at low velocities, but with increases in velocity the material beneath the impact surface is crushed and the theory can no longer be used to estimate the damage zone characteristics.

In general, the damage patterns at higher velocity impacts (>50 m/s) were found to be dramatically different from those of the static indentation. High-velocity impacts introduce dynamic (inertial) effects, wave propagation, nucleation of multiple microcracks, dynamic crack growth, and comminution of material ahead of the penetrating projectile. Chaudhri and Walley (1978) conducted ball impact experiments on SL and BS glasses and observed that radial and lateral cracks were formed during loading and unloading (rebound), respectively, up to a velocity of 180 m/s. Ring cracks and cone cracks formed immediately upon impact, and lateral crack formation was enhanced at higher velocities up to 340 m/s. Eventually, cone cracks turn up, leading to further lateral cracking. Interestingly, the cone cracks grew at terminal crack velocities, and the apical angle of the cone crack was significantly lower than that observed in quasi-static indentations (Knight et al. 1977). At impact velocities above 150 m/s, BS glass and fused silica revealed cone cracks, but SL glass exhibited splinter cracks (Knight et al. 1977). SL glass sheets impacted by steel and nylon balls displayed similar behavior (Kirchner and Gruver 1977). Significant crushing was also observed beneath the impact surface, and the maximum impact damage radius was larger than the Hertzian crack radii. Therefore, Knight et al. (1977) suggested that the traditional Hertzian analysis previously conducted by Evans (1973) is not applicable for these impact velocities. However, the radius of the innermost ring crack was found to be the same as that during the static indentation experiments on these glasses, suggesting that a quasi-static approach may be applicable for crack length calculations.

Tensile wave reflections also play a role in the evolved damage patterns during high-velocity impacts on plates and rods. In these indentation and impact experiments, the radial stress is compressive, and the circumferential stress is tensile (Chaudhri and Walley 1978; Chaudhri and Walley 1978; Chaudhri and Walley 1978). However, unlike indentation tests, the high velocity ball impact tests do not have a well-defined unloading phase such as retraction of the indenter; instead, the projectile continues to penetrate the target until its velocity is reduced to zero or the target is fully penetrated. The continued penetration of impactor further complicates the analysis of deformation response under ball impact as compared to static indentation. In-depth studies of damage progression at the end-of-penetration phase investigated by Anderson et al. (2009) and Bavdekar et al. (2019) were briefly reviewed in Chapter 3 and will be discussed further in this chapter.

7.4.2 Ball Impact of Chemically Strengthened Glass Rods

Recall that a chemically strengthened glass exhibits a high surface compressive stress extending to a small depth followed by interior tension in the bulk of the material (see Figures 7.2 and 7.3). Thus, there is large amount of stored elastic energy (1/2 × stress × strain) distributed throughout the specimen material. Although the material becomes less sensitive to surface flaws due to surface compressive stress, the available total elastic energy can have a profound influence on the damage evolution characteristics because once the initial damage (due to impact or otherwise) extends to a depth beyond the surface compressive layer, the stored elastic energy in the tensile zone is released in an explosive manner. As the cracks enter the inner tensile zone, they are aided by the local tensile stress until the entire specimen is fully consumed by the growth of cracks. The damage grows in an uncontrolled fashion with damage growth velocity higher than that in a stress-free material.

One of the first systematic studies on strengthened glass were conducted by Chaudhri and Liangyi (1986) where the surface compressive stress was approximately 200 MPa, which decreased to zero within a depth of 2 mm. The maximum interior tensile stress was around 100 MPa at the center plane of the specimen. Crack velocities as high as 1800 m/s were measured in the chemically strengthened glass, whereas in a typical glass, the velocity was only 1500 m/s. Jannotti et al. (2014) measured crack velocities in chemically strengthened (ion-exchanged) glass up to 1950 m/s. This material exhibited surface compressive stresses up to 1 GPa, and the average tensile stress in the interior bulk was only 45 MPa (see Figures 7.2 and 7.3). Clearly, the increased density of atoms (or increased compressive stresses) in the chemically strengthened glass aid in increasing the crack velocities.

Figure 7.9 compares the induced damage in an unstrengthened glass rod and a chemically strengthened glass rod at a ball impact velocity of about 300 m/s

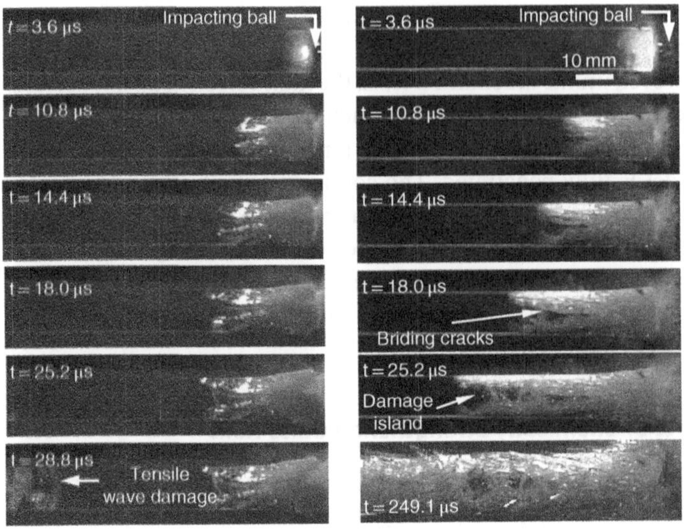

Figure 7.9 Comparison of damage propagation in unstrengthened glass bars (a) and chemically strengthened glass bars (b) impacted by steel balls (Jannotti et al. 2014). Source: Reproduced with permission from Jannotti et al. (2014).

(Jannotti et al. 2014). Upon impact, the end surfaces of both the rods are severely damaged, producing a mist-like appearance. The kinetic energy (KE) of the ball impact is absorbed by the glass rod through wave propagation, fracture, fragmentation, comminution, ejecta formation, and KE of fragments. Fine debris ejects off from the impact end in the opposite direction of the incoming projectile. Impact on a chemically strengthened glass rod creates a damage front (also referred to as FF) consisting of multiple cracks at the impact location, and this damage front propagates along the length of the rod with an increase in time. On the other hand, the damage growth in the unstrengthened glass becomes less severe at distances farther away from the impact location. Many of the initial damage features are common to both the unstrengthened and strengthened glasses, but distinct differences emerge with time and distance from the impact surface of the rod.

In the unstrengthened glass, the propagating damage is arrested after a short distance from the impact end, but additional damage evolves from the opposite end of the rod and travels toward the arrested damage. Interestingly, the nature of damage in these two damage zones is different. At the impact end of the rod, the cracks within the damage front grow parallel to the length of the rod forming needle-like fragments parallel to the direction of the failure wave propagation (or length of the rod). The cracks on the rear side of the rod are oriented perpendicular to the length of the rod. These two different damage modes can be rationalized by recalling the nature of damage evolved under compressive and tensile loads. Upon impact by the ball, a compressive wave develops in the rod from the impact location and as long as its amplitude is greater than the compressive strength of the material, compression cracks evolve parallel to the loading direction, and the damage propagates along the length of the rod. The damage growth velocity is well below the longitudinal stress wave propagation velocity in the rod due to the inertia associated with the crack growth process. However, in shock experiments, crack (damage) growth can occur at velocities comparable to that of the shock wave if the amplitude of the shock overwhelms the material, as will be seen in Section 7.8 later. Once the compressive stress amplitude falls below the compressive failure strength of the material, the damage growth stops but the impact generated compressive stress wave continues to propagate to the other end of the rod and reflects as a tensile wave (due to the free end). Recall that most brittle materials are weaker in tension than in compression, and hence, tensile damage initiates at the rear end in a direction perpendicular to the tensile wave propagation direction. The energy in the stress wave is being consumed to create this tensile damage. Once the stress wave amplitude falls below the tensile strength of the material, the damage growth stops.

The evolution of columnar fragments at the impact-end indicates that the impact-induced stress state in the rod is almost uniaxial compression (see discussion on uniaxial compression-induced fracture in Chapter 4), which causes axial splitting. The cracks developed at the other end of the rod are perpendicular to the length, and the stress state is akin to uniaxial tension. Both these fracture patterns confirm that in ball impact experiments on long rods, as long as the diameters of the ball and rod are comparable, the induced damage pattern follows the principles of damage growth under uniaxial stress.

On the other hand, the damage growth characteristics in the strengthened glass are dramatically different. Recall that due to the high surface compression and low-magnitude interior tension, every material point in the rod has stored elastic strain energy. Upon impact at sufficiently high velocity, a compressive stress wave is generated which travels slightly faster along the outer surface than the interior of the rod, due to the higher number of heavier atoms packed closer together along the periphery. The impact also generates cracks that penetrate beyond the depth of the surface compressive zone and creates a damage front that propagates along the length of the rod. The damage front propagates in a self-sustained fashion: it feeds off of the stored elastic strain energy available at every material point along the length of the rod. Periodic bridging cracks propagate transversely, creating partially damaged spherical islands at the center. The interior tensile stress aids the damage growth process to grow in a self-sustained manner by allowing a rapid release of the stored elastic energy. Consequently, the highly stressed atoms relax in the surrounding vicinity of the newly generated cracks. This rapid relaxation causes new cracks to be initiated in the next neighborhood and relax the stress at that location. This process continues until the rod is completely consumed by the damage. The result is a domino effect that manifests as a damage front propagating at

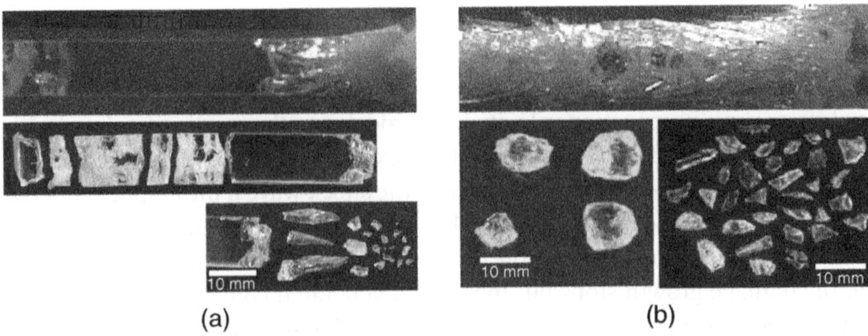

Figure 7.10 Fragmentation resulting from ball impact of (a) normal glass and (b) chemically strengthened glass. Source: Reproduced with permission from Jannotti et al. (2014).

a characteristic velocity based on the magnitude of the surface compressive stress and properties of the material. Interestingly, no tensile wave reflection or associated damage can be observed at the rear end of the bar. The reasons for this behavior are discussed later in this section.

Analysis of the fragments from the unstrengthened glass and strengthened glass rods, shown in Figure 7.10 (Jannotti et al. 2014), attests to the nature of the stress state along the length of the rod. Close to the impact surface, the dominance of impact energy pulverizes the material and forms fine debris which ejects off in the opposite direction of the ball (see images on the right of Figure 7.9). Away from this surface, columnar (needle-like) fragments are formed due to the dominance of compressive stress imposed by the ball impact. In the case of the unstrengthened rod, the stress amplitude diminishes with time while the stress wave travels to the other end of the bar and reflects as a tensile wave causing cracks perpendicular to the wave propagation direction. A considerable amount of tensile damage is seen in each fragment in Figure 7.10a. As the energy in the reflected wave is continuously dissipated by creating tensile fracture, the magnitude of the stress decreases further, and the tensile fracture eventually stops, leaving behind an intact material in the central section of the rod. The development of cracks parallel to the bar length at the impact end and cracks perpendicular to the bar axis at the rear end confirm the development of a nearly uniaxial stress state during the wave propagation in the cylindrical rod. On the other hand, in the chemically strengthened rod, the stress wave amplitude does not diminish as there is continuous availability of energy at every material point and the damage propagates unhindered until all the material is fully disintegrated. The fragments are needle-like in the compressive zone and almost spherical in the central tensile zone as seen in Figure 7.10b.

From the high-speed video images of the impact and damage propagation events, one can determine the velocity of the FF as shown in Figure 7.11. For

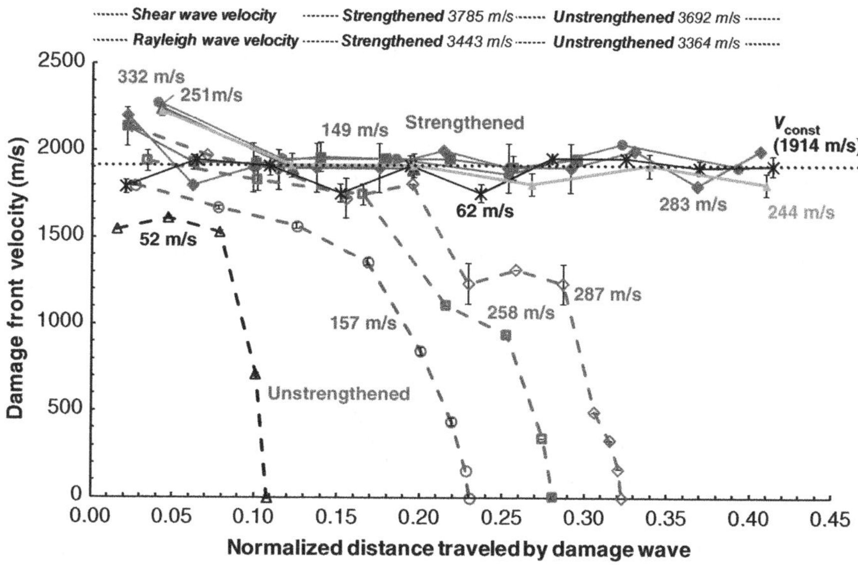

Figure 7.11 Measurement of damage front velocity versus the normalized distance of damage front. Note that there is no decline in damage front velocity with distance for the strengthened glass. Source: Jannotti et al. (2014). © 2014, John Wiley & Sons.

the unstrengthened glass rod, the initial damage front velocity increases with increasing impact velocity but quickly slows down and stops within a short distance. Greater impact velocities result in greater depths of damage in the rod. The tensile fracture at the rear end is not considered in this analysis because it is not a part of the damage resulting directly from the initial impact. On the contrary, for the strengthened glass, the damage-front velocity immediately upon impact increases with increasing impact velocity but once the influence of the initial impact (KE) subsides, the damage front propagates at a constant velocity of approximately 1914 ± 20 m/s, irrespective of the impact velocity. Thus, the stored elastic energy is released at a constant rate creating a self-sustaining damage front at a constant velocity. The steady-state velocity of the FF is purely a function of the magnitude of the stored elastic energy and material properties. While the chemical strengthening provides a higher strength and lower sensitivity to surface flaws, once damage is initiated it propagates in an uncontrolled manner until the target is fully consumed.

The question now is why is there no tensile damage at the rear end of the strengthened rod due to the wave reflection? This is because the amplitude of the reflected tensile stress wave is insufficient to cause the tensile cracking in the strengthened rod. Recall that the large surface compressive stress in the rod prevents surface flaws from opening, meaning that the tensile strength of the chemically strengthened rod is be significantly higher than that of the unstrengthened rod. Even if damage initiates from bulk defects in the interior tensile zone, this zone is confined within the outer compressive layer, increasing the driving force required for the crack growth into the compressive layer.

The above sustained FF in the strengthened glass has been a fascinating and intriguing phenomenon from the Middle Ages. The fracture of a Prince Rupert tear drop[1] (Chandrasekar and Chaudhri 1994; Chaudhri 1998) illustrates the damage propagation of tempered glass. This tear-shaped glass has attracted the attention of royalty and common people alike since the 1660s. Venetian glass makers created glass beads in the form of a tear drop by dropping hot molten glass into a cold-water bath. When the bead was struck hard by a hammer it resisted fracture, but by simply nipping the tail of the tear drop, the entire bead explodes, leaving behind fine rubble. This mystery can be unraveled from the knowledge of the fracture behavior of strengthened glass. When the hot molten glass is dropped into the cold-water bath, the outer layer of the glass quickly quenches, leaving behind the inner glass unable to cool (and shrink) in unison with the outer layer of glass. As the interior cools slowly, it too tries to shrink, but the outer layer resists this process because it is already in a solid state. Thus, the outer layer remains under high compressive stresses (due to shrinkage), and the interior of the glass remains in a state of tensile stress. Thus, the head of the tear-drop has a thick layer of high compressive stress on the surface and resists fracture when stuck with a hammer, but the narrow tail can easily be nipped, causing the material on that cross section to relax and release the stored internal elastic energy instantaneously. This relaxation on one plane initiates stress relaxation in the adjacent regions, causing additional fracture and the process continues in a self-sustaining manner which is seen as a moving damage-front in high-speed video images until the entire tear drop explodes.

A similar phenomenon occurs when a car window is punctured with a sharp tool. The window is made of tempered glass with a thin surface layer of compressive stress. Once the sharp tool penetrates beyond the compressive layer, fracture initiates in the tensile zone, and the entire glass collapses into equisized fine fragments. The shape and size of fragments can be controlled by proper design of the compressive and tensile stress profiles and the thickness of the glass sheet.

7.4.3 Ball Impact of Transparent Crystalline Ceramics

Transparent crystalline ceramics (aluminum oxynitride – AlON or $Al_{23}O_{27}N_5$), spinels (magnesium aluminate spinel – $MgAl_2O_4$)), and sapphire (single crystal alumina – Al_2O_3)) are much stronger and have better impact resistance than

1 The reader is recommended to watch video clips on YouTube® to get acquainted with this dynamic phenomenon. One such recommended video is https://youtu.be/xe-f4gokRBs.

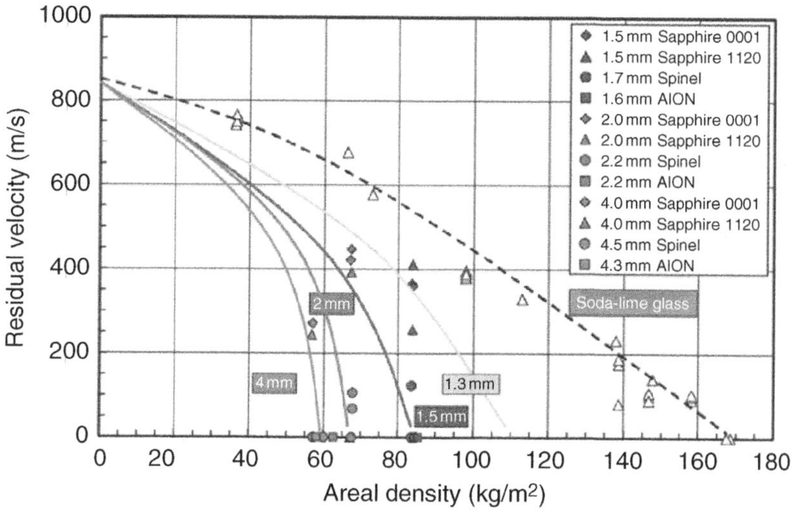

Figure 7.12 Residual velocity versus areal density for various transparent material targets. Source: Strassburger (2009). © 2009, Elsevier.

glasses. Evidence of their superior ballistic performance has been reported by various research groups. Swab et al. (2000) and Patterson et al. (2009) noted that the V_{50} performance (i.e. a projectile velocity at which there is a 50% probability that it penetrates a target) of Mg-spinel target and AlON were superior to those of sapphire and glass for a range of area densities. Similarly, in Figure 7.12, the residual velocity of a bullet (exit velocity after complete penetration of a target) is the lowest for AlON followed by spinel, sapphire, and then glass (Strassburger 2009). On the contrary, comparison of the properties of these materials in Table 7.2 reveals that aside from the compressive strength, all other properties of sapphire are superior to spinel. This contradiction of superior performance of transparent ceramics (spinel) in ballistic tests over sapphire despite spinel's inferior properties questions our fundamental dogma that superior mechanical properties always result in superior performance in the field tests.

To investigate the reasons for the superior ballistic resistance of spinel over sapphire, Haney and Subhash (2013) conducted a series of ball impact experiments on cylindrical rods of magnesium aluminate spinel (Mg-spinel) and sapphire and analyzed their fracture characteristics. In both cases, severe damage evolved close to the impact site, and the damage depth increased with impact velocity. The damage front velocity was a function of impact velocity but was significantly less than the shear wave velocity. Close to the impact site, damage evolves in the form of fine fragments in both materials. However, the damage modes in the rods are distinctly different.

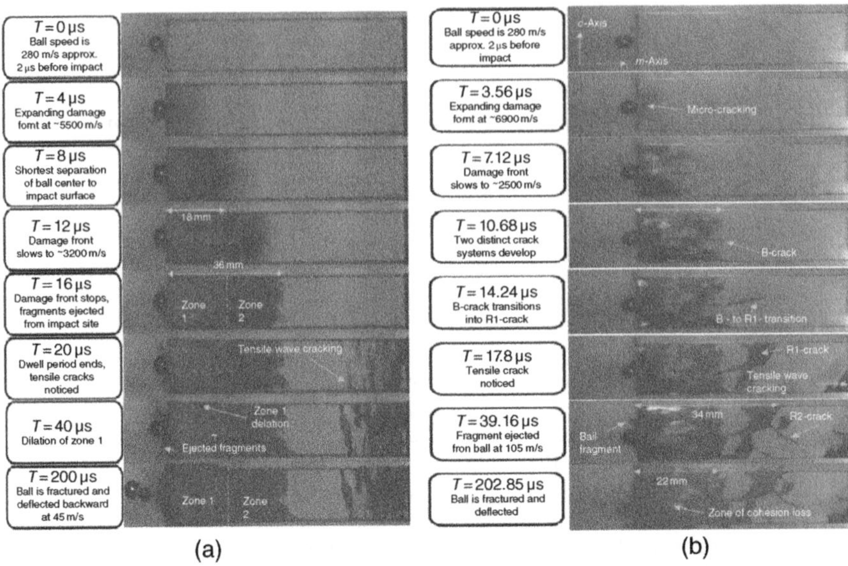

Figure 7.13 Evolution of damage due to ball impact in (a) spinel (left) and (b) sapphire (right). Source: Reproduced with permission from Haney and Subhash (2013).

Figure 7.13 compares the ball-impact-induced fracture patterns in Mg-Spinel (Haney and Subhash 2013) and sapphire rods. In both the rods, extensive inter- and intra-granular fractures occur close to the impact site in response to the large incident KE. With distance from the impact site, this mixed-mode fracture (Zone 1) in Mg-spinel transitions solely into intergranular cracking (Zone 2) as shown in Figure 7.13a. Eventually, the damage propagation ceases as the compressive stress amplitude diminishes (due to the energy dissipated in fracture creation), but tensile damage emerges at the back end of the rod due to the stress wave reflection. Once again, these tensile cracks are perpendicular to the axis of the rod, reflecting the uniaxial nature of stress wave loading. As the ball tries to penetrate the rod, the fine fragments created at the impact site flow in the direction opposite to that of impact and are ejected off the rod. This ejecta can be seen in Figure 7.13a at the time $T = 12\,\mu s$ and beyond. The ejecta also opposes the incoming motion of the ball, not only decelerating it but also deflecting it back. In the process, the ball is fractured into two fragments as seen in the images at time $T = 200\,\mu s$.

The observed fracture modes in Zone 1 and Zone 2 can be rationalized based on the available impact energy. Due to the high-impact energy, the strain rate of deformation is high in Zone 1, and the damage is primarily in the form of extensive pulverization because of combined inter- and intragranular fracture. Away from

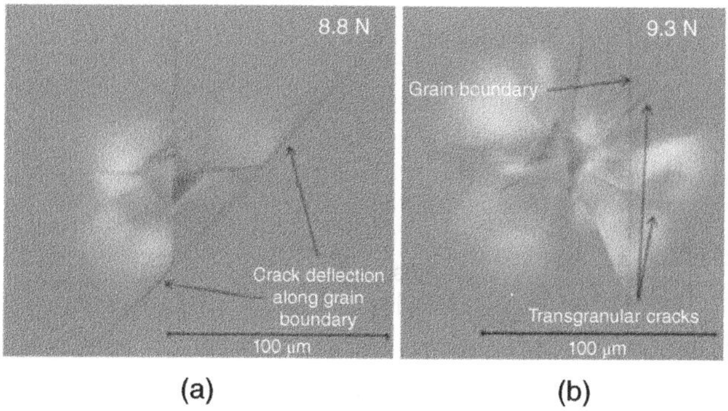

Figure 7.14 Induced damage in spinel due to (a) static indentation and (b) dynamic indentation. Source: Reproduced with permission from Haney and Subhash (2011b).

the impact site in Zone 2, the strain rate of deformation has diminished significantly, and the damage is mostly intergranular fracture. This result is consistent with the fracture modes observed under static and dynamic indentations by Haney and Subhash (2011b). Cracks developed due to a static indentation within a grain are either arrested at the grain boundary or deflected to propagate along the grain boundary, but similar indentations at dynamic rates induce cracks that cut across the grain boundary as seen in Figure 7.14 (Haney and Subhash 2011b). Thus, these fundamental crack modes reveal that at high strain rates (where the rate of incident energy is high) the fracture is predominantly mixed mode and at low strain rates, it is mostly intergranular.

Contrary to the damage developed in spinel, the crack pattern in sapphire for similar ball impact velocities is dramatically different as evident in Figure 7.13b. Since sapphire is a single crystal, the cracks emerge along specific crystallographic planes. Before we delve into understanding the above impact-induced damage patterns, we will review the crystal structure of sapphire and identify dominant cleavage planes because the orientation of these planes with respect to the impact direction plays a critical role in fracture evolution (Haney and Subhash 2011a). A cutaway schematic of an indentation placed on the α-plane of sapphire is shown in Figure 7.15. Compare this figure with Figure 7.5b to get a better understanding of the orientation of fracture planes. In the α-sapphire, three dominant cleavage planes exist: two rhombohedral planes (R1 and R2) and a basal plane (B). Distinct cracks evolve along these planes. The R1 crack plane is angled at 43° relative to the α-plane, while both B and R2 cracks are perpendicular (Haney and Subhash 2011a) to this plane. The angular relationships of these crack planes with the indentation surface can be used to identify the specific cracks along rhombohedral and basal planes (Haney and Subhash 2011c) during the impact process.

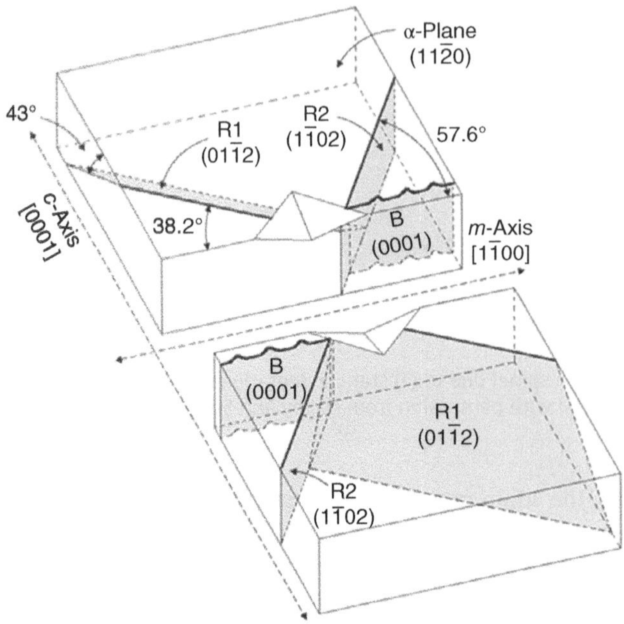

Figure 7.15 Schematic diagram of indentation-induced fracture planes in α-sapphire. Source: Haney and Subhash (2011a). © 2011, Elsevier.

Wavy sets of cracks running parallel to *m*-axis are designated as B-cracks, while the "'X"-pattern cracks are designated as R1- and R2-cracks.

Using sequential Vickers indentations (one indentation followed by another in close proximity) at different orientations, Haney and Subhash (2011a) analyzed induced crack orientations in a single-crystal sapphire. They varied the angle, θ, between two indents from 0 to 135° and identified the unique fracture modes as depicted in Figure 7.16. First an indentation was made at one location on the surface, and a second indentation was then placed at a specific orientation at a suitable distance to promote crack interaction. At $\theta = 0°$, the indent diagonals are aligned, and a new material removal mode emerges, where R1 and R2 cracks from each indent join beneath the surface to form a weakened zone. An elongated prism shaped fragment will eventually separate from the parent materials (Haney and Subhash 2011a). At $\theta = 45°$, the R2 cracks display a unique "pass-by" behavior as the crack planes are parallel to each other. At $\theta = 90°$, B-cracks from each indentation join each other. Finally, at $\theta = 135°$ the cracks from the two indentations once again appear to pass-by on the surface but they interact in the subsurface region. All these crack interaction behaviors are summarized in Figure 7.16. With this knowledge of crack interactions and crack

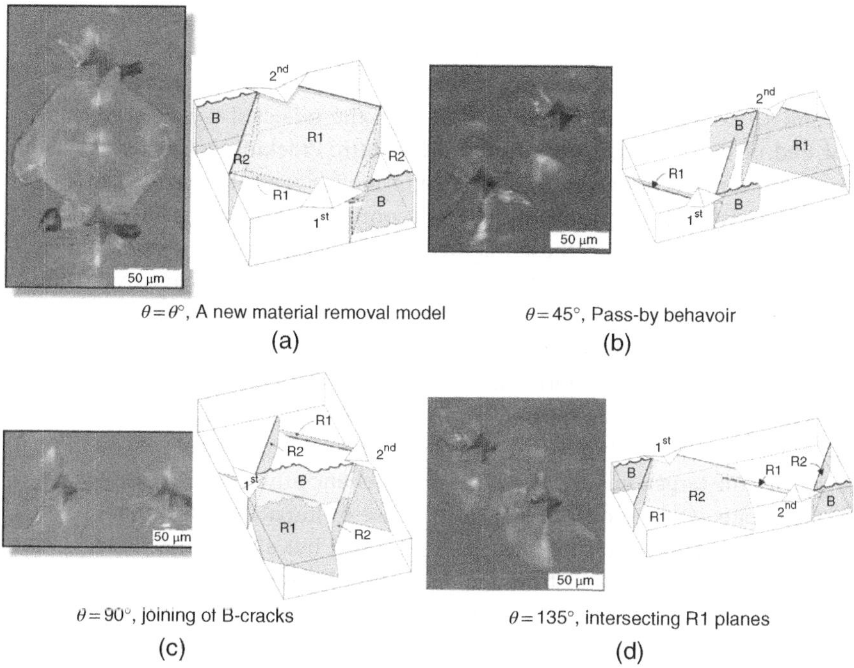

Figure 7.16 Evolved crack patterns due to sequential indentation. For each orientation, optical images of crack patterns are shown on the left and schematics of crack interactions and indentation sequence are shown to the right of the image. Source: Haney and Subhash (2011a). © 2011, Elsevier.

plane orientations, one can now rationalize the damage patterns observed in Figure 7.13b where the impact occurs along the *m*-axis and the surface being viewed is the *a*-plane. At the impact site, numerous fine microcracks evolve as the impact energy overpowers the resistance of the material and no preferential damage evolves along any specific plane. At a short distance away from the impact site, damage occurs preferentially along specific crystallographic planes, and a network of cracks evolve along the basal-plane (B-crack) and rhombohedral (both R1 and R2) planes. These cracks propagate at different velocities along these planes. Although not specifically identified, McCauley et al. (2013) measured crack velocities on various planes to range between 4500 and 5000 m/s. The orientation of these evolving cracks can be identified by comparing the crack patterns to those shown in Figure 7.15. Note from the same figure that both the B-plane and R2 planes are perpendicular to the *a*-plane, and hence, these cracks appear as thin dark lines in the images when viewed through the *a*-plane. On the other hand, the R1 plane is inclined at an angle to the *a*-plane and appears much wider in Figure 7.13b. With time, the crack growth slows and fewer cracks evolve, and hence, these cracks can be identified more easily.

In general, it is noted that rhombohedral cracking dominates in sapphire. Around 15 μs after the impact, cracking from the reflected tensile waves at the rear end of the rod begins to emerge. The entire cracking process stops about 40 μs after impact. Interestingly, no ejecta formation was observed at the impact site in these images. It appears that the ball continues to penetrate the rod as large fragments are formed and these fragments move away from the projectile path, allowing deeper penetration of the ball. The impacting ball is recovered intact (due to the lack of erosion from ejecta) unlike the ball that fractured upon impact on the spinel rod (see Figure 7.13a) (Haney and Subhash 2013).

If the properties of a material are the predictors of material's performance in applications, these results contradict this assertion. When the fragments formed are large, they are easily pushed away (due to lack of confinement) as the ball penetrates the target, and the projectile does not encounter much resistance from the fragments. On the other hand, in the case of impact on the spinel rod, the fine ejecta generated not only decelerate the ball but also erodes it (mass loss) as the particles travel against the incoming ball. Erosion marks are clearly seen on impacted balls, which is discussed in the next section (Haney and Subhash 2013). Such erosion marks were also observed on steel core projectiles used in impact experiments on spinels and AlON by Strassburger et al. (2013) Thus, the fine ejecta moving in the opposite direction of the ball is far superior in resisting the penetration than large fragment formation, which simply allows the fragments to separate from the material and move away from the incoming projectile path. In addition, spinel provides many more energy absorption mechanisms when compared to sapphire. Spinel exhibits both intergranular and transgranular fractures at the grain scale, and the energy is absorbed in the creation of numerous new surfaces that contribute to fine ejecta formation and rapid dissipation of the incident KE. Away from the impact site, only intergranular fracture occurs, and the rod maintains structural integrity because this fracture mode is limited to only a few grain boundaries. On the other hand, the preferential crack planes in sapphire allow large fragments to form, which easily move away from the path of the projectile due to the lack of lateral confinement. This mechanism-based description can provide a rational explanation for the superior performance of spinel when compared to sapphire in the observations of ballistic impact results discussed in Figure 7.12.

7.4.4 Energy Partition During Impact

Having discussed the damage mechanisms in glasses, spinel, and sapphire, it is now appropriate to compare the energy dissipation in various deformation modes in each of these materials. Understanding the energy partition in various deformation and fracture processes from a projectile impact on a target helps to clarify the relationship between the stored energy (in case of strengthened glasses), incident KE, and the mechanisms of deformation (fragmentation process, plasticity, ejecta formation, dilation of the target material due to fracture, and erosion and fracture of the projectile).

First, let us consider the energy dissipated in the elastic stress wave propagation in a solid. Immediately upon impact of a projectile onto a target a compressive stress wave is generated. The fraction (λ) of the KE of the projectile consumed in the wave propagation can be estimated from the following equation proposed by Reed (1985) as

$$\lambda = 7.267\frac{\delta}{\rho_2 C_o^3}(1+\nu_2)\left(\frac{1-\nu_2^2}{1-2\nu_2}\right)^{1/2}\rho_1^{-1/5}K^{-6/5}V^{3/5} \tag{7.2}$$

where the subscript 1 refers to the projectile and 2 refers to the target (glass). The variable δ is a dimensionless constant, V is the impact velocity, ρ is the density of the material, ν is the Poisson's ratio, C_o is the longitudinal wave velocity in the target, and K is given by:

$$K = \frac{4}{3}\left(\frac{1-\nu_1^2}{E_1}+\frac{1-\nu_2^2}{E_2}\right)^{-1} \tag{7.3}$$

where E is the modulus of elasticity. In general, this energy is a major fraction of the incident KE of the ball in these experiments.

Upon impact, material at the impact site is pulverized and ejected off the target. By assuming the deformed volume due to impact to be an ellipsoid of certain depth h and contact radius a, the volume of material that is fully pulverized can be calculated (Haney and Subhash 2012; Jannotti et al. 2015). The average ejecta velocity can be approximated from video images. Thus, the total mass of the ejecta and hence the KE absorbed by the ejecta particles can be calculated. Next, the impact generates fracture and fragmentation in the target. One common question raised here is the relationship between the stored energy and the fragmentation size. By collecting the fragments, measuring their distribution, and assuming a reasonable shape for each fragment, the total fracture surface area of the fragments (A_{FS}) can be estimated. Knowing the surface energy (γ) for the target material, one can estimate the total energy consumed in the fragmentation process (U_{FS}) using:

$$U_{FS} = A_{FS}\gamma \tag{7.4}$$

The fracture process creates new surfaces, and the material dilates. Energy is also consumed in the dilation process. The video images of the impact-induced failure process allow estimation of the dilation volume and its expansion rate. For

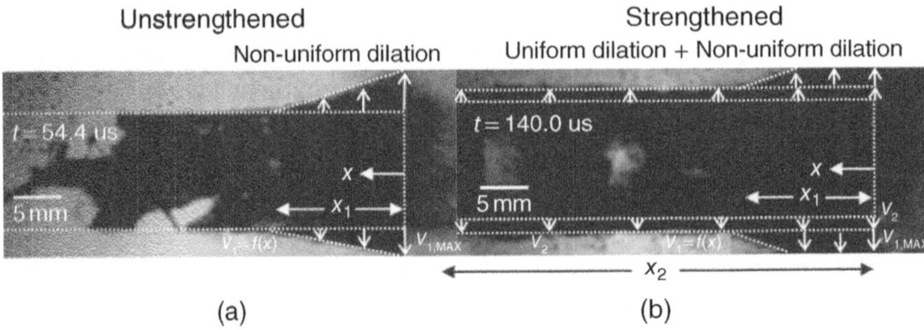

Figure 7.17 (a) Nonuniform dilation of unstrengthened glass rod and (b) and uniform dilation in the rod and nonuniform dilation at the impact end of chemically strengthened glass at comparable impact velocities of 287 and 283 m/s, respectively. Source: Reproduced with permission from Jannotti et al. 2015.

example, in Figure 7.17, the fracture process for the strengthened glass creates a nonuniform dilation at the impact-end of the rod and almost a uniform dilation along the rest of the rod. By knowing the expansion rate and volume of the material, one can estimate the KE associated with this dilation process (Jannotti et al. 2015).

Finally, the energy consumed in friction between the target and the spherical projectile during the impact can be analyzed by considering the interaction area between the two surfaces. Due to the high-impact velocity, significant heat is generated, and the material beneath the impact surface melts and adheres to the ball as seen in Figure 7.18.

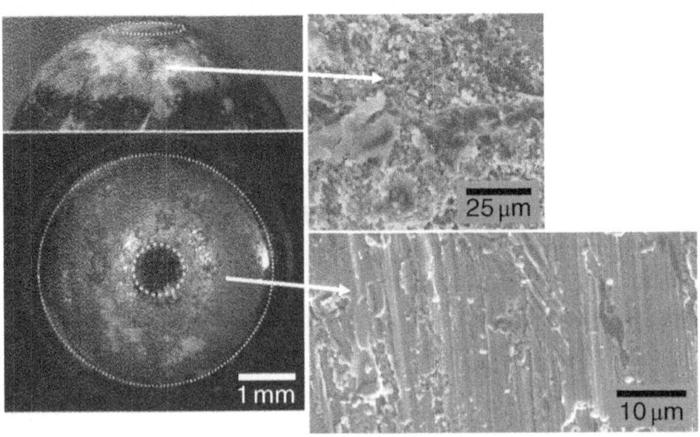

Figure 7.18 Damage to ball from impact with chemically strengthened glass. Note that there are glass particles adhered to the ball and gouges in the steel from high velocity ejecta particles. Source: Reproduced with permission from Jannotti et al. (2015).

Simultaneously, the material that is fragmented and pulverized in the adjacent regions is also ejected. The sharp edges of the ejecta scratch the surface of the projectile ball and cause erosion marks. This frictional energy U_{FR} is a function of the average force associated with deceleration of the ball ($\mathrm{d}V/\mathrm{d}t$), friction coefficient (μ) between the contacting surfaces, and the arc distance (d) over which the ejecta interacts with the incoming projectile (see Figure 7.18). From this information, we can now estimate the energy consumed in the friction process as (Haney and Subhash 2012; Jannotti et al. 2015):

$$U_{FR} = F_{avg}\mu d \tag{7.5}$$

where

$$F_{avg} = m\frac{\mathrm{d}V}{\mathrm{d}t} \tag{7.6}$$

and m is the mass of the ball. In some instances, the ball (projectile) may plastically deform and fracture, which should also be considered. With the above framework, the energy partition in various deformation and fracture mechanisms of spinel and sapphire were estimated and are summarized in Table 7.3.

In the case of strengthened glass, additional stored energy must also be considered. For the residual stress profile shown in Figure 7.2, the stored elastic energy

Table 7.3 Contribution of energy consumed in from various deformation modes during ball impact on slender rods.

	Spinel		*m*-Axis sapphire	
Energy mechanism	**Energy (J)**	**Fractional energy (%)**	**Energy (J)**	**Fractional energy (%)**
Input				
Ball kinetic energy	17.4	100	15.9	100
Dissipation				
Elastic wave energy (UEW)	7.4	43	5.7	36
Intergranular fracture (WOF)	2.8	16	0	0
Intragranular fracture (WOF)	0.8	5	1.7	11
KE of ejecta	1.2	7	0.1	1
Friction	1.5	9	1.4	9
KE of rod dilation	1.1	6	4.3	27
KE of deflected ball	0.6	3	0.8	5
Ball deformation	1.4	8	1.2	8
Tensile fracture (WOF)	0.6	3	0.7	4

Source: Haney and Subhash (2012). © 2012, Elsevier.

is given by

$$U = \frac{\sigma^2}{2E} \tag{7.7}$$

For the compressive and tensile regions, the stored elastic energies were 0.74 and 0.04 J, respectively. Note from Figure 7.17 that the strengthened glass contains not only the nonuniform dilation at the impact end but also uniform dilation along the rest of the bar. The KE of fragments from uniform dilation region is 0.7 J, which is about the same as the stored elastic energy of the compressive region (Jannotti et al. 2015). This analysis reveals that the stored compressive energy contributes to the explosive nature of fragmentation along the length of the chemically strengthened rod, seen in Figure 7.9.

What about the tensile stored energy (0.04 J) in the interior of the strengthened rod? To answer this question, the fragments formed from the impact process of the strengthened and unstrengthened rods were analyzed. The fragment size distributions as a function of size and aspect ratio are shown in Figure 7.19 (Jannotti et al. 2015). The area between the two plots is a measure of the total stored elastic energy. By separating glass fragments into macro-fragments (>1 mm size) and micro-fragments (micron-sized) and approximating their surface area, one can estimate the energy dissipated in fragmentation using:

$$U_s = A_s \gamma \tag{7.8}$$

where the surface energy (γ) is assumed to be 3.7 J/m^2 for glass (West et al. 1999). The energy associated with micro-fragmentation was found to be around 0.35 J and 0.31 J in the unstrengthened and strengthened glasses, respectively, and the corresponding energies in macro-fragmentation were 0.07 and 0.13 J, respectively (Jannotti et al. 2015). Thus, the difference between the energies of macro-fragmentation (0.06 J) is close to around 0.04 J, which is the stored energy due to tensile stress in the strengthened rod. Thus, it is concluded that the compressive stored energy contributes to the explosive nature of the rod

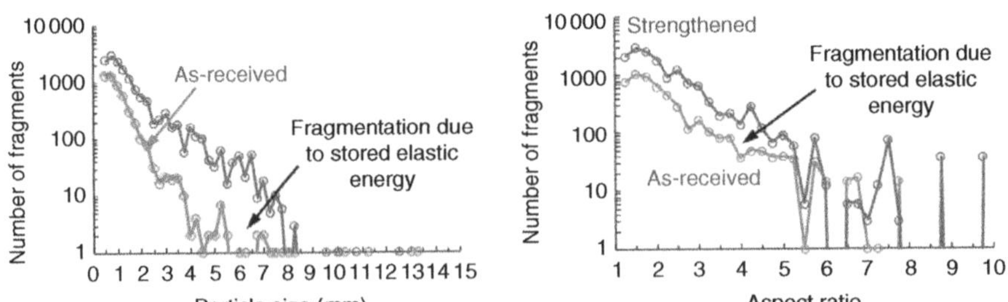

Figure 7.19 Particle size distribution of fragments from impact of unstrengthened (as-received) and strengthened glass. The aspect ratio (indicating sphericity) is shown in the right graph. Source: Jannotti et al. (2015). © 2015, Elsevier.

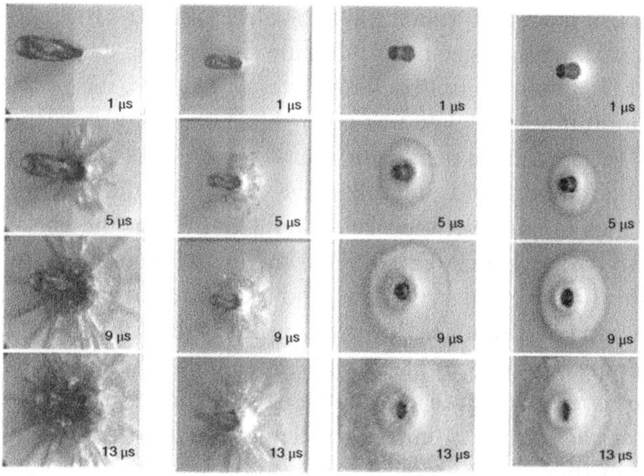

Figure 7.20 Time-lapsed images of damage patterns on different transparent targets. Source: Reproduced with permission from McCauley et al. (2013).

and the tensile strain energy causes additional macro-fragmentation of the strengthened rod.

7.4.5 Normal Impact of Projectiles on Plates

While the impact on a long rod by a spherical projectile illustrates the damage propagation along the length of the rod, the projectile impact on thin plates allows visualization of damage propagation on the impact surface of a target. Figure 7.20 shows time-lapsed images of different axisymmetric damage patterns on different types of glass targets (McCauley et al. 2013). A few transparent glasses such as SL glass and fused silica develop radially growing circular damage fronts with the damage primarily in the form of fine fragments. Others such as AlON develop both radial cracks and circumferential damage growth. Such differences in damage patterns emerge due to differences in the microstructure of the material. Glasses are open structures with a large defect density at the atomic level and have no preferential planes. Hence, damage propensity is the same throughout the structure. This results in fine fragmentation extending radially equidistance from the impact location, whereas ALON is a ceramic with a propensity for radial cracking and forms both radial and circumferential cracking.

Strassburger et al. (2013) evaluated grain size effects in spinel at two different velocities. Spinel with smaller grain sizes (0.6 and 1.6 μm) as well as a bimodal spinel with large grains (∼100–200 μm) dispersed in a fine-grained matrix of

5–20 μm were tested. Their impact performance at two velocities (850–1100 m/s) were compared to AlON with a grain size of 150–200 μm. In all these cases, the ceramics were supported by an aluminum backing plate. At 850 m/s, the penetration depth into the aluminum back plate was negligible, but at 1100 m/s, the residual penetration into the composite plate with AlON was 45% less compared to the spinels. Both circumferential and radial damage evolved in these plates with average crack velocities ranging between 2900 and 3700 m/s for the above materials. The spinel with 0.6 μm grain size exhibited the lowest crack velocity, followed by the spinel with 1.6 μm grain size, AlON, and then bimodal spinel.

Experiments of this type are very limited in the literature due to the difficulty and cost associated with obtaining big sheets or plates of these materials. Additionally, the evolved fracture patterns are difficult to interpret due to the complex interplay of waves reflecting from the back surface. Hence, we will not delve into this specific type of test geometry any further.

7.5 Edge-on Impact of Glass Targets

Previous studies on the impact response of rods by a ball (seen in Figure 7.13) provide visual illustration of damage evolution in a 1D setting. The effects of lateral damage growth and the influence of material lateral confinement are not captured in this target geometry. Impact normal to a thin plate (see Figure 7.20), on the other hand, mainly reveals damage evolution on the impact surface. A lack of sufficient thickness in the impact direction precludes observing evolving damage in the interior of the bulk solid. Visualization of damage growth in the bulk solid is essential for understanding the mechanics of impact-induced damage during a projectile penetration. To this end, edge-on-impact (EOI) experiments were developed which allow visualization of damage propagation in a 2D framework. A schematic of the setup was earlier shown in Figure 2.20. The edge of a plate is impacted with either a spherical or a blunt cylindrical projectile in the plane of the plate, and the stress wave propagation and damage evolution are captured with a polariscope and a high-speed camera, respectively. Through this technique, the induced macroscopic fracture patterns, crack velocities, and damage-front-propagation characteristics can be investigated. Due to the finite dimension of the plate and the high velocity of the projectile, wave effects become significant in the analysis of results at times well after the impact. For a sound foundation on wave propagation in a finite body, the reader is referred to the publications by Schardin (1950), Vollkommer (1967), and Beinert (1974).

Figure 7.21 shows spherically expanding longitudinal and shear wave fronts developed due to edge-on impact on a monolithic Starphire® glass (a low-iron glass, also called water-clear glass) by a steel sphere at 440 m/s (McCauley et al.

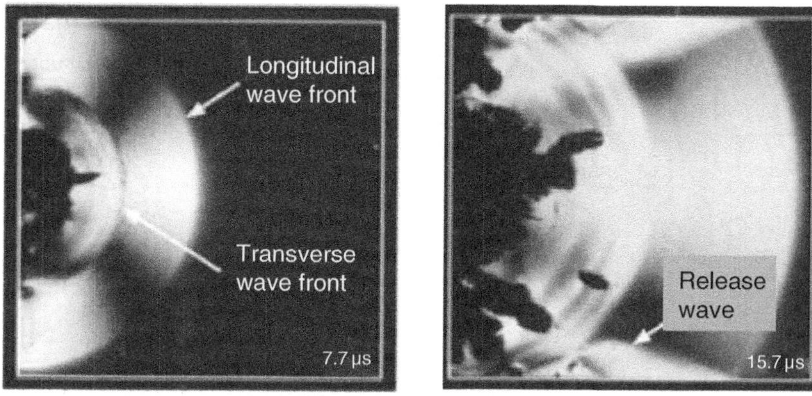

Figure 7.21 Shadowgraph images of wave propagation and damage front evolution in Starphire® glass due to a steel sphere impact at 440 m/s. (a) Longitudinal and shear wave fronts are seen immediately after the impact and (b) damage evolution and void growth behind the shear wave as well as release wave from lateral edges are seen. Source: Reproduced with permission from McCauley et al. (2013).

2013). The longitudinal and shear wave velocities were calculated to be about 5780 and 3270 m/s, respectively. Initially, damage nucleates at discrete locations just behind the shear wave. The major axis of these isolated damage zones is aligned radially, and these zones eventually merge with the incoming damage front. Interestingly, with an increase in the impact velocity, the damage front growth rate was found to coincide with that of the shear wave velocity, implying that the shear wave amplitude was sufficiently high to create instantaneous damage as it arrives.

Figure 7.22 illustrates wave profiles and damage patterns generated during the EOI test (Strassburger et al. 2008). Upon impact, a longitudinal compressive wave emanates first from the impact location followed by a shear wave, both expanding in a spherical fashion into the bulk material at velocities C_L and C_s, respectively. Of interest are the waves generated at the edges of the projectile and their influence on fracture generation. Due to the finite thickness of the plate and the bluntness of the projectile, two additional shear waves evolve, one at the top-edge and another at the bottom-edge of the cylindrical projectile. The angles formed by these cylindrical wave fronts with the growing edge waves are given by

$$\alpha = \sin^{-1}\left(\frac{C_s}{C_L}\right) \tag{7.9}$$

$$\beta = \sin^{-1}\frac{C_D}{C_s} \tag{7.10}$$

where C_D is the damage-front velocity.

Strassburger and Senf (1995) noted that at low-impact velocities, the damage initially grows from the edge of the impacted cylindrical on a plane perpendicular to

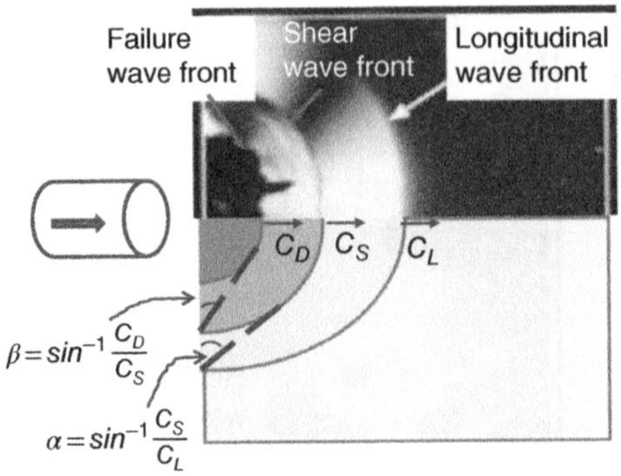

Figure 7.22 Illustration of evolution of longitudinal, shear, and failure waves in a transparent material upon edge-on impact. Source: McCauley et al. (2013). © 2013, Springer Nature.

the line of impact and then curves into the bulk material. This damage evolves due to intense shear stress along the edge of the impact zone and grows at the shear wave velocity of the material. However, the individual cracks within the damage zone have been found to grow at terminal velocity, which is characteristic of a material and typically ranges between 0.55 and 0.65 times the shear wave velocity. At higher impact velocities (>200 m/s), damage nucleates and evolves faster ahead of the projectile in a spherical zone because enough shear has already been attained for crack growth. At even higher projectile velocities, (>400 m/s), extensive damage evolves. If the plate dimensions are small, reflected tensile waves can create additional damage and spalling along the edges. Also note from Figure 7.21b that isolated voids emerge ahead of the damage front but this damage lags behind the shear wave. These voids are first noted to nucleate just behind the longitudinal wave of sufficiently high magnitude, implying that damage is activated (nucleated) at the preexisting defects upon arrival of the longitudinal wave but their growth is delayed due to the time required for crack opening (inertia). Severe damage growth occurs upon the arrival of the shear wave. Following the shear front is the damage front consisting of severely fragmented and comminuted zone emanating from the impact site, where the high density of energy deposition has occurred. In polycrystalline ceramics and glasses, the EOI experiments activate more bulk defects than surface defects. Two types of interior damage are noted (Senf et al. 1994): (i) needle-like cracks whose plane is perpendicular to the viewing surface and (ii) shell-shaped cracks whose plane is almost parallel to the viewing surface and curve toward the surface at distances farther away from the impact site.

EOI experiments on single-crystal sapphire induced damage patterns dramatically different from those in the glasses described above. Fracture in sapphire evolves along preferred cleavage planes as presented in Figure 7.23a,b. At impact velocity below 400 m/s, cracks originate from these planes and grow at their characteristic speed. Figure 7.23c presents crack growth rates along few of these

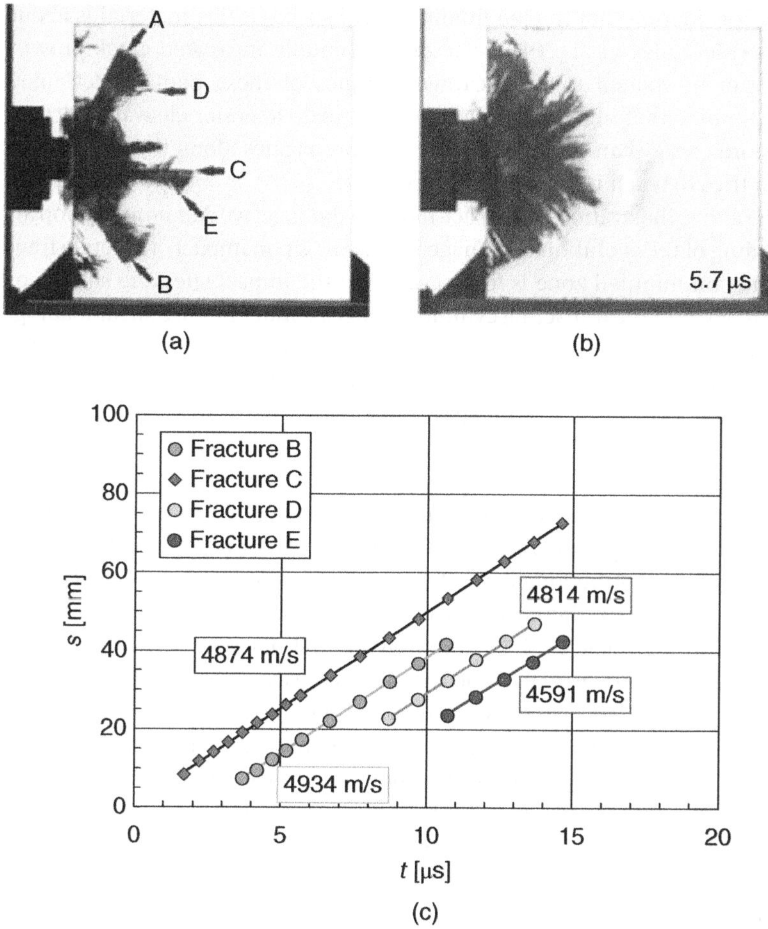

Figure 7.23 (a) Damage evolution on a *c*-plane sapphire due to edge-on impact along a-axis, (b) plot of damage growth rates along selected fracture planes, and (c) spherical damage front evolution at higher impact velocities. Source: (a) Reproduced with permission from McCauley et al. (2013), (c) McCauley et al. (2013). © 2013, Springer Nature.

planes on a *c*-plane sapphire when impacted along *a*-axis. It is seen that the growth rates of these cracks vary depending on the cleavage plane (McCauley et al. 2013). Interestingly, at higher velocities (>450 m/s) the difference between these fracture growth rates diminishes due to excessive energy input. A dense, semi-circular fracture zone can be seen in Figure 7.23b. The average crack growth speeds increase well in excess of 5200 m/s and the damage front velocity reaches as high as 8137 m/s at the center of the front. This high velocity of damage is surprising considering that the longitudinal wave speed in this material is about 11 450 m/s (McCauley et al. 2013). The experimentally measured crack growth velocities can be related to the cleavage energies of these planes. McCauley et al. (2013) noted that higher cleavage energies relate to easier cleavage fracture because more energy can be expended if crack propagates along these planes as the system tries to reach its minimum energy state.

From the above illustrations of impact-induced damage, we can now develop an understanding of the evolution of damage in glasses. Upon impact, a severely fragmented and comminuted zone is formed close to the impact site. The surface of glass in contact with the impacting cylindrical edge is under intense shear loading, whereas the material directly ahead of the impact surface is under a compressive stress state. Cracks can originate more easily under shear as compared to compression, and hence edge cracks generate first and grow faster to large size as noted in the high-speed images reported by Strassburger and Senf (1995). However, bulk defects are formed readily in the compressive zone. When the amplitude of compressive stress is low, cracks often nucleate at preexisting defects when the shear wave arrives. At high enough compressive stress levels (or at high-impact velocities), the amplitude of the longitudinal wave is sufficient to cause high shear at that location. In between these two waves, isolated damage is initiated ahead of the damage front (Grujicic et al. 2008), implying that damage is activated at the preexisting defects upon arrival of the longitudinal wave. A slight delay is often observed due to the inertia associated with crack opening (see Figure 7.21). Upon arrival of a shear wave, severe damage evolves. When the dimensions of the plate are small, the longitudinal wave reflects from the free edges and generates release waves, which are both tensile and shear waves. When these waves reach potential crack nuclei, further damage evolves which may eventually merge with the incoming damage front. Once the intensity of stress wave diminishes, the damage ceases to grow.

7.6 Damage in Multilayered Glass Composites

Thus far, we have discussed damage evolution in a single glass specimen of rod or plate geometry impacted either edge-on or along normal to its major plane. While these studies provide valuable information on fundamental damage evolution modes, many structures made of transparent materials consist of multiple layers of glass to provide adequate protection against high-velocity impacts. Transparent materials of different properties are stacked together to obtain the desired performance in the final composite. For example, high hardness outer layer to achieve impact resistance and rapid erosion of the projectile, and high fracture toughness inner layers to prevent shattering of glass. Upon impact, damage created in one layer may propagate to the next layer, and understanding of this transfer mechanism through an interface is fundamental to the design of better protective composites.

Earlier studies on the low-velocity impact of polymers and glasses (e.g. homalite and PMMA) by Suresh et al. (1992) revealed that cracks may arrest at an interface, but at higher velocity impacts, cracks extend into the next layer. It has been observed that when the crack reaches the interface, it pauses for some time and then propagates into the adjacent material (Theocaris and Milios 1980a,b, 1981a,b). If the adhesive layer has low strength, then the crack propagates along the interface and separates the two materials (Theocaris and Demakos 1986; Theocaris and Milios 1981a; Xu et al. 2003; Xu and Rosakis 2002a,b, 2003a,b, 2005). Chai and Ravichandran (2007) performed systematic experiments on a bonded soda lime glass and polycarbonate composite multilayer system using sharp and blunt projectiles to investigate the damage propagation characteristics and threshold energy for initiation of unstable fracture. It was noted that median cracks that originate at the impact surface from the sharp tip of a projectile are arrested at the interface and spread the damage in the impacted glass along its interface.

Fundamental studies of crack propagation through an interface between two transparent plates were investigated by Park and Chen (2011) and Parab and Chen (2014). The specimen design consisted of two rectangular thin plates either kept next to each other edge-on to create an interface with no adhesive or the interface between them was bonded with a polymer adhesive of varying thicknesses. To ensure that a crack is initiated at a desired location and arrived normal to the interface, a V-notch was created on the front plate. The sharp tip of the V-notch faces the interface, and the wide mouth of the V-notch faces a cylindrical projectile launched from a gas gun, as shown in the schematic of Figure 7.24 (Parab and Chen 2014). The impact results in opening of the V-notch, initiating a mode-I tensile crack which grows and arrives perpendicular to the interface. The dimensions of the two plates were carefully chosen to ensure that the release waves from the impact site or specimen edges do not interfere with the region of interest along the crack path during the observation time. Using a PU projectile and adhesives of different strengths and thicknesses, impact experiments were conducted on composite specimens of two plates in the configuration shown in Figure 7.24 and the induced crack velocities were measured to be up to 1500 m/s (Parab and Chen 2014).

When the two glass plates were simply in edge-on contact without any adhesive, the crack did not propagate across the interface. This implies that even a small gap may impede the energy transfer between the two plates. On the other hand, when the interface was bonded with a thin (presumably zero-thickness) layer, the crack effectively entered the second glass with short delay at the interface, as noted in the top images of Figure 7.25. There was no change in the crack speed or direction.

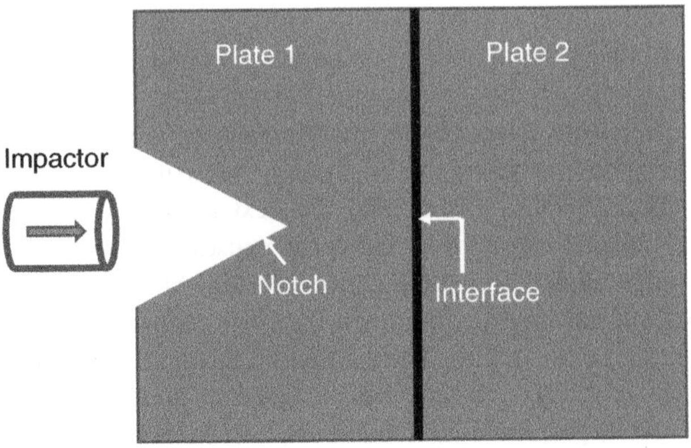

Figure 7.24 Schematic of a test specimen for studying the interaction between a dynamic crack and an interface between two transparent glass plates. Source: Modified from Parab and Chen 2014.

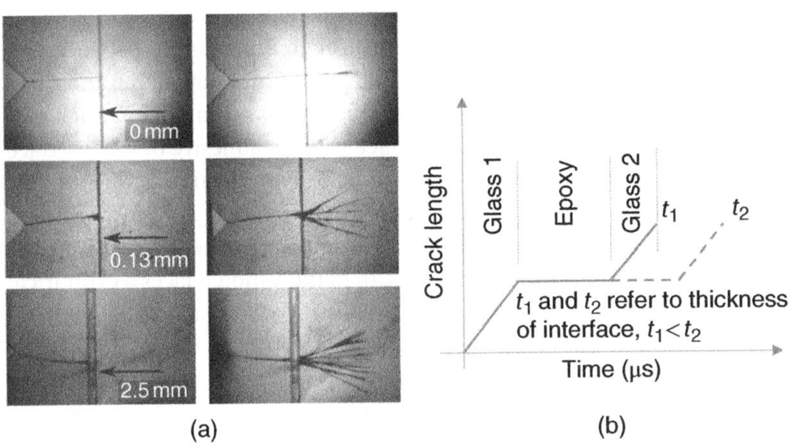

Figure 7.25 Influence of interface thickness on the Interaction of a crack tip with a polymer interface between two glass plates. The images (a) are at the time when crack reaches the interface and those on the right are obtained after the crack emerges from the interface. Source: Reproduced with permission from Park and Chen (2011). © 2011, American Society of Mechanical Engineers. The schematic in (b) illustrates the crack growth characteristics with time in the composite with epoxy interlayer.

When the adhesive thickness was increased to 0.13 mm there was a 20-μs delay in the crack propagation at the interface. Upon arrival of the crack at the interface, the crack growth is arrested, and the energy accumulates in the low-impedance epoxy layer. The stress builds-up until a threshold is reached and then released suddenly to reinitiate damage in the form of multiple cracks in the second plate in a 50° cone (middle image in Figure 7.25).

Interestingly, all cracks grew at the same velocity. Thus, even a thin layer of adhesive influences the interface behavior compared to no adhesive. Increasing the interface thickness to 2.5 mm arrests the crack for a longer duration in the epoxy interface, as shown in the schematic of Figure 7.25. The crack reemerged in the second glass after a 45-μs delay with multiple branches emanating in a cone with apex angle as large as 65°. The delay allowed accumulation of large amount of energy at the interface. While the time to cross the thickness of the adhesive layer increases, the time to reinitiate the crack does not change once the crack reaches the second glass (Parab and Chen 2014).

Motivated by these experimental studies, (Sundaram and Tippur 2016a,b) conducted in-depth mechanistic investigations on the influence of the interface strength of a two-layer PMMA specimen bonded with an acrylic adhesive. A V-notch induced crack was dynamically propagated in mode-I configuration in layer 1 toward the interface, similar to the experiments described above. The images are given in Figure 7.26. A mechanistic analysis of the competition among crack penetration, deflection, and crack branching at the interface was conducted utilizing the digital gradient sensing (DGS) technique in conjunction with high-speed photography. Time-resolved optical measurements of crack-tip deformations, crack-tip velocity, and stress intensity factor (SIF) histories were obtained. The interface was designed to be either weak or strong based on their crack initiation toughness values (Sundaram and Tippur 2016a). The analysis revealed that a low-velocity mode-I crack with a low SIF penetrates the interface, whereas a high-velocity mode-I crack at higher SIF branches out along the interface in opposite directions for some distance and then emerges in the second layer as two daughter cracks with equal stress intensity but in a mixed mode fashion. When the interface was strong, the distance over which the crack

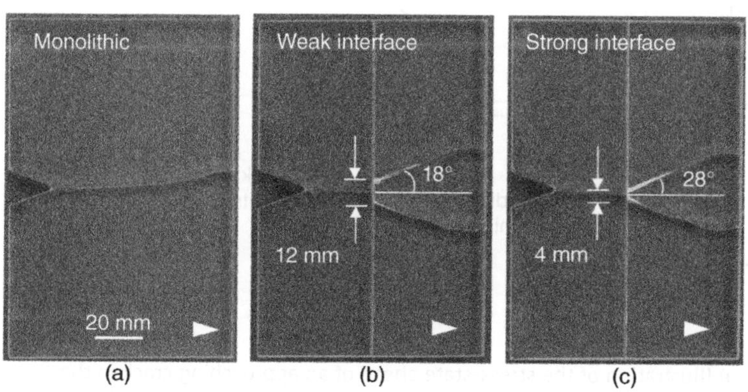

Figure 7.26 Mode-I crack propagation characteristics in (a) monolithic PMMA glass and (b and c) in bi-layer glasses with weak and strong interfaces. Source: Reproduced with permission from Sundaram and Tippur (2016b).

branches propagated along the interface was shorter than the distance traveled in a weaker interface, but the angle between the emerging daughter cracks in Layer 2 was smaller when the interface was weaker; see Figure 7.26. In both cases, the interface influences the mode-I crack propagation by branching and deflecting the crack to create a greater surface area of the crack than that would occur in a monolithic glass.

Interestingly, the crack-tip velocity measurements indicated that as the mode-I crack approached the interface, its velocity increased due to the reflected tensile wave from the interface, aiding crack growth and increasing the crack tip driving force. This increase in crack growth rate was interpreted as a result of the information received from the reflected wave to the crack-tip of the approaching weaker polymer interface material (Sundaram and Tippur 2016b). The crack velocity is also greater while growing along the interface compared to that in monolithic material, with higher crack velocities being registered along a weaker interface which is reflected as a greater separation between the daughter cracks compared to that in a stronger interface. Based on experimentation and fracture analysis, Sundaram and Tippur (2016b) concluded that, the proximity of the interface to crack tip and its strength dictate the crack penetration and branching characteristics, which are summarized in Figure 7.27. When the interface is strong (strength greater than the tensile stress induced by the approaching crack), the mother crack penetrates the interface and enters into the second plate in mode-I fashion. When the interface is weak, the tensile stress across the interface induced by the approaching mother crack causes debonding of the interface, allowing crack

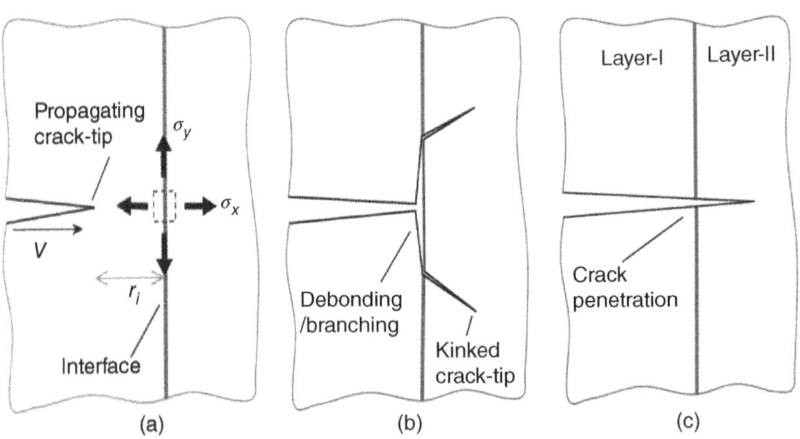

Figure 7.27 (a) Illustration of the stress-state ahead of an approaching crack in the vicinity of an interface, (b) crack branching along a weak interface, and (c) crack penetration when the interface is strong. Source: Sundaram and Tippur (2016b). © 2016, Elsevier.

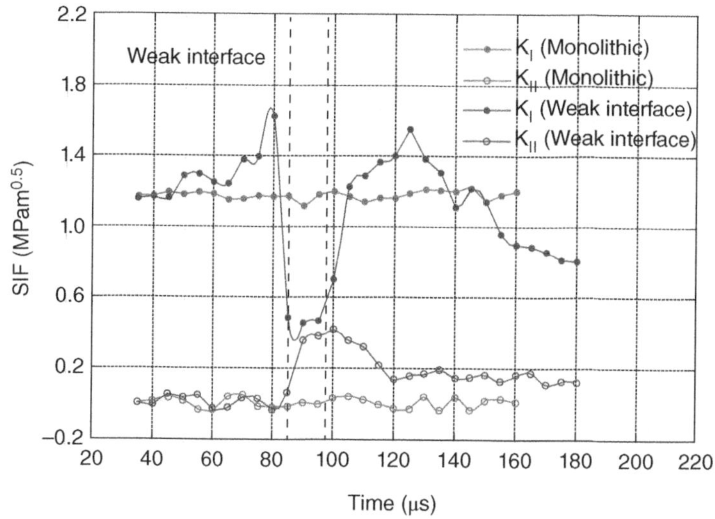

Figure 7.28 Stress intensity factor histories for crack growing into the interface and emerging in Layer 2. Source: Sundaram and Tippur (2016a). © 2016, Springer Nature.

branching and growth of two cracks along the interface in opposite directions. The competition between the mixed-mode crack growth along the interface and that into the second layer dictates when the crack enters the second layer.

The histories of SIFs for both mother and daughter cracks, shown in Figure 7.28, reveal that as the crack approaches the interface, its K_I increases but upon entering the interface, K_I drops rapidly to a low value, whereas the K_{II} increases as the daughter cracks grow along the interface, indicating the shear dominance during this phase. As the daughter cracks reemerge in Layer 2, K_I jumps back to the original value as in Layer 1 and K_{II} settles down to a high value (compared to that in Layer 1). This behavior indicates the dominance of mixed mode fracture during their growth. Overall, both K_I and K_{II} values are higher for a specimen with weak interface than for a strong interface. Analysis of individual SIFs indicated that the two daughter cracks interact during their growth process, i.e. when the top crack experiences a high K_I value, the bottom crack experiences a low value and vice versa.

The studies described above provide fundamental mechanistic information on crack interaction with an interface. However, the damage propagation under dynamic loads between two glass layers is much more complex. Strassburger et al. (2007) and McCauley et al. (2013) studied damage propagation in laminated glass composites by conducting EOI experiments. Test specimens of two or three Starphire® glass laminates bonded with different polymer interlayer thickness (0.64, 1.27, 2.54 and 5.08 mm) were impacted by a cylindrical rod at velocities up

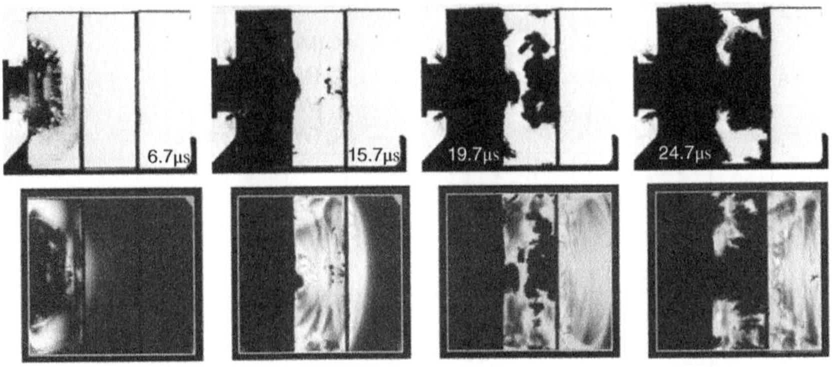

Figure 7.29 Images of edge-on impact of a three-layer-laminate transparent armor composed of Starphire glass and polyurethane. Source: Reproduced with permission from McCauley et al. (2013).

to 400 m/s. The stress wave transfer and damage propagation across the interface were captured using a high-speed Cranz-Schardin camera and photoelasticity. Figure 7.29 illustrates damage growth in the top row of images and stress wave propagation in the bottom row of images in a three-layer-laminate specimen. Upon impact, damage is initiated in the first laminate and after accumulation of significant damage, the stress wave transfers to the second laminate, which can be seen in both top and bottom images. However, damage occurs initially on the backside of the second plate due to the transmitted compression wave, which reflects as a tensile wave and then the damage grows with time. Similarly, stress wave then travels to the third laminate and the process repeats.

In Figure 7.30, the effect of interlaminar polymer thickness is illustrated. Note that for thin interlayers, the stress wave is quickly transferred to the second laminate and reflected off the free surface to cause tensile damage at the rear end of the second laminate. In contrast, when the polymer interlayer thickness is large (5.08 mm), the stress wave transfer into the adjacent layer is delayed. The delay corresponded to the time required for the longitudinal wave to travel the PU interlayer at a velocity of 2 km/s. Similar results were also found by Choi and Kim (2016) who evaluated laminates composed of glass with a rubber interlayer. However, the relationship between the severity of damage in the second layer and the thickness of interface were not discussed in this study, probably due to limited thickness of the layers. Shorter thickness of layers allows stress wave interactions (from reflected waves from the boundaries) and prevents observation of damage evolution due to impact alone.

In summary, these studies imply that a polymer layer at the interface of a laminate can only delay the time for energy transfer and, upon reaching a threshold, will release that energy to the adjacent layer in a single burst to initiate multiple

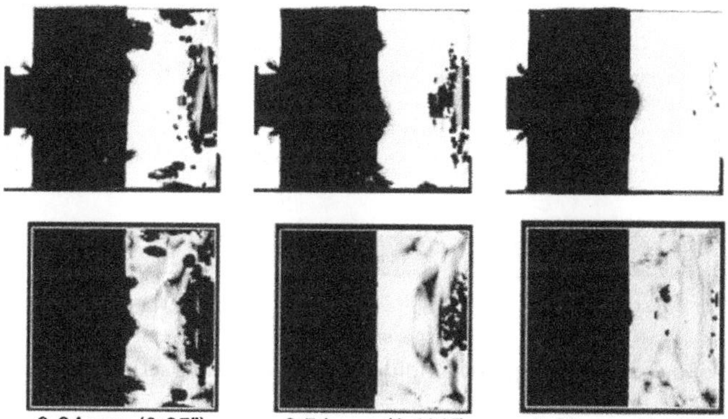

Figure 7.30 Illustration of the effect of interlaminar polyurethane layer thickness between a two-layer Starphire glass laminate due to edge-on impact. Source: Reproduced with permission from McCauley et al. (2013).

cracks. When the amplitude of the incident energy is low, the interface acts as a cushion and does not allow damage to propagate into the next layer, but if the amplitude of the incoming stress is high, the wave energy transfers to the next layer causing severe damage at the entry in the next layer. Even when there is no damage at the entry in the second layer, additional damage is possible at the rear side of that layer due to reflected tensile wave.

7.7 Rod-on-Rod Impact (Taylor Impact) Experiment

Early experimental investigations on glasses subjected to shock loading revealed a unique particle velocity signature just behind the shock front: the shock front was followed by another high amplitude wave whose velocity increased with increased impact velocity. Researchers interpreted the appearance of this feature as an indication of a propagating damage wave as they were dealing with shock wave profiles, and they termed this phenomenon a "failure wave" propagation (Bourne et al. 1996a; Brar 2000; Espinosa and Xu 1997; Rosenberg et al. 1996). However, it was unclear if the damage growth followed typical wave characteristics, i.e. once initiated, the waves travel at a steady state with no additional energy input. In case of impact by a penetrator, this phenomenon, if exists, should yield a damage front propagation even after the driving force from the impactor is fully subsided (or penetrator has stopped or fully eroded). The fundamental mechanisms of damage evolution were not understood. Two test methods, as illustrated in Figure 7.31, were considered to resolve this issue: (i) Rod impact against a rigid planar surface or anvil (referred to as a Taylor impact) and (ii) symmetric rod-on-rod axial impact

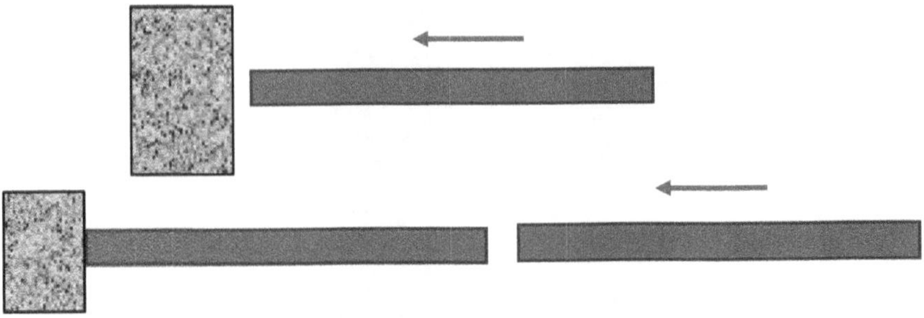

Figure 7.31 Schematic of a rod impact against a rigid surface (Taylor impact) and symmetric rod-on-rod impact.

where one rod is rigidly held while another rod of similar lateral dimensions is launched against it to induce damage under 1D-stress loading in both rods. High-speed photography was utilized to observe and analyze the evolving damage characteristics.

The traditional Taylor test was originally developed by Taylor (1948) to investigate plastic deformation characteristics of metals at large strains and high strain rates and for validating constitutive models. However, friction along the impact surface was found to influence the deformation behavior of the specimens. Therefore, symmetric rod-on-rod tests were employed (Bless and Brar 1994; Brar et al. 1991a; Erlich and Seaman 1981).

In Taylor impact experiments on glass rods (Radford et al. 2004; Willmott and Radford 2005), a shock wave is initially generated upon impact and is followed by a failure wave. A typical result on fused quartz is shown Figure 7.32, and similar observations were made on SL and BS glasses (Radford et al. 2004; Willmott and Radford 2005). Initially, the stress state in the rod is 1D-strain until the release wave

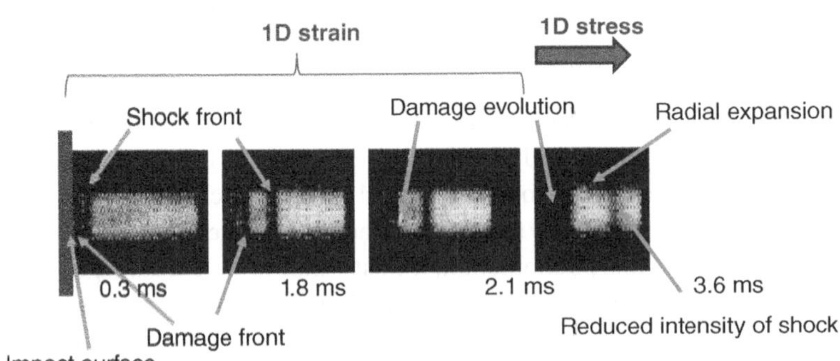

Figure 7.32 Evolution of shock front and damage front during Taylor impact of a glass rod. Source: Reproduced with permission from Radford et al. (2004).

from the radial boundary arrives at the center of the rod, and then the damage wave travels in a 1D-stress state. The early stage longitudinal wave velocity is given by Eq. (2.1), and the latter 1D-stress wave velocity is given by Eq. (2.2). The transition seems to occur around five diameters from the impact face (Brar et al. 1991), and is accompanied by two effects: (i) a loss in shock velocity from the uniaxial strain condition to the uniaxial stress condition and (ii) gradual loss in the visible intensity of the shock front due to a reduced pressure gradient across the shock front in the 1D-stress state (see Figure 7.32). The appearance of the shock wave in these high-velocity long-rod impact experiments is associated with densification that occurs in the shock front (Rosenberg et al. 1996; Bourne et al. 1996a). Glass densification up to 11% has been recorded in such Taylor impact shock experiments by Okuno et al. (1999) due to a reduction in the Si–O–Si angle and the collapse of the ring cavities in the glassy structure. This densification causes changes in the refractive index of the glass (Reynard et al. 1999; Okuno et al. 1999), which is reflected as a dark band in the above images. Doyoyo (2002) developed a theoretical framework where the densification causes roughening of the shock front and initiates the damage. This damage starts to grow and eventually emerges as a failure wave. These theoretical analyses will be discussed later in this chapter. Behind this shock front, significant radial expansion of material is seen in Figure 7.32, which corresponds to the evolving internal damage. Such features were also reported by Murray et al. (1998) and by Jannotti et al. (2015).

The damage evolution features due to the symmetrical rod impact in two types of glasses were investigated by Murray et al. (1998) and Wilmott and Radford (2005). Upon impact, a damage front initiates at the impact plane in both rods and propagates into the rods in opposite directions. The damage can be detected as loss of transparency (dark region) in each rod. The elastic wave travels in SL glass at 5400 m/s and at 5700 m/s in BS glass. The damage front, however, travels at increasing velocities with increasing impact stress level. In this specific case, it travels at 4500 ± 400 m/s in SL at a stress level of 2.0 GPa and 4900 ± 400 m/s in BS glass at 2.5 GPa. Interestingly, the radial damage expansion in BS glass occurs with a smooth profile, but that in SL occurs in an irregular fashion, as shown in Figure 7.33, where periodic radial jets appear to spike in the profile (Willmott and Radford 2005). The authors speculated that these morphological features may relate to the fracture process and final shape of the comminutia evolved in each material. The SL forms rough large size fragments, whereas the BS glass forms finer fragments.

In these rod-on-rod experiments (Radford et al. 2004; Willmott and Radford 2005), the FF velocities exceeded $\sqrt{2}\,C_s$ and sometimes reached close to that of the longitudinal wave speed when the impact pressure exceeded 2 GPa. From a classical dynamic fracture mechanics perspective, the maximum velocity of a single Mode-I crack is limited to the Rayleigh wave velocity, C_R, which is approximately

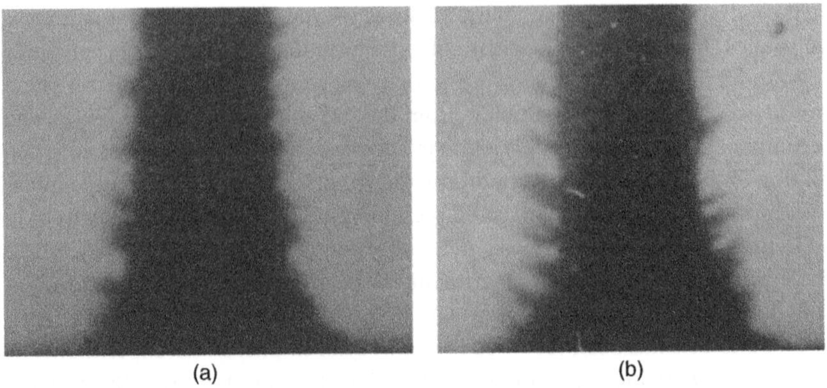

Figure 7.33 Damage due to symmetrical impact of (a) borosilicate glass revealing smooth profile and (b) soda-lime glass with periodic needle like damage profile. Source: Reproduced with permission from Willmott and Radford (2005).

90% of C_s (Freund 1998). It is also well-established that a single crack in shear mode can propagate at speeds close to $\sqrt{2}\,C_s$ (Rosakis et al. 1999). So why does the damage growth in these rods occur at velocity in excess of $\sqrt{2}C_s$? We will answer this question in Section 7.10 after gaining further insight into the damage evolution due to shock loading in the next section.

7.8 Shock-wave Experiments on Glasses

Unlike ceramics, the open structure of glass allows densification under high pressure, and this densification behavior has been argued to initiate damage (Clifton 1993). To understand densification, damage propagation, and the shock response of glasses, normal plate-impact experiments were conducted by instrumenting the specimen with longitudinal and lateral strain (or stress) gages. Schematics of two such specimen configurations are shown in Figure 7.34. Embedded manganin gages or gages mounted at the back of the specimen and supported by a thick PMMA block, as shown in Figure 7.34a, capture the shock waves from which the material response was inferred (Bourne et al. 1996a). Split-target confinement is used (Figure 7.34b) to measure stress in both σ_x (longitudinal) and σ_y (lateral) directions, so that the shear stress ($(\sigma_x - \sigma_y)/2$) can be calculated during the wave propagation process. By bonding two strain gages at different locations along the path of stress wave propagation, one can infer if there are any differences in the velocities of shock propagation and damage propagation. A variety of transparent materials were tested using metallic (Al or Cu) or ceramic (WC) flyer plates (Bourne et al. 1996a). In the following, these results are summarized separately for

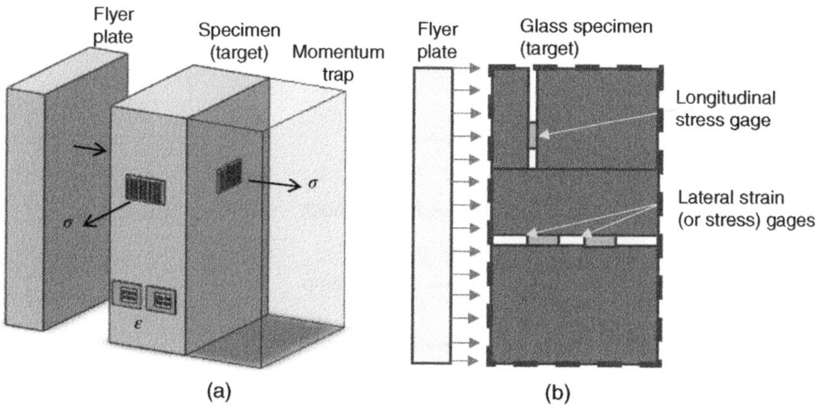

Figure 7.34 Schematic of flyer plate specimen geometries with (a) embedded gages and transparent PMMA momentum trap and (b) split specimen configuration to measure longitudinal and shear stresses during the experiment.

low-density and high-density glasses due to their different responses under shock loading.

7.8.1 Low-density Glass

The density of glass plays a major role in its impact response. In contrast to the shock response of ceramics, glasses reveal a softer response in plate impact experiments because of their network structure, which collapses and densifies the glass at higher stress levels. Recall from Chapter 5 that for ceramics, the particle velocity history during shock loading reveals a sharp rise until the HEL is reached. This sharp rise indicates that the shock state is attained in the material immediately upon flyer plate impact. For low-density glasses, however, the shock state is not immediately achieved upon impact, but the stress gradually builds to a certain level up (called an elastic ramp), which is then followed by a steep rise (elastic shock) until the HEL is reached (Bourne et al. 1996b). A typical particle velocity measurement at the interface between a glass specimen and a PMMA window during shock loading is shown in Figure 7.35 (Alexander et al. 2008). The initial ramp implies loss in stiffness of the glass due to the collapse of its open-network. As the glass gets denser during the loading, a steep rise in particle velocity occurs (elastic shock) until the HEL. For example, low-density glasses such as fused quartz and BS have a ramp up to 2.5 GPa, followed by a steep rise to the HEL of 6–9 GPa (Bourne et al. 1996b). However, the HEL is not a sharp point but a gradual change over a pressure range. To describe this deformation behavior, Bourne (2008) defined lower (HEL_L) and upper (HEL_U) HEL limits, each corresponding, respectively, to fracture initiation at favorable locations and

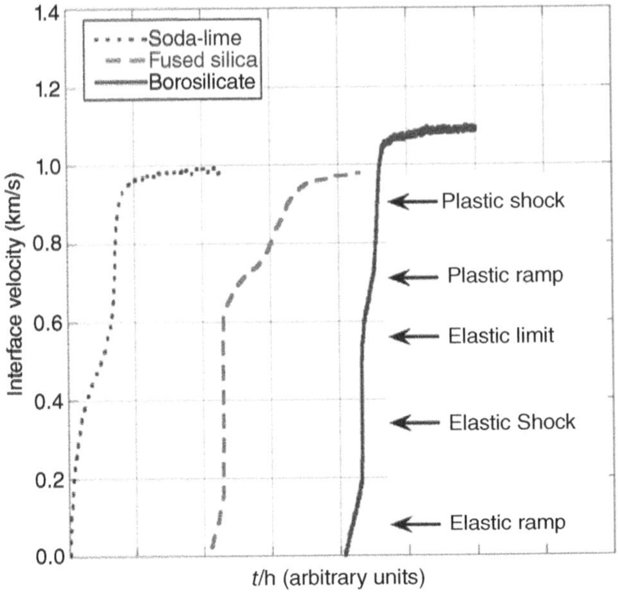

Figure 7.35 Shock profiles revealing various phases of deformation in low-density glasses. All deformation phases occur gradually rather than in an abrupt fashion as is typically observed in ceramics. Source: Alexander et al. (2008). © 2008, Elsevier.

spread of damage to the entire specimen. During this phase, the curve transitions from a convex to a concave rise. The extensive failure at the end of HEL_U manifests into an FF at higher stress levels, which has been captured during the flyer plate impact experiment using transverse stress signals as discussed later.

Beyond the HEL, a plastic ramp is observed followed by a plastic shock. Alexander et al. (2008) speculated this plastic state to be densification, phase transformation, or even failure wave generation. The plastic ramp indicates that the plastic deformation does not occur instantaneously but is stretched over a period where softer and/or defective regions first undergo inelastic deformation. As the stress level increases, other regions gradually start to undergo permanent deformation. By suitably positioning lateral strain gages in the specimen, one can track the propagation characteristics of the plastic (failure) shock wave. These shock wave profiles are shown for BS and sodalime glasses in Figure 7.36 (Alexander et al. 2008). Note that with an increase in shock amplitude, the plastic ramp becomes steeper and eventually catches up with the elastic ramp, indicating the increasing velocity of plastic deformation wave at higher shock amplitudes.

The densification process during the elastic and plastic responses of glasses has been a subject of considerable interest. The HEL in glasses is often defined by fracture and comminution of the material, leading to a loss in the longitudinal

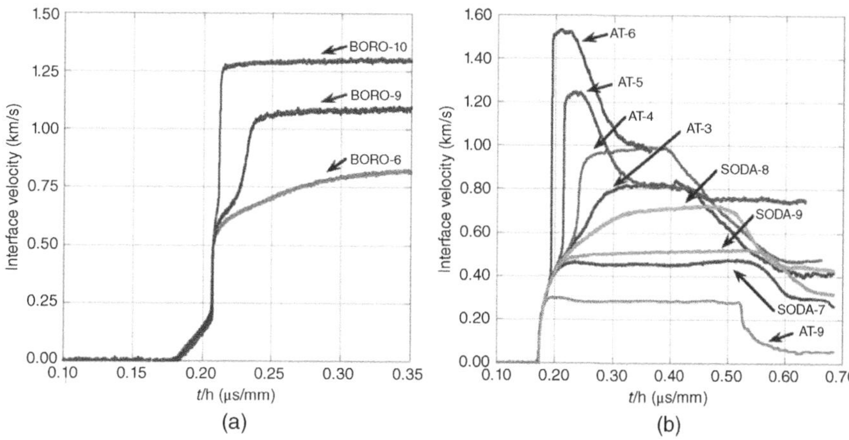

Figure 7.36 Interface velocity as measured between the glass specimen and PMMA window for (a) borosilicate and (b) soda-lime glass. (Alexander et al. 2008). Source: Alexander et al. (2008). © 2008, Elsevier.

modulus. During this phase, the release wave velocity falls below the elastic wave speed (Bourne et al. 1996a). The longitudinal stress slowly relaxes behind the wave front. To illustrate occurrence of permanent densification during shock loading, two example plots of strain versus time are shown in Figure 7.37: one for low-density SL glass (Rosenberg et al. 1996) and the other for high-density lead-based glass (Bourne et al. 1996a). For each material two curves are provided, one below the stress amplitude corresponding to the HEL and the other above the HEL. For SL glass (HEL is 6.0 GPa and the corresponding strain is 0.06), the response at two amplitudes of shock stress (4.3 and 7.7 GPa) are shown in orange (Rosenberg et al. 1996). While the slopes of the elastic responses are similar at both the shock stress amplitudes, the lower amplitude (i.e. at 4.3 GPa) response below the HEL reveals a constant peak stress followed by a complete release of the stress, resulting in no residual strain. The higher amplitude response at 7.7 GPa (i.e. peak stress above HEL) on the other hand, exhibits an elastic strain of 0.06 up to HEL followed by a hardening response until a maximum strain value close to 0.08 is reached, and then the elastic unloading where the material partially recovers. The unloading slope changes after this point where slow relaxation (recovery) may occur eventually leaving behind a residual strain of 0.02.

Similar observations were made for DEDF glass except that the HEL and residual strain values are different. The small value of residual strain in these glasses is attributed to densification but not to failure or phase transformation. These measurements provide clear evidence for densification as the primary elastic–plastic deformation mode during the shock loading of glasses. Reynard et al. (1999)

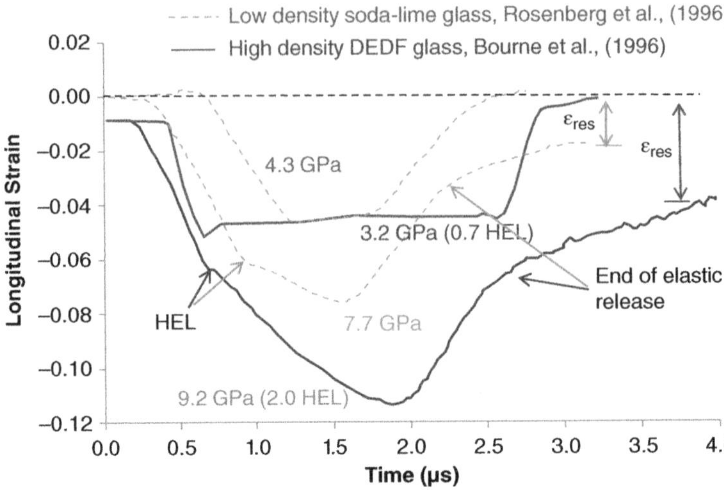

Figure 7.37 Longitudinal strain measurements unraveling the densification behavior of glass during shock loading. Sources: Rosenberg et al. 1996; Bourne et al. 1996a.

and Okuno et al. (1999) have shown that a low-amplitude shock wave causes an increase in density (d_p/d_{po}) up to 2.2% for a shock pressure of 24 GPa and up to 11% increase in silica glass at comparable shock pressures. Concurrently, there is a significant increase in refractive index of glass as shown in Figure 7.38. At higher shock pressures, the high temperature induced by the post-shock process causes annealing or particle recovery which brings the refractive index and density close to their original values. In some glasses such as fused quartz, permanent densification has been observed at stress levels as high as 16 GPa (Cagnoux and Longy 1988). The tetrahedral crystal structure of SiO_2 remains unaltered until this pressure is reached but is distorted beyond this pressure up to 22 GPa. On the contrary, permanent densification in Pyrex® glass was observed until 17 GPa, but at pressures above this value, the densification was less. This behavior was attributed to annealing of material due to the high temperature generated upon reaching the Hugoniot state. For fused silica, temperatures as high as 710 K at 9.5 GPa and 1920 K at 22 GPa were predicted. As a result, the material softened (reduction in shear modulus), and the deviatoric stress remained constant around 3.8 GPa during the densification regime (Cagnoux and Longy 1988).

7.8.2 Dense Glass

Contrary to low-density glasses, the shock response of lead-based, high-density DEDF glass does not show the early elastic ramping behavior, but similar to

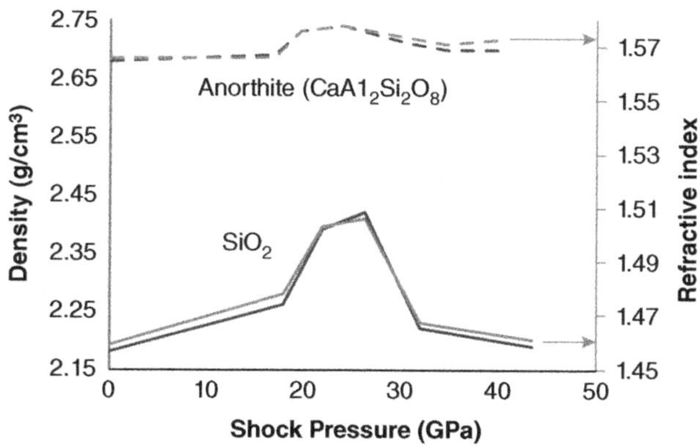

Figure 7.38 Change in density and refractive index of glasses as a function of shock pressure (Reynard et al. 1999; Okuno et al. 1999). Sources: Reynard et al. 1999; Okuno et al. 1999.

Figure 7.39 Lateral stress at various levels of shock amplitude in DEDF glass (Bourne et al. 1996a). Delayed damage evolves at stresses below the HEL; the damage front follows the elastic stress wave. The delay decreases with increasing shock amplitudes. Source: Bourne et al. (1996a). © 1996, AIP Publishing.

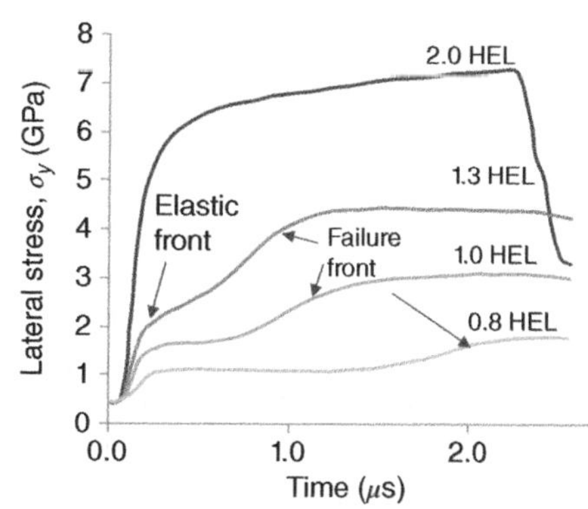

ceramics it exhibits a well-pronounced HEL of about 4.3 to 4.6 ± 0.1 GPa with a pre-cursor decay. It shows the typical elastic–plastic Hugoniot response (Bourne et al. 1996a), as shown in Figure 7.39. The absence of an initial ramp is attributed to the filled microstructure of lead glass. Due to the high initial density, at sufficiently high amplitudes beyond HEL, the material strength is lost immediately, similar to a high-strength ceramic. The inelastic deformation in the form of either phase transformation wave or damage can be observed when the imposed stress exceeds a threshold value. Figure 7.39 illustrates the behavior of DEDF glass at shock amplitudes from 0.8 times HEL to twice the HEL value. The recorded signal from lateral stress gages (see Figure 7.34) is plotted to illustrate the inelastic changes during the shock wave propagation.

At a low stress level (0.8 HEL), the failure does not initiate immediately upon reaching the imposed stress level but is delayed; the elastic front traverses in the intact material, which is then followed by a FF after a delay of almost 1.7 μs. This delay which decreases with increasing shock pressure may indicate incubation period and relate to the inertia associated with damage nucleation. Eventually the FF catches up with the shock wave at twice the HEL stress. At these sufficiently high pressures, failure occurs simultaneously along with the shock in the material due to the over-driven condition. The occurrence of the FF (at low stress levels) well after the HEL is reached indicates that the mechanism that causes the HEL is different from that causes failure. The HEL may indicate temporary densification (Bourne et al. 1996a) or onset of inelastic deformation processes including phase transformation.

7.9 Phase Transformation and Damage Nucleation Under Shock Loading

Shock loading imposes large compressive stresses on the material. One may wonder how damage (cracking) can initiate under such high compressive stress states. Clifton (1993) postulated that at these high pressures, glasses undergo a phase transformation, and the associated volumetric strain is considerably larger than that possible from densification alone. This large volume change leads to microcracking and could manifest as another wave traveling behind the elastic front through the specimen. This volume-change induced microcracking and subsequent comminution were argued to be responsible for the observed lack of spall strength (Raiser and Clifton 1994).

Evidence for phase transformation in glasses can also be inferred from the $U_S - u_P$ relationship such as $U_S = C_o + Su_P$ (Eq. (5.15)). For low-density glasses, the dimensionless parameter S was found to be between 1.60 and 1.69 (Alexander et al. 2008). Figure 7.40 presents data for two glasses (SL and fused silica) along with that of a ceramic (Al_2O_3) for comparison purposes (Reinhart and Chhabildas 2003). Note that for the ceramic material, the experimental data extrapolates to the bulk wave velocity, which is around 7.45 km/s. However, for the selected two glasses, the extended line gives a velocity far below the bulk wave velocity, indicated by arrows (the two arrows almost overlap). The bulk wave velocities for soda lime glass and fused silica are 4.175 and 4.07 km/s, respectively (Alexander et al. 2008), whereas the experimental data extrapolates to 2.18 and 0.95 km/s, respectively. The under-prediction of bulk wave speeds has been speculated to suggest phase transformation in these glasses. Many glasses such as quartz (Chhabildas and Miller 1985) and fused silica (Chhabildas and Grady 1984) are known to exhibit phase transformation, and the above experimental results may point to phase transformation in these materials.

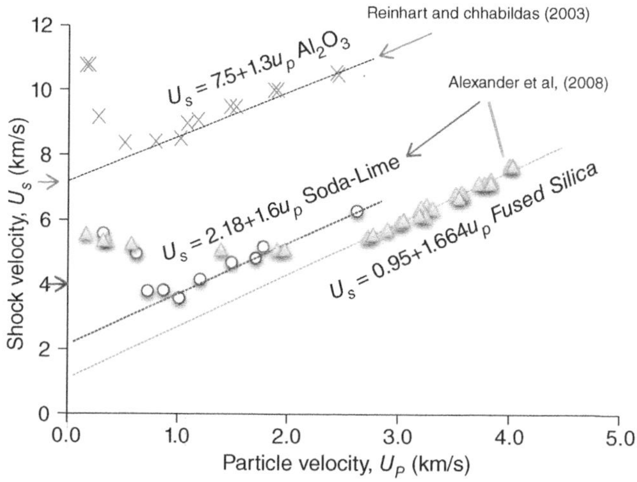

Figure 7.40 Particle velocity versus shock velocity for soda-lime glass, fused silica, and alumina (Reinhart and Chhabildas 2003). For alumina, the data can be extrapolated to the bulk wave velocity (around 7.45 km/s). The extrapolated line for the glasses is much less than the bulk wave velocity (4.175 km/s for soda lime glass and 4.07 km/s for fused silica (Alexander et al. 2008) indicating a possible phase transformation in the glass. Sources: Reinhart and Chhabildas 2003; Alexander et al. 2008.

7.10 Strength of Damaged Glasses During the Shock-wave Propagation

In Figure 7.39, we have seen that at stress levels close to and above the HEL values, experiments on glasses have revealed a unique lateral stress signature where an increase in stress is seen just behind the elastic precursor. This bump has been associated with failure (damage) generated in glasses immediately after the shock loading. Cracks occupy additional volume and the associated strain increase is captured by the lateral strain gages. The boundary between the shocked material and the damaged material was defined as the FF. Due to its damaged state, the material behind the FF has a lower impedance and complete loss of its tensile strength, but only partial loss of shear strength as it is confined by the surrounding intact material. The increase in lateral stress is also an indication of the loss of shear strength in the material as can be seen from the equation $\tau = (\sigma_x - \sigma_y)/2$. From the measured lateral stress signals, the shear strength was calculated before the FF arrived (shocked state of intact material) and after the FF had passed (strength of damaged material), and plotted in Figure 7.41 for DEDF and SL glasses (Bourne et al. 1999). Ahead of the front the shear stress increased linearly with longitudinal stress, but behind the FF (in the damaged material), the stress initially decreased and then increases with increasing impact stress. This increase is due to the mechanical

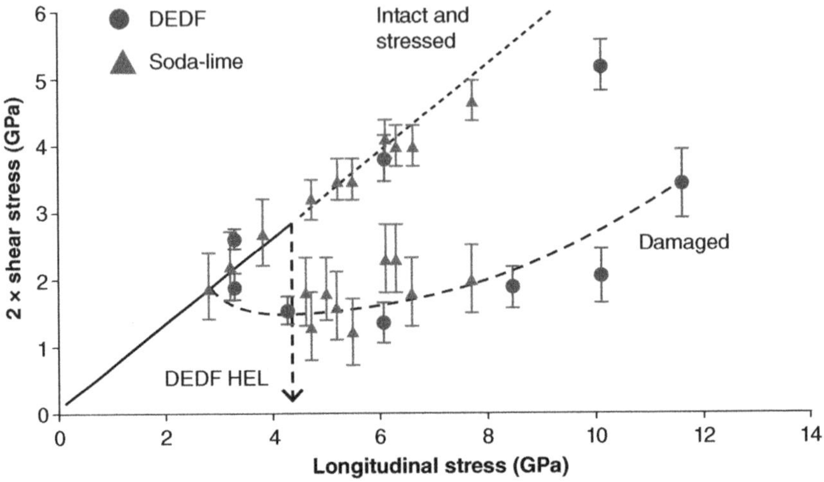

Figure 7.41 Shear strength of glasses in intact and damaged states measured before and after the shock front. Source: Bourne et al. (1999). © 1999, The Royal Society.

resistance of the comminuted material at high confinement pressures. The failure initiated at a stress well below the HEL value. The shear strength of the material behind the FF is almost half that of the intact material. Interestingly, the slope of the intact stressed material strength is 2/3 (for $\nu = 0.25$) for both the glass materials despite the large difference in density (2.5 versus 5.2 g/cm^3).

The reasons for the partial loss in the shear strength of glasses during shock loading were investigated by Clifton et al. (1998) by conducting pressure-shear plate-impact experiments. They noted a loss of shear resistance (as indicated by transverse velocity measurements) at sufficiently high shear stresses, confirming the above observations by Bourne et al. (1999, 1996a). From interferometric traces obtained during the experiments, it was noted that the unloading response of the glass in uniaxial strain is same as the elastic response of an undeformed glass. Later, Sundaram and Clifton (1998) investigated the behavior of SL glass and quantified the loss in shear strength. They argued that the primary mechanism responsible for strength loss is associated with the switching of strong, directional covalent Si–O bonds at large shear strains.

7.11 Do Failure Waves Develop in Glass Under Shock?

In Figure 7.39, it is seen that there is an increase in lateral stress behind the shock front and this increase has been associated with delayed damage evolution due to the applied shock stress. With an increase in the applied shock amplitude, the time delay between the elastic precursor and the damage signal decreases. This phenomenon has been interpreted earlier as a failure wave being generated in glasses immediately following the elastic precursor. However, whether it is appropriate to call it a "failure wave" or what features define this failure wave have been debated in literature. A failure wave was defined as a characteristic damage mode that evolves in rods and plates of transparent materials impacted by a high-velocity projectile (Brar et al. 1991b; Haney and Subhash 2013; Jannotti et al. 2015; McCauley et al. 2013; Haney and Subhash 2013; Jannotti et al. 2015; McCauley et al. 2013). The material ahead of the FF has high shear strength, but the material behind the failure wave is comminuted and exhibits loss of shear strength, as seen in Figure 7.41. Raiser and Clifton (1994) demonstrated that material behind the failure wave has no spall strength.

Let us now try to answer the question posed at the end of Section 7.6. Why does the damage in glass rods propagate at velocity above $\sqrt{2}\ C_s$ when the classical dynamic fracture mechanics limits the maximum velocity of a single Mode-I crack to the Rayleigh wave velocity? This is because there is no single growing crack or propagating damage wave, but only nucleation of damage behind a propagating longitudinal shock wave. As the stress wave arrives at a given location, cracks (or voids) nucleate from the preexisting defects either due to densification-caused crushing or local shear, and they grow to larger size behind the wave. The open structure of glass allows the large density of cracks to nucleate. These cracks or voids neither move with the wave to cause a moving boundary nor do they transfer energy as the material particles do during the wave motion. Hence, there is no failure "'wave," but only a FF that encompasses the nucleation and growth of new cracks as the wave arrives at a point. The continuous growth of new cracks behind a shock wave manifests as a moving boundary, but this boundary has no features that characterize a wave. Thus, the failure wave is not a propagating boundary (such as a longitudinal wave), but it is the profile of the growing damage zone with time-dependent nucleation of multiple (shear-induced) new microcracks behind the shock front. From this perspective, the name "failure wave" is a misnomer in that it has no characteristics of a wave. The failure developed under shock requires continuous driving force to sustain its growth, where as a propagating wave requires no driving force because once initiated it propagates at steady state. However, the boundary of this growing damage can only be called a "FF" of "damage front" but not a "failure wave."

While the aforementioned shock impact experiments on glasses provided indirect evidence for the nature of damage evolved behind a shock wave in the form of lateral stress (or strain) increase, Chocron et al. (2016) made a direct observation of damage nucleation that forms the FF behind a shock wave. They conducted symmetric plate impact experiments by utilizing BS glass projectiles and BS glass targets at impact velocities from 116 to 351 m/s (or stress levels from 0.7 to 2.15 GPa) and optically observed the damage front propagation during the shock wave loading. For a low shock amplitude (figure not shown), a dark band developed a few nanoseconds after the impact and propagated in the glass. This band is indicative of the shock front as a result of intense densification and change in the refractive index of the material. The band propagates to the back end of the specimen and reflects as a tensile wave, releasing the densified material. After a short incubation time, tensile damage emerged and resulted in a loss of optical transparency. With time the damage expanded and fully encompassed the glass specimen. The main conclusions were that there is no failure wave behind the initial shock front at this low shock amplitude and there appears to be a delay between the reflected tensile wave and damage initiation.

At higher shock amplitudes, shown in Figure 7.42, a shock front (a dark vertical band) can be seen at a time of 0.6 μs, and damage evolves at 1.0 μs along a parallel plane slightly behind the shock front (Chocron et al. 2016). The damage plane consists of several bead-like features at equal distances. This behavior suggests that the nucleation of damage occurs at regular intervals along a plane at some characteristic distance behind the shock wave. Interestingly, the damage front does not move but grows in severity at the same location. After the shock front has moved farther away from the damage plane, the driving force for the damage at this location subsides, and the damage severity stalls. As the shock front moves forward, another damage plane with similar bead-like appearance nucleates and grows in severity behind the shock front. Once again, this second damage plane

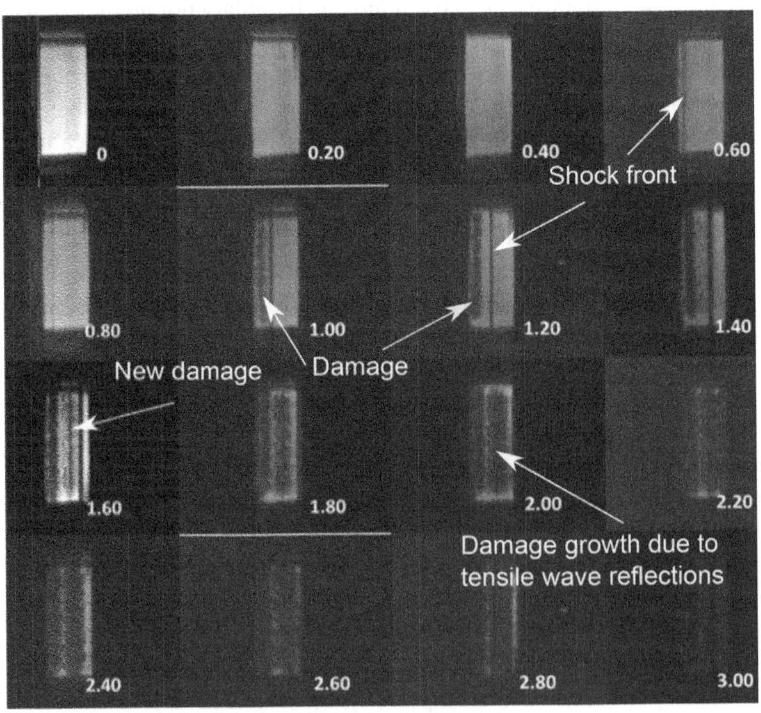

Figure 7.42 Time resolved damage evolution behind a shock front. Source: Reproduced with permission from Chocron et al. (2016).

seems to be at the same distance behind the shock front as the first one (compare the images at $t = 1.0\,\mu s$ and $t = 1.6\,\mu s$ in Figure 7.42). The shock front then reaches the back end of the specimen at about 1.8 μs and reflects as a tensile wave, which then accelerates the growth of damage along the second plane and eventual complete loss of transparency.

These experiments conclusively determine that the so called "failure wave" is not a propagating boundary of damage, but it is instead a process consisting of time-dependent nucleation of multiple (shear-induced) microcracks that evolve behind the propagating shock front. The increase in lateral stress (behind the elastic precursor) seen in Figure 7.39 is due to the damage that evolves after some incubation time. The fact that the lateral strain gages reveal a decreasing separation between the shock front and the FF at higher shock amplitudes may indicate that the damage may evolve immediately with shorter incubation time at high amplitudes behind the shock front and at even shorter spacing.

Contrary to this observation of periodic damage in shock loading, the damage characteristics in ball-on-rod (Figures 7.9 and 7.13), EOI experiments (Figures 7.21 and 7.22), and rod-on-rod impact experiments (Figures 7.32 and 7.33) are quite different. In these latter set of experiments, no periodicity in the failure pattern nor formation of parallel damage planes were noted. So, how do we reconcile this difference? The simple answer is that the damage patterns depend on the state of stress, geometry of the specimen, and the duration of loading. In the flyer plate impact experiments, a uniaxial strain state is achieved in a short duration loading of only a few hundred nanoseconds, and hence the damage evolves along a plane parallel to the shock front at regular intervals. At short pulse durations, the evolved damage quickly relieves the neighborhood of the available energy as the stress wave moves away, and there is no more driving force left at this location for initiated damage to grow. As the shock wave moves farther away from the first damage plane, stress rebuilds once again behind the shock front and away from the first damage plane to initiate damage along a new plane, and the process continues. The periodicity of damage in both longitudinal and lateral directions suggests that there is a threshold stress needed for damage nucleation as well as a critical spacing. This threshold stress is highly strain rate dependent where higher stress levels are required at higher strain rates. On the contrary, the damage front in rods and EOI plates is noticeably different for several reasons. First, the stress state is not as well-defined as in plate impact experiments and second, the duration of loading (and hence the developed compressive wave duration) is significantly longer (hundreds of microseconds). Once initiated, the damage not only continues to grow but new damage also initiates immediately behind the moving compressive front. The driving force for damage to grow persists well after the compressive wave front has passed by. Thus, the initiated damage grows in an uncontrolled manner. In both cases, the crack nucleation and growth is a time-bound process. Upon reaching the required threshold stress level, cracks are nucleated behind the shock front, but it takes time for them to grow to a critical size and coalesce.

With an increase in the shock stress level, the process speeds up and a connected network of cracks is formed, resulting in a FF.

7.12 Spall Strength of Glass

Early studies on the spall strength of glasses were conducted by loading the specimens to a pressure below the HEL. These studies revealed that spall strength for window glass was greater than 3.5 GPa and that of fused silica was around 3.1 GPa (Molodets and Dremin 1977; Rosenberg et al. 1985). Glasses have been found to retain high spall strength even when shocked above the HEL, which could be due to fusing of glass fragments as high temperatures are generated during shock loading (Rosenberg et al. 1985). Cagnoux and Longy (1988) determined the spall strength of BS glass to be around 1.8 GPa. Pyrex® reveals an increase in spall strength up to 2.3 GPa as the shock amplitude is increased to 5 GPa (Cagnoux and Longy 1988). The spall strength of K8 glass (HEL 7.5–8 GPa) was about 3 GPa at 20 GPa shock stress (Kanel et al. 1992). In general, the high spall strength of glasses has been attributed to their high internal tensile strength. Rosenberg et al. (1985) measured the spall strength of SL glass to be 3.8 GPa when shock stress was below the HEL value (6.4 GPa) and zero when the shock stress was above the HEL. Cagnoux and Longy (1988) reported that the spall strength for glass increases with impact stress below the HEL but decreases with increasing Hugoniot stress. The increase below the HEL was due to the increased velocity of the projectile and the resulting high-strain-rate loading. Higher strain rates allow less time for the activation of flaws, resulting in higher strength values. On the other hand, in the Hugoniot range above the HEL, a greater intensity of damage, was noted. This high level of damage results in negligible spall strength. In some glasses such as K-8, a high stress above the HEL causes densification and simultaneous heat generation, which gives finite but lower spall strength than that when shocked below the HEL (Cagnoux and Longy 1988).

The above inconsistencies in spall strength for glasses can be rationalized based on the discussion of FF. The spall strength depends on whether the spall plane is behind or ahead of the FF. Recall that the material ahead of the FF is intact and that behind the FF is comminuted. By proper selection of the thickness of the flyer plate, the release wave can be made to arrive at different locations (or times) in the specimen and the spall plane location can be made to fall either in the intact portion or comminuted portion of the target as shown in Figure 7.43 (Espinosa et al. 1997).

The particle velocity versus time plots at two flyer velocities for DEDF glass and SL glass are shown in Figure 7.44 (Chocron et al. 2016). No spall signature was noted for the DEDF glass, and a well-defined spall signature was observed for the SL glass. When the spall plane lies ahead of the FF (see Figure 7.43a), the spall

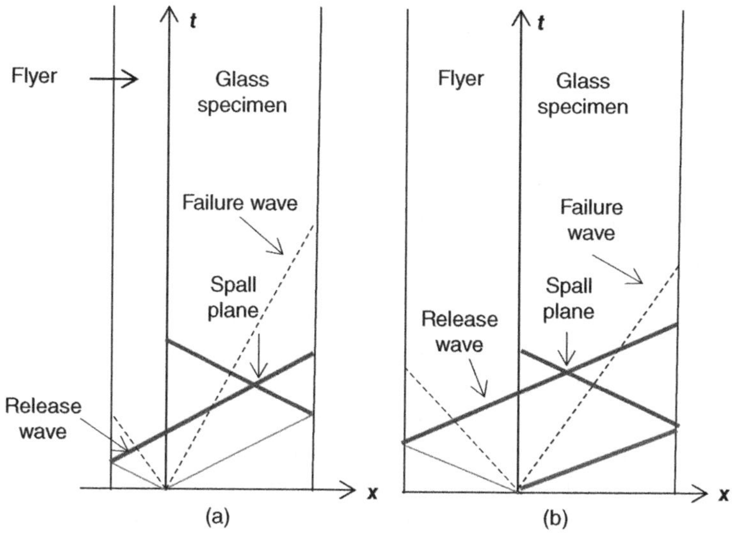

Figure 7.43 x–t plot illustrating the location of spall (a) ahead of the failure front and (b) behind the failure front. Source: Espinosa et al. (1997).

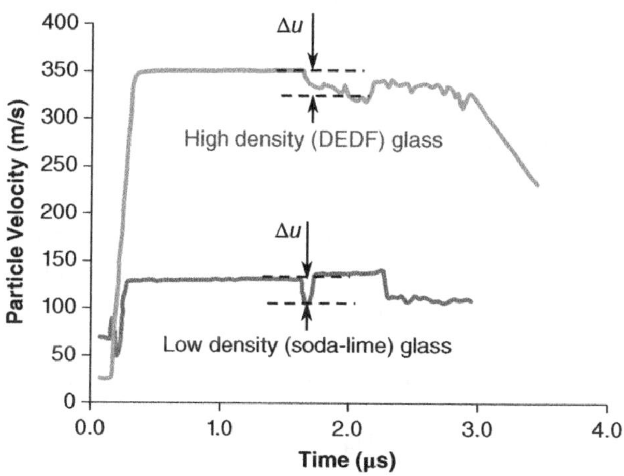

Figure 7.44 Particle velocity measurements on glass specimens with a spall signal in soda-lime glass and without a well-defined spall signal in DEDF glass. Source: Chocron et al. (2016). © 2016, Springer Nature.

signature is strong, and a finite spall strength can be measured. On the other hand, when the spall plane is behind the FF (see Figure 7.43b), the spall signature is almost nonexistent, and the strength of this damaged material is almost zero. The spall strength is calculated using

$$\sigma_{spall} = \frac{1}{2}\rho_o c\,(\Delta u) \tag{7.11}$$

where Δu is the jump in the particle velocity in the spall pull back signal as in Figure 7.44. A spall strength as high as 3.0 GPa was measured on the same SL glass in the elastic range where no failure wave was noted in the gage signals (Espinosa and Xu 1997).

7.13 Penetration (Reverse Ballistic) Experiments on Glass

Having established that a failure or damage front develops in glasses when impacted at a high velocity, we will now explore the implications of the FF on the projectile penetration mechanics. Projectile penetration experiments into bulk glasses are a useful means to visualize the damage ahead of a penetrator and the interaction of the damaged material with the penetrator. Orphal et al. (2006) evaluated the impact of gold rods on borofloat glass and a DEDF glass with high PbO content. The choice of a gold penetrator was to eliminate the effect of rod rigidity so as to investigate the FF velocity (V_F) and failure growth characteristics (Orphal et al. 2007). The FF velocity was determined from high-speed images. It was observed that the V_F value, so determined, was 3.18 km/s which falls between $\sqrt{2}\,C_s$ (2.86 km/s) and the longitudinal wave velocity of 3.52 km/s for this glass (Orphal et al. 2006). This high value of FF velocity above the shear wave velocity indicates that damage is initiated by the high magnitude of initial longitudinal wave and grows behind the wave as it propagates. Figure 7.45 illustrates a schematic of the induced damage ahead of a projectile in the target (Strassburger 2013) where the projectile impact velocity V_p, failure-front velocity (V_F), and the penetration velocity u of the projectile are indicated. Figure 7.46a presents the experimental results by Orphal et al. (2007) on DEDF and borofloat glasses at various impact velocities. The FF velocity increases almost linearly with projectile impact velocity in both glasses, but it was always lower than the shear wave velocity, C_s. Experiments by Behner et al. (2008) with gold rods against

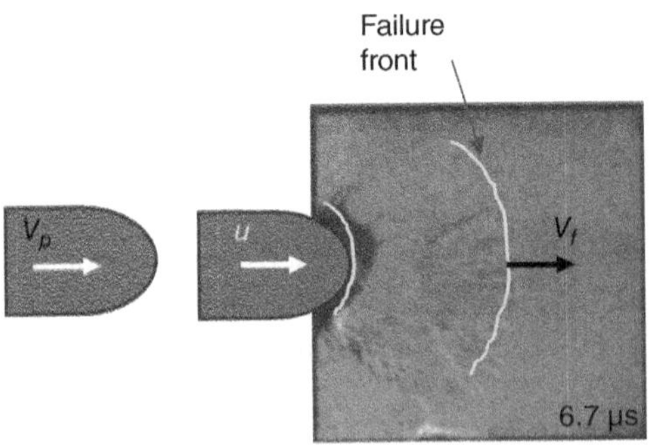

Figure 7.45 Illustration of penetrator interaction with damaged material in the target when the damage front velocity is greater than the projectile penetration velocity. Source: Reproduced with permission from Strassburger (2013).

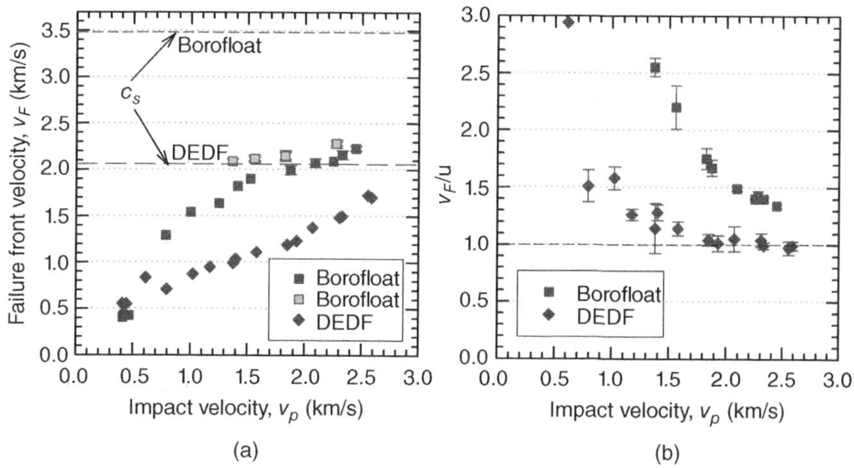

Figure 7.46 (a) Plot of failure front velocity versus impact velocity for two glasses. (b) The ratio of V_F and u reaches unity at high impact velocities, indicating that the penetrator is in contact with intact material at higher impact velocities. Source: Orphal et al. (2007). © 2007, AIP Publishing.

DEDF glass confirmed similar results. Immediately upon impact, the FF velocity is high (close to that of the shear wave velocity), and then it decreases to a lower steady-state value which is shown in the plot.

The penetration velocity and rod consumption velocity vary linearly with impact velocity. However, with an increase in impact velocity, the initial V_F can exceed the shear wave velocity. The steady-state velocity also reaches a high value close to that of the shear wave velocity. For all the impact velocities considered in Figure 7.46b, the rod penetration velocity (u) in the bulk solid was observed to be lower than the FF velocity, implying that significant damage evolved in the bulk target ahead of the projectile but definitely well behind the longitudinal wave generated by the impact process. This is because the KE of the penetrator is being dissipated to create damage in the target. At sufficiently high-impact velocities (V_p), the penetration velocity of the rod in the bulk material increases, and the ratio V_F/u approaches unity. For steady-state penetration, the distance between the FF and the tip of the penetrating rod is about 1.5 times the diameter of the rod. For high-density DEDF glass, the FF velocity almost reaches C_s at impact velocities around 3 km/s (Orphal et al. 2007). At higher impact velocities, the penetration velocity exceeds the FF velocity and hence the rod now penetrates the intact material (inside the bulk solid), and this situation is possible at a velocity of 3.5 km/s for borofloat glass and around 2.7 km/s for DEDF. Thus, with increasing V_p, the resistance for penetration increases, which implies that the Tate term R_t

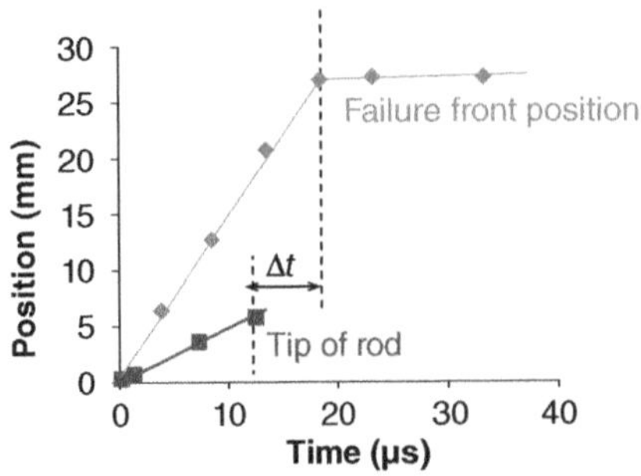

Figure 7.47 Plot of failure front position and the penetrator tip position as a function of time, indicating that the failure front extends for a finite time after the penetrator ceases to propagate. Source: Anderson et al. (2009). © 2009, Elsevier.

(target resistance, see Chapter 3) increases with V_p and eventually converges to the maximum value of intact material (Behner et al. 2008).

Anderson et al. (2009), Orphal et al. (2009), and Behner et al. (2008) investigated failure kinetics during the terminal phase of penetration by impacting short gold rods onto a BS glass target. The critical question was "Does the FF continue to propagate even after the projectile stops in the target material after the driving force by the penetrator has been fully expended in the penetration process?" By tracking the position of the tip of the penetrator and the FF as a function of time, shown in Figure 7.47, it is seen that the FF continues to propagate for few microseconds even after the rod was fully eroded (no more driving force), and then abruptly ceases to propagate. The extra driving force for the FF even after the full erosion of the penetrator comes from the release waves from the projectile-target interface that reach the FF and eventually unload the target (i.e. release the compression in the volume of material ahead of the projectile). Thus, the release wave effectively communicates with the FF to indicate that there is no more driving force for continued damage propagation or there is no overstress driving the process.

From the earlier result that the impactor always penetrates the damaged target (not the intact material) and that the FF ceases to propagate once the impactor is fully eroded, Orphal et al. (2009) inferred that the FF progression cannot be described by a wave equation. This is because the FF does not propagate at a steady state once the driving force from the impactor is fully subsided (penetrator stopped or fully eroded), and the FF stops abruptly without any deceleration. On the other hand, a typical stress wave continues to propagate at a steady velocity without any additional driving force, even well after the impactor comes to rest or is fully eroded. To further validate the above argument, they conducted additional experiments where they noted that a second collinear impactor at the same location reinitiated the FF. The abrupt change in FF propagation velocity immediately following complete erosion of the bullet indicates that the FF is not governed by diffusion-like phenomena but rather is a nucleation-and-growth-dominated phenomenon, possibly with some densification. From the data provided in

Figure 7.47, the communication that the rod has eroded and the time that the FF should stop, occurs at the bulk sound speed of the undamaged glass, even though this information travels through damaged material. This phenomenon implies that the wave speed of the damaged material is the same as that of the undamaged material. This is counter intuitive because the cracks in the material should slow the wave velocity. However, recall that the damaged material is under high compressive stress as it is confined between the incoming projectile and the bulk solid. This compressive stress closes cracks in the damaged material and allows wave propagation uninterrupted as if it were an intact solid. This inference is consistent with the earlier results discussed in Section 4.4.1 that a failed ceramic under confinement exhibits a stiffness similar to that of an intact ceramic.

7.14 Laser Shock

Aside from physical impact, shock can also be induced by imparting laser energy onto a glass surface. Compared to plate-impact experiments, laser shock experiments take less time to setup and are also less expensive. Another advantage of the pulsed laser technique to induce shock is that shock amplitudes that are 100–1000 times the yield strength of the material can be applied in a nanosecond duration, resulting strain rates in excess of $10^6/s^{-1}$ (Li et al. 1998). The main disadvantage is that no diagnostic measurements (e.g. strain gages) of wave profiles are often possible. Wong et al. (2006) used UV laser pulses to create shock on a fused silica surface. In addition to a central crater that forms due to the thermal explosion of molten silica at the center, molten nodules and fibrous morphology were seen in the crater, which is surrounded by radial cracks. Concentric fragments surrounded this core region. By analyzing the compacted layers (10-μm thick) at the bottom of the crater, Wong et al. (2006) found that fused silica is highly compressible up to 30 GPa, where a reduction in volumetric strain of 50% was observed. The compacted layer was fully amorphous.

Laser shock experiments with a pulse duration of 2.4 ns on SL glass of three different thicknesses were conducted to investigate spallation behavior at the rear surface (de Resśguier and Cottet 1995). When the stress amplitude exceeded the tensile strength of glass, a spall fracture was created in the form of a damage layer and material removal near the back surface. It was found that microcracks originated perpendicular to the laser direction, and the depth of this damaged zone from the rear surface initially increased with incident peak pressure up to 30 GPa. It then remained almost constant at around 0.8–0.9 mm for peak pressures up to 60 GPa. No spallation was observed for pressures below 8 GPa. Neither the thickness of the specimen (1–3 mm) nor the pulse duration (0.6–27 ns) had any measurable influence on the depth of damage zone. This behavior is due to a rapid hydrodynamic decay of the pressure pulse generated by the laser impulse at the front surface. This decay occurs within a short distance from the surface, and then a steady elastic pulse of constant amplitude propagates. Unlike plate impact experiments where the applied stress increases with impact velocity, in laser experiments, the elastic pulse amplitude increases with incident laser energy but beyond a threshold, the amplitude remains constant. However, the pulse duration has significant effect. At larger pulse durations (27 ns compared to 0.6 ns), other parameters being approximately the same, damage was so large that no irradiated material could be recovered.

The depth of damage was modeled by considering a damage parameter D defined by the volume fraction of microvoids. By considering a damage evolution law, they related the maximum tensile (σ_c) strength to spall strength (σ_0) as

$$\sigma_c = \sigma_0(1 - D)^b \tag{7.12}$$

where b is a material constant. Assuming a maximum tensile strength of 0.8 GPa and $b = 3.0$, de Resśguier and Cottet (1995) were able to capture the experimentally measured spall depth.

Experiments on glasses due to laser impact were also conducted to detect damage patterns and determine the internal strength of glass. Since the failure strength of glasses is controlled by the surface defects, it is difficult to determine the internal tensile strength of glasses. Holmquist and Wereszczak (2014) utilized laser shock experiments coupled with numerical analysis to estimate the tensile strength of BS glass. Unlike the previous works of Song et al. (2011) and de Resśeguier and Cottet (1995) who used a uniaxial strain condition to determine the nature of surface damage or the spall strength, Holmquist and Wereszczak (2014) determined the internal strength by focusing the laser beam on an opaque material (black tape) which is sandwiched between the target glass and a sheet of water. The laser evaporates the tape, creating a plasma plume, and the rapid expansion of the plasma creates a stress pulse of 4-ns rise time and 20-ns duration in the glass over a loading area of 5 mm. No cracks or damage were seen when the amplitude of pressure was below 4 GPa. At a pressure of 8.1 GPa, steep cone cracks emerged from the edges of the loaded surface, potentially from the surface flaws, as seen in Figure 7.48. But when the stress amplitude was slightly lower, the damage evolved at a depth between 7 and 11 mm beneath the glass surface as shown in Figure 7.49, which when viewed from the impact side or rear side revealed ring cracks. No surface damage was detected in this experiment, and hence the damage was deemed from the bulk defects. By analyzing the wave interactions, it was also deemed that the damage is not from the spall condition.

The loading and unloading of the laser shock produce complex stress wave interactions within a thick specimen. A compressive wave is generated at the center of the loaded region, but tensile release waves also generate simultaneously along

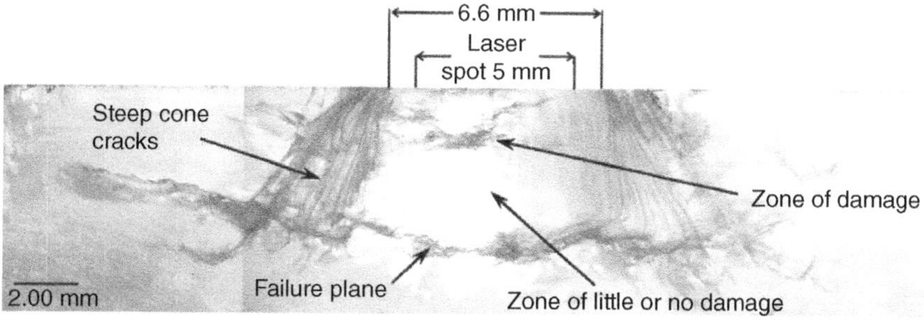

Figure 7.48 Damage in borosilicate glass due to a stress wave generated from laser-induced plasma plume expansion on the surface. Source: Reproduced with permission from Holmquist and Wereszczak (2014).

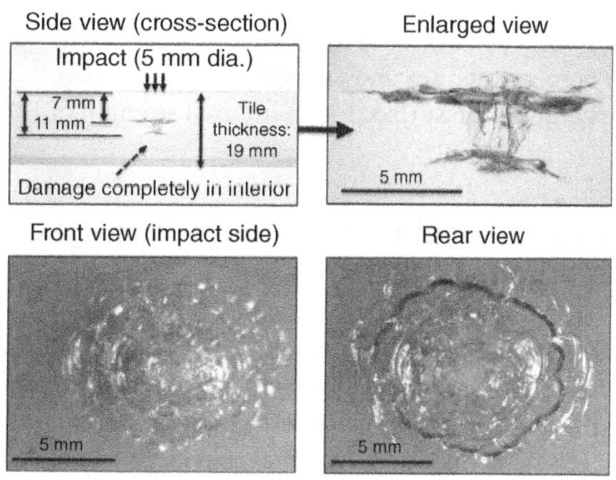

Figure 7.49 Damage induced due to bulk defects in laser shock experiments. Source: Reproduced with permission from Holmquist and Wereszczak (2014).

the circumferential edges of the expanding plasma core responsible for the cone cracks in Figure 7.48. Upon completion of the plasma expansion, a tensile release wave is also generated, which trails the leading compression wave. These wave interactions with defects cause complex damage patterns depending on the stress amplitude.

To gain insight into the observed damage and stress wave interactions, Holmquist and Wereszczak (2014) conducted finite element analysis of this loading event. They noted that the tension produced at the periphery of the loading zone amplifies the tensile stress in the material (when superposed with the trailing tension) but reduces the amplitude of the compression wave. Hence, the amplitude of the reflected tension from the rear surface was found to be insufficient to cause spall. It was noted that no surface or interior damage occurs when the internal strength was set at 1.5 GPa but interior damage evolves at a depth of 7 mm when the strength was set at 1.2 GPa due to the maximum tensile

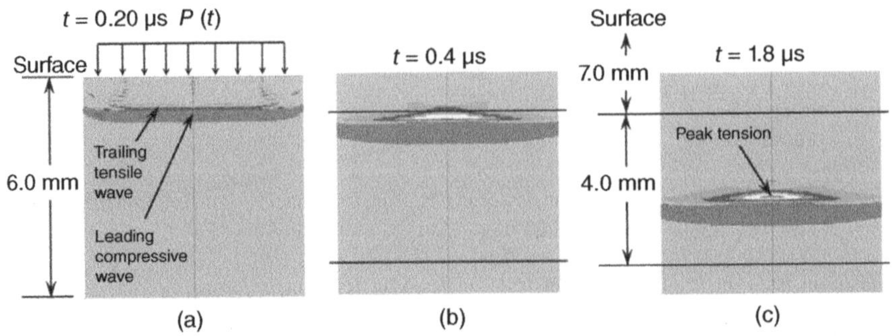

Figure 7.50 Simulations of wave interactions revealing the location of maximum tensile stress in the material. Source: Holmquist and Wereszczak (2014). © 2014, John Wiley and Sons.

stress occurring at a location consistent with the experimental observations. These results are shown in Figure 7.50. Also, it was shown that the damage extends to the surface for an internal strength of 1.0 GPa.

7.15 Modeling of Damage Propagation under Shock Loading

Evolution of damage under shock loading is a complex phenomenon. Why does glass fracture under such a high level of compression? Raiser and Clifton (1994) rationalized that glasses undergo phase transformation under high stress levels, and the large volume change associated with this transformation can cause heterogeneous nucleation of cracks from these transformed regions. Subsequently, a phase transformation wave, where a dramatic change in volume strain is followed by concurrent damage evolution. The high level of damage has also been speculated to be the main cause for loss of spall strength at high stress levels (Clifton 1993). Bourne et al. (1996a) subscribed to this argument and stated that in low-density glasses, the observed failure was due to structural collapse that accompanies phase transformation, and the resulting microcracking was due to the differences in the volumetric strains of the pre- and post-transformed structures. Shear cracks that grow from these microcracks at velocities as high as the shear wave velocity manifest as the damage front. On the other hand, Feng (2000) argued that for such a transformation wave to travel at the failure wave velocity, the longitudinal stress must increase slowly, and no such experimental evidence was available. Espinosa and Xu (1997) and Espinosa et al. (1997) theorized that the failure wave evolution is a process consisting of shear-activated inhomogeneous plastic flow leading to microcracking. They provided a micromechanical description of failure wave generation by utilizing a multiplane crack evolution model that captured the kinetics of the failure process, reduction in spall strength and progressive attenuation of stress behind the FF. Hong-Liang et al. (1997) hypothesized the failure wave formation to be a result of microcrack nucleation from the impact surface due to the rearrangement of SiO_2 tetrahedra in the region undergoing permanent densification behind the shock wave. In the following, few of these models will be described with an intent to provide a broad overview of various theories on this subject. The focus of these approaches has been to capture the experimentally observed lateral strain increase during damage front evolution and estimation of loss of shear strength behind the FF.

Partom (1998) used a simple damage growth rate condition coupled with shear modulus degradation and implemented this model in a 1-D Lagrangian code. The damage rate, $\dot{D}$, was assumed to be a function of the damage velocity (U_D), and the gradient of damage (dD/dx) in the neighboring region. The degradation in shear strength behind FF was modeled by considering the degradation of shear modulus as:

$$G = G_o\left[1 + \left(\sigma_x/\sigma_G\right)^2\right] \tag{7.13}$$

where G_o and σ_G are parameters, and σ_x is the longitudinal stress. Using this model, Partom (1998) predicted the loss of shear strength of the material after the FF has arrived and was able to capture the measured lateral strain profiles successfully as depicted in Figure 7.51.

Assuming a nonlinear elastic response ahead of the FF, no change in field quantities in the longitudinal direction across the FF, and neglecting thermal effects

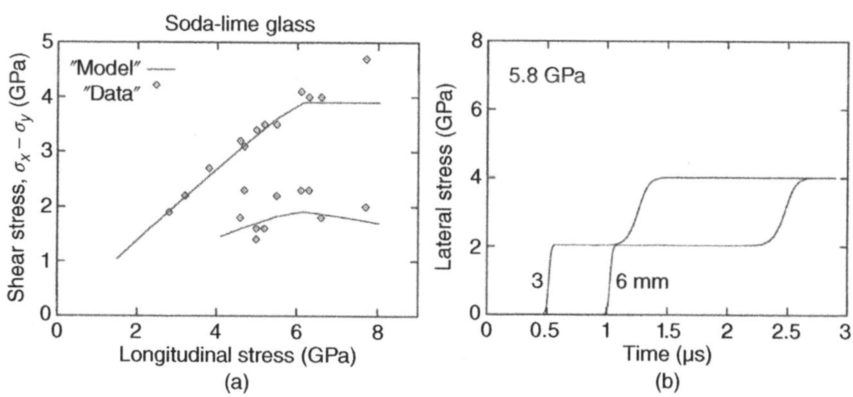

Figure 7.51 (a) Comparison of model results and experimental data of shear strength before and after the failure front and (b) lateral stress profiles upon arrival of failure front during shock wave loading in glasses. Source: Partom (1998). © 1998, Elsevier.

(Feng 2000) developed a theory for FF propagation in shocked glasses. The FF was considered to be a progression of microfissures into the bulk material. The microfissures originate at local defects on the impact surface due to the imposed high shear stress (deviatoric energy) upon shock loading. These cracks result in dilation in the material. Thus, the shock-induced deviatoric energy in the intact material is converted to volumetric potential energy in the comminuted material behind the shock front. The failure wave propagation is modeled as a nonlinear time-dependent coupled diffusion and evolution of dilatant volume. Recall that experimental evidence clearly suggests a loss of shear strength behind the FF, which means that the lateral stress must increase while the longitudinal stress remains constant at the applied shock amplitude. This lateral stress increase comes from dilation of the material in the lateral direction, and the overall mean stress must increase. Thus, the model assumes lateral percolation of damage which is modeled as a diffusion process.

The governing equation for the evolution of the dilated volume ($V_d(x, t)$), which is a measure of damage, is written as:

$$\frac{\partial V_d}{\partial t} = \frac{\partial}{\partial x}\left[D(x,t)\frac{\partial V_d}{\partial x}\right] + f(x,t) \tag{7.14}$$

where the damage evolution function $f(x, t)$ is given by:

$$f(x,t) = s(x,t)\frac{V_d - V_{do}}{T_d} \geq 0 \tag{7.15}$$

In this equation, t is time and x is the longitudinal coordinate. The evolution of the damage diffusivity, $D(x, t)$, is defined by:

$$D(x,t) = \lambda s(x,t) \tag{7.16}$$

where λ is the diffusion coefficient and

$$s(x,t) = \begin{cases} 0 \text{ if } Y \leq Y_{THD} \text{ and } V_d \leq V_{do} \\ \dfrac{Y(x,t) - Y_F}{Y_{HEL} - Y_F} \geq 0 \text{ otherwise} \end{cases} \tag{7.17}$$

where Y is a measure of the stress deviator defined as the difference between the axial and lateral stresses:

$$Y = \left|\sigma_{xx} - \sigma_{yy}\right| \tag{7.18}$$

The parameters Y_{HEL} and Y_F are the stress deviators in the intact material at the HEL and in the damage material upon complete damage, respectively. The use of threshold values Y_{THD} and V_{do} allows diffusion to be inactive below these values and drives the failure process above these values. Also, $Y < Y_F$ indicates unloading. Assuming an initial value of V_{do}/V_o as 10^{-3} and some experimental parameters, the above equations were numerically solved (Feng 2000), and a good comparison between the lateral strain measurements of (Kanel et al. 1992) and the model were obtained.

Doyoyo (2002), on the other hand, hypothesized that densification-induced fragmentation of glass was responsible for FF generation under shock loading. The damage was assumed to originate in an unstable manner along the interface between densified and unstressed material, and this damage front was assumed to move with certain velocity. Thus, the failure wave was correctly interpreted as a moving fracture boundary generated by the densification-induced interface roughness and cracking. Considering the total energy of the densified and damaged volumes and interfacial surface area, and using the relationships of time-dependent perturbations of the fluctuation dissipation theorem (Goldenfeld 2018), Doyoyo (2002) derived the interface roughness and damage fragmentation size for various stress states. The fragmentation size was found to be inversely proportional to the jump in strain energy density across the FF. The fragmentation size for a shear stress state is given by:

$$\lambda_{cr} = 2\pi\gamma\mu_o\left[\sigma_{12}\right]^{-2} \tag{7.19}$$

and the jump in shear stress over the FF is:

$$\left[\sigma_{12}\right] = \frac{1-2\nu_o}{2-2\nu_o}\sigma_{22} - k \tag{7.20}$$

where k is a proportional limit in shear which can be obtained from the shear strength values reported for glass behind the FF. For example, $k = 0.9$ GPa for BS glass (Bourne et al. 1998). Fragmentation size under tension is derived as:

$$\lambda_{cr} = 8\pi\gamma\mu_o\frac{\left(1-\nu_o\right)}{\left(3-4\nu_o\right)}\left[\sigma_{11}\right]^{-2} \tag{7.21}$$

and the σ_{11} is calculated to be:

$$\left[\sigma_{11}\right] = \left(\frac{\nu_o}{1-\nu_o} - \frac{1-f}{1+f}\right)\sigma_{22} \tag{7.22}$$

where f is the internal friction coefficient of the Mescall zone and γ is the densification interface energy, which is unknown, and k is the proportional limit in shear, ν_o is the Poisson's ratio, and μ_o is the shear modulus. The dependence of fragmentation size on interface energy for shear and tensile modes of cracking is shown in Figure 7.52. The fragmentation size is almost 5–10 times larger in shear than in tension. Thus, the fragment size will be larger in plate impact experiments (where a higher shear stress persists) than under penetration by a long rod projectile.

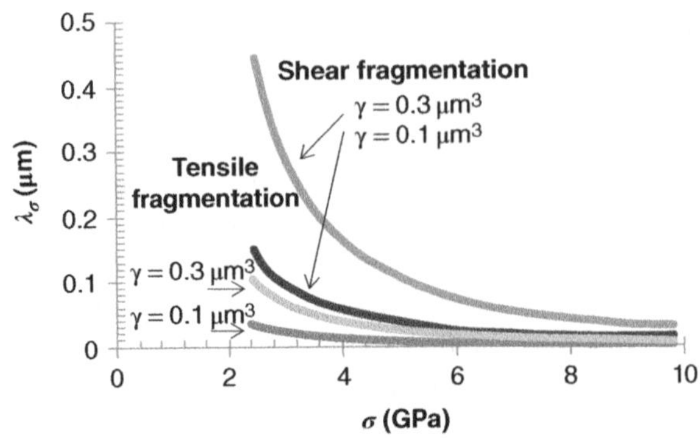

Figure 7.52 Estimated fragmentation size under tensile and shear stress states. Note that tensile fragmentation size is almost 10 times smaller than that in shear mode. Source: Doyoyo (2002). © 2002, Elsevier.

7.16 Summary and Conclusions

We conclude this chapter by emphasizing that experiments on transparent materials provide a greater insight into the damage nucleation and growth characteristics than what could be achieved from conducting similar experiments on opaque ceramics. Similar to the inherent defects in ceramics, glasses also have inherent defects in the form of an open chain structure, which enable investigations of damage initiation and growth under a range of dynamic loading and stress states. By summarizing damage evolution and propagation behaviors in rods, plates, and laminates at a range of velocities, this chapter has provided a broad, yet fundamental, picture of the impact characteristics of a variety of transparent materials. As noted here, the damage evolution mechanisms are complex, and combined experimental, theoretical, and computational approaches are vital for in-depth understanding. The literature on impact-induced damage in these materials is broad and we have only illustrated few relevant examples with the hope of familiarizing the reader to plethora of issues in this field.

References

Alexander, C., Chhabildas, L., Reinhart, W., and Templeton, D. (2008). Changes to the shock response of fused quartz due to glass modification. *International Journal of Impact Engineering* 35: 1376–1385.

Anderson, C., Orphal, D., Behner, T., and Templeton, D. (2009). Failure and penetration response of borosilicate glass during short-rod impact. *International Journal of Impact Engineering* 36: 789–798.

Andrews, J. (1931). Experiments on impact. *Proceedings of the Physical Society* 43: 8–17.

Auerbach, F. (1891). Measurement of hardness. *Annals of Physical Chemistry* 43: 61.

Bavdekar, S., Subhash, G., and Satapathy, S. (2019). A unified model for dwell and penetration during long rod impact on thick ceramic targets. *International Journal of Impact Engineering* 131: 304–316.

Behner, T., Anderson, C., Orphal, D. et al. (2008). Penetration and failure of lead and borosilicate glass against rod impact. *International Journal of Impact Engineering* 35: 447–456.

Beinert, J. (1974). Schlierenoptische Untersuchungen zur Ausbreitung kurzer elastischer Impulse in Platten.

Bless, S.J. and Brar, N.S. (1994). Impact induced fracture of glass bars. *American Institute of Physics Conference Series* 309: 1813–1816.

Bourne, N. (2008). The relation of failure under 1D shock to the ballistic performance of brittle materials. *International Journal of Impact Engineering* 35: 674–683.

Bourne, N., Millett, J., and Rosenberg, Z. (1996a). Failure in a shocked high-density glass. *Journal of Applied Physics* 80: 4328–4331.

Bourne, N., Rosenberg, Z., and Ginzburg, A. (1996b). The ramping of shock waves in three glasses. *Proceedings of the Royal Society a-Mathematical Physical and Engineering Sciences* 452: 1491–1496.

Bourne, N., Millett, J., Rosenberg, Z., and Murray, N. (1998). On the shock induced failure of brittle solids. *Journal of the Mechanics and Physics of Solids* 46: 1887–1908.

Bourne, N., Millett, J., and Field, J. (1999). On the strength of shocked glasses. Proceedings of the Royal Society of London A: Mathematical, Physical and Engineering Sciences, pp. 1275–1282. The Royal Society.

Brar, N.S. (2000). Failure waves in glass and ceramics under shock compression. AIP Conference Proceedings, 601–606. American Institute of Physics.

Brar, N.S., Bless, S.J., and Rosenberg, Z. (1991). Impact-induced failure waves in glass bars and plates. *Applied Physics Letters* 59: 3396.

Brar, N., Bless, S., and Rosenberg, Z. (1991a). Impact-induced failure waves in glass bars and plates. *Applied Physics Letters* 59: 3396–3398.

Brar, N., Rosenberg, Z., and Bless, S. (1991b). Spall strength and failure waves in glass. *Journal de Physique III* 1: 639–644.

Bridgman, P.W. (1948). The compression of 39 substances to 100,000 kg/cm^2. 55–70.

Bridgman, P. and Simon, I. (1953). Effects of very high pressures on glass. *Journal of Applied Physics* 24: 405–413.

Cagnoux, J. and Longy, F. (1988). Spallation and shock-wave behavior of some ceramics. *Journal de Physique* 49: 3–10.

Chai, H. and Ravichandran, G. (2007). On the mechanics of surface and side-wall chipping from line-wedge indentation. *International Journal of Fracture* 148: 221–231.

Chandrasekar, S. and Chaudhri, M. (1994). The explosive disintegration of Prince Rupert's drops. *Philosophical Magazine B* 70: 1195–1218.

Chaudhri, M.M. (1998). Crack bifurcation in disintegrating Prince Rupert's drops. *Philosophical Magazine Letters* 78: 153–158.

Chaudhri, M. and Liangyi, C. (1986). The catastrophic failure of thermally tempered glass caused by small-particle impact. *Nature* 320: 48–50.

Chaudhri, M. and Walley, S. (1978). Damage to glass surfaces by impact of small glass and steel spheres. *Philosophical Magazine A-Physics of Condensed Matter Structure Defects and Mechanical Properties* 37: 153–165.

Cherepanov, G.P., De Wit, R., and Cooley, W. (1979). *Mechanics of Brittle Fracture*. New York, NY: McGraw-Hill International Book Co.

Chhabildas, L.C. and Grady, D.E. (1984). Chapter IV: 6 – shock loading behaviour of fused quartz. In: *Shock Waves in Condensed Matter* (eds. J.R. Asay, R.A. Graham and G.K. Straub), 175–178. Amsterdam: Elsevier.

Chhabildas, L. and Miller, J. (1985). *Release-adiabat Measurements in Crystalline Quartz*. Albuquerque, NM: Sandia National Labs.

Chocron, S., Barnette, D.D., Holmquist, T.J. et al. (2016). Damage threshold of borosilicate glass under plate impact | SpringerLink. *Journal of Dynamic Behavior of Materials* 2: 167–180.

Choi, J. and Kim, D. (2016). Crack and shock propagation through the interlayer in soda lime glass under detonation loading. *International Journal of Applied Glass Science* 7: 104–117.

Clifton, R.J. (1993). Analysis of failure waves in glasses. *Applied Mechanics Reviews* 46: 540–546.

Clifton, R., Mello, M., and Brar, N. (1998). Effect of shear on failure waves in soda lime glass. AIP Conference Proceedings, 521–524. AIP.

Cooper, S. and Swiler, D. (2019). Glass innovation in the grocery store. *American Ceramic Society Bulletin*. The American Ceramic Society, 18–23.

Dally, J.W. and Riley, W.F. (1991). *Experimental Stress Analysis*. McGraw-Hill.

Doyoyo, M. (2002). A theory of the densification-induced fragmentation in glasses and ceramics under dynamic compression. *International Journal of Solids and Structures* 39: 1833–1843.

Erlich, D. and Seaman, L. (1981). Symmetric rod impact technique for dynamic yield determination. In: *Bulletin of the American Physical Society*, 658. Woodbury, NY: Amer Inst Physics Circulation Fulfillment Div.

Espinosa, H., Xu, Y., and Brar, N. (1997). Micromechanics of failure waves in glass .1. Experiments. *Journal of the American Ceramic Society* 80: 2061–2073.

Espinosa, H., Xu, Y., and Brar, N. (1997). Micromechanics of failure waves in glass .2. Modeling. *Journal of the American Ceramic Society* 80: 2074–2085.

Evans, A. (1973). Strength degradation by projectile impacts. *Journal of the American Ceramic Society* 56: 405–409.

Evans, A. and Wilshaw, T. (1976). Impact damage in brittle materials. *American Ceramic Society Bulletin* 55: 400.

Evans, A. and Wilshaw, T. (1977). Dynamic solid particle damage in brittle materials – appraisal. *Journal of Materials Science* 12: 97–116.

Feng, R. (2000). Formation and propagation of failure in shocked glasses. *Journal of Applied Physics* 87: 1693–1700.

Fischer-Cripps, A. (1997). Elastic–plastic behaviour in materials loaded with a spherical indenter. *Journal of Materials Science* 32: 727–736.

Freund, L.B. (1998). *Dynamic Fracture Mechanics*. Cambridge University Press.

Goldenfeld, N. (2018). *Lectures on Phase Transitions and the Renormalization Group*. CRC Press.

Grujicic, M., Pandurangan, B., Coutris, N. et al. (2008). A ballistic material model for starphire (R), a soda-lime transparent-armor glass. *Materials Science & Engineering, A: Structural Materials: Properties, Microstructure and Processing* 491: 397–411.

Haney, E. and Subhash, G. (2011a). Analysis of interacting cracks due to sequential indentations on sapphire. *Acta Materialia* 59: 3528–3536.

Haney, E. and Subhash, G. (2011b). Rate sensitive indentation response of a coarse-grained magnesium aluminate spinel. *Journal of the American Ceramic Society* 94: 3960–3966.

Haney, E. and Subhash, G. (2011c). Static and dynamic indentation response of basal and prism plane sapphire. *Journal of the European Ceramic Society* 31: 1713–1721.

Haney, E. and Subhash, G. (2012). Edge-on-impact response of a coarse-grained magnesium aluminate spinel rod. *International Journal of Impact Engineering* 40–41: 26–34.

Haney, E. and Subhash, G. (2013). Damage mechanisms perspective on superior ballistic performance of spinel over sapphire. *Experimental Mechanics* 53: 31–46.

Holmquist, T.J. and Wereszczak, A.A. (2014). The internal tensile strength of a borosilicate glass determined from laser shock experiments and computational analysis. *International Journal of Applied Glass Science* 5: 345–352.

Hong-liang, H., Xiao-gang, J., Fu-qian, J., and Kanel, G. (1997). Dynamic failure of shock-loaded glass. *Chinese Physics Letters* 14: 538.

Jannotti, P., Subhash, G., Ifju, P. et al. (2011). Photoelastic measurement of high stress profiles in ion-exchanged glass. *International Journal of Applied Glass Science* 2: 275–281.

Jannotti, P., Subhash, G., Ifju, P. et al. (2012). Influence of ultra-high residual compressive stress on the static and dynamic indentation response of a chemically strengthened glass. *Journal of the European Ceramic Society* 32: 1551–1559.

Jannotti, P., Subhash, G., and Varshneya, A. (2014). Ball impact response of unstrengthened and chemically strengthened glass bars. *Journal of the American Ceramic Society* 97: 189–197.

Jannotti, P., Subhash, G., and Varshneya, A. (2015). Impact-induced deformation mechanisms in unstrengthened and chemically strengthened glass bars. *International Journal of Impact Engineering* 75: 53–64.

Kanel, G., Rasorenov, S., and Fortov, V. (1992). The failure waves and spallations in homogeneous brittle materials. *Shock Compression of Condensed Matter–1991*, Elsevier,451–454.

Kingery, W.D., Bowen, H.K., Uhlmann, D.R., and Bowen, H.K. (1976). *Introduction to Ceramics*, 2e. New York, NY: Wiley.

Kirchner, H. and Gruver, R. (1977). Localized impact damage in glass. *Materials Science and Engineering* 28: 153–160.

Knight, C., Swain, M., and Chaudhri, M. (1977). Impact of small steel spheres on glass surfaces. *Journal of Materials Science* 12: 1573–1586.

Lawn, B. and Swain, M. (1975). Microfracture beneath point indentations in brittle solids. *Journal of Materials Science* 10: 113–122.

Lawn, B. and Wilshaw, R. (1975). Indentation fracture – principles and applications. *Journal of Materials Science* 10: 1049–1081.

Lawn, B. and Wilshaw, T.R. (1993). *Fracture of Brittle Solids*. Cambridge University Press.

Lawn, B., Wiederhorn, S., and Johnson, H. (1975). Strength degradation of brittle surfaces – blunt indenters. *Journal of the American Ceramic Society* 58: 428–432.

Lawn, B., Fuller, E., and Wiederhorn, S. (1976). Strength degradation of brittle surfaces – sharp indenters. *Journal of the American Ceramic Society* 59: 193–197.

Li, X.-Z., Nakano, M., Yamauchi, Y. et al. (1998). Microcracks, spall and fracture in glass: a study using short pulsed laser shock waves. *Journal of Applied Physics* 83: 3583–3594.

Longchambon, L. (1934). Mechanical properties of glasses. *Comptes Rendus Hebdomadaires Des Seances De L Academie Des Sciences* 199: 1381–1383.

Mackenzie, J. (1963). High-pressure effects on oxide glasses: I, densification in rigid state. *Journal of the American Ceramic Society* 46: 461–470.

McCauley, J., Patel, P., Chen, M. et al. (2009). AlON: a brief history of its emergence and evolution. *Journal of the European Ceramic Society* 29: 223–236.

McCauley, J.W., Crowson, A., Gooch, W.A. Jr., et al. (2012). *Ceramic Armor Materials by Design*. Wiley.

McCauley, J., Strassburger, E., Patel, P. et al. (2013). Experimental observations on dynamic response of selected transparent armor materials. *Experimental Mechanics* 53: 3–29.

Molodets, A. and Dremin, A. (1977). Investigation of singularities of glass strain under intense compression waves. *Combustion, Explosion and Shock Waves* 13: 772–777.

Murray, N.H., Bourne, N.K., Field, J.E., and Rosenberg, Z. (1998). Symmetrical Taylor impact of glass bars. *American Institute of Physics Conference Series* 429: 533–536.

Nie, X., Chen, W., Wereszczak, A., and Templeton, D. (2009). Effect of loading rate and surface conditions on the flexural strength of borosilicate glass. *Journal of the American Ceramic Society* 92: 1287–1295.

Okuno, M., Reynard, B., Shimada, Y. et al. (1999). A Raman spectroscopic study of shock-wave densification of vitreous silica. *Physics and Chemistry of Minerals* 26: 304–311.

Orphal, D.L., Behner, T., Hohler, V. et al. (2006). Failure wave in DEDF and soda-lime glass during rod impact. *AIP Conference Proceedings* 845: 1391–1394.

Orphal, D.L., Anderson, C.E., Behner, T. et al. (2007). Failure kinetics in borosilicate glass during rod impact. *AIP Conference Proceedings* 955: 759–762.

Orphal, D., Anderson, C., Behner, T., and Templeton, D. (2009). Failure and penetration response of borosilicate glass during multiple short-rod impact. *International Journal of Impact Engineering* 36: 1173–1181.

Parab, N. and Chen, W. (2014). Crack propagation through interfaces in a borosilicate glass and a glass ceramic. *International Journal of Applied Glass Science* 5: 353–362.

Park, H. and Chen, W.W. (2011). Experimental investigation on dynamic crack propagating perpendicularly through interface in glass. *Journal of Applied Mechanics* 78 (5): 051013.

Partom, Y. (1998). Modeling failure waves in glass. *International Journal of Impact Engineering* 21: 791–799.

Patterson, M.C.L., DiGiovanni, A.A., Gilde, G., and Roy, D.W. (2009). Spinel armor - clearly the way to go. Annual Cocoa Beach Conference on Advanced Ceramics and Composites: A: Ceramic Engineering and Science Proceedings, 441–446.

Persson, J., Breder, K., and Rowcliffe, D. (1993). Loading rate effects during indentation and impact on glass with small spheres. *Journal of Materials Science* 28: 6484–6489.

Radford, D.D., Willmott, G.R., and Field, J.E. (2004). The effect of structure on failure front velocities in glass rods. *AIP Conference Proceedings* 706: 755–758.

Raiser, G. and Clifton, R.J. (1994). Failure waves in uniaxial compression of an aluminosilicate glass. *American Institute of Physics Conference Series* 309: 1039–1042.

Reed, J. (1985). Energy-losses due to elastic wave-propagation during an elastic impact. *Journal of Physics D: Applied Physics* 18: 2329–2337.

Reinhart, W. and Chhabildas, L. (2003). Strength properties of Coors AD995 alumina in the shocked state. *International Journal of Impact Engineering* 29: 601–619.

de Resśguier, T. and Cottet, F. (1995). Experimental and numerical study of laser induced spallation in glass. *Journal of Applied Physics* 77: 3756–3761.

Reynard, B., Okuno, M., Shimada, Y. et al. (1999). A Raman spectroscopic study of shock-wave densification of anorthite ($CaAl_2Si_2O_8$) glass. *Physics and Chemistry of Minerals* 26: 432–436.

Rioux, J., Jones, C., Mandelartz, M., and Pluen, V. (2007). Transparent armor. *Advanced Materials & Processes* 165: 31–33.

Rosakis, A., Samudrala, O., and Coker, D. (1999). Cracks faster than the shear wave speed. *Science* 284: 1337–1340.

Rosenberg, Z., Yaziv, D., and Bless, S. (1985). Spall strength of shock-loaded glass. *Journal of Applied Physics* 58: 3249–3251.

Rosenberg, Z., Bourne, N., and Millett, J. (1996). Direct measurements of strain in shock-loaded glass specimens. *Journal of Applied Physics* 79: 3971–3974.

Rouxel, T., Ji, H., Hammouda, T., and Moréac, A. (2008). Poisson's ratio and the densification of glass under high pressure. *Physical Review Letters* 100: 225501.

Sakka, S. and Mackenzie, J. (1969). High pressure effects on glass. *Journal of Non-Crystalline Solids* 1: 107–142.

Sands, J., Fountzoulas, C., Gilde, G., and Patel, P. (2009). Modelling transparent ceramics to improve military armour. *Journal of the European Ceramic Society* 29: 261–266.

Schardin, H. (1950). Results of cinematographic investigation of the fracture process in glass. *Glastechnische Berichte (Reports on Glass Technology)* 23: 325–336.

Senf, H., Strassburger, E., and Rothenhausler, H. (1994). Stress wave-induced damage and fracture in impacted glasses. *Journal de Physique IV* 4: 741–746.

Song, Y., Yu, G., Jiang, L. et al. (2011). Shock induced damage and damage threshold of optical K9 glass investigated by laser-driven shock wave. *Journal of Applied Physics* 109: 073103.

Strassburger, E. (2009). Ballistic testing of transparent armour ceramics. *Journal of the European Ceramic Society* 29: 267–273.

Strassburger, E. (2013). Edge-on impact investigation of fracture propagation in boron carbide. *Advances in Ceramic Armor IX*. 15–24.

Strassburger, E. and Senf, H. (1995). Experimental investigations of wave and fracture phenomena in impacted ceramics and glasses. Fraunhofer-Inst Fuer Kurzzeitdynamik-Ernst-Mach-Inst Weil Am Rhein, Germany.

Strassburger, E., Patel, P., Mccauley, J.W., and Templeton, D.W. (2007). Wave propagation and impact damage in transparent laminates. 23rd International Symposium on Ballistics, Tarragona, Spain, 1381–1388.

Strassburger, E., Patel, P., McCauley, J.W. et al. (2008). High-speed transmission shadowgraphic and dynamic photoelasticity study of stress wave and impact damage propagation in transparent materials and laminates using the edge-on impact (EOI) method. 1–16. US Army Research Laboratories.

Strassburger, E., Hunzinger, M., Patel, P., and McCauley, J.W. (2013). Analysis of the fragmentation of AlON and spinel under ballistic impact. *Journal of Applied Mechanics* 80.

Sundaram, S. and Clifton, R.J. (1998). Flow behavior of soda-lime glass at high pressures and high shear rates. *American Institute of Physics Conference Series* 429: 517–520.

Sundaram, B. and Tippur, H. (2016a). Dynamic crack growth normal to an interface in bi-layered materials: an experimental study using digital gradient sensing technique. *Experimental Mechanics* 56: 37–57.

Sundaram, B.M. and Tippur, H.V. (2016b). Dynamics of crack penetration vs. branching at a weak interface: an experimental study. *Journal of the Mechanics and Physics of Solids* 96: 312–332.

Suresh, S., Sugimura, Y., and Tschegg, E. (1992). The growth of a fatigue crack approaching a perpendicularly-oriented, bimaterial interface. *Scripta Metallurgica et Materialia* 27: 1189–1194.

Swab, J.J., Gilde, G.A., Patel, P.J. et al. (2000). Fracture analysis of transparent armor ceramics. In: *Fractography of Glasses and Ceramics IV: Proceedings of the Fourth Alfred Conference on the Fractography of Glasses and Ceramics* (eds. J.R. Varner and G.D. Quinn), 489–508. Alfred, NY: Alfred University.

Swain, M. and Hagan, J. (1976). Indentation plasticity and ensuing fracture of glass. *Journal of Physics D: Applied Physics* 9: 2201.

Taylor, G.I. (1948). The use of flat-ended projectiles for determining dynamic yield stress I. Theoretical considerations. *Proceedings of the Royal Society of London. Series A: Mathematical and Physical Sciences* 194: 289–299.

Theocaris, P. and Demakos, C. (1986). Crack bifurcation modes in composite plates under impact. *International Journal of Fracture* 32: 71–92.

Theocaris, P. and Milios, J. (1980a). Crack-propagation velocities in bi-phase plates under static and dynamic loading. *Engineering Fracture Mechanics* 13: 599–609.

Theocaris, P. and Milios, J. (1980b). Dynamic crack-propagation in composites. *International Journal of Fracture* 16: 31–51.

Theocaris, P. and Milios, J. (1981a). Crack-arrest at a bimaterial interface. *International Journal of Solids and Structures* 17: 217–230.

Theocaris, P. and Milios, J. (1981b). The process of the momentary arrest of a moving crack approaching a material discontinuity. *International Journal of Mechanical Sciences* 23: 423–436.

Tillett, J. (1956). Fracture of glass by spherical indenters. *Proceedings of the Physical Society. Section B* 69: 47–54.

Tsai, Y. and Kolsky, H. (1967). A study of fractures produced in glass blocks by impact. *Journal of the Mechanics and Physics of Solids* 15: 263.

Uhlmann, D. (1973). Densification of alkali silicate glasses at high pressure. *Journal of Non-Crystalline Solids* 13: 89–99.

Varshneya, A.K. and Spinelli, I.M. (2009). High-strength, large-case-depth chemically strengthened lithium aluminosilicate glass. *American Ceramic Society Bulletin* 88: 27–33.

Vollkommer, H. (1967). *Schlierenoptische Untersuchung des Schneidenstoßes auf eine Plattenkante*. Ernst-Mach-Inst.

Warren, B. (1934). The diffraction of x-rays in glass. *Physical Review* 45: 0657–0661.

West, J.K., Mecholsky, J.J. Jr., and Hench, L.L. (1999). The application of fractal and quantum geometry to brittle fracture. *Journal of Non-Crystalline Solids* 260: 99–108.

Willmott, G. and Radford, D. (2005). Taylor impact of glass rods. *Journal of Applied Physics* 97: 093522.

Wong, J., Ferriera, J., Lindsey, E. et al. (2006). Morphology and microstructure in fused silica induced by high fluence ultraviolet 3ω (355 nm) laser pulses. *Journal of Non-Crystalline Solids* 352: 255–272.

Xu, L. and Rosakis, A. (2002a). Impact failure characteristics in sandwich structures part I: basic failure mode selection. *International Journal of Solids and Structures* 39: 4215–4235.

Xu, L. and Rosakis, A. (2002b). Impact failure characteristics in sandwich structures part II: effects of impact speed and interfacial strength. *International Journal of Solids and Structures* 39: 4237–4248.

Xu, L. and Rosakis, A. (2003a). An experimental study of impact-induced failure events in homogeneous layered materials using dynamic photoelasticity and high-speed photography. *Optics and Lasers in Engineering* 40: 263–288.

Xu, L. and Rosakis, A. (2003b). Real-time experimental investigation of dynamic crack branching using high-speed optical diagnostics. *Experimental Techniques* 27: 23–26.

Xu, L. and Rosakis, A. (2005). Impact damage visualization of heterogeneous two-layer materials subjected to low-speed impact. *International Journal of Damage Mechanics* 14: 215–233.

Xu, L., Huang, Y., and Rosakis, A. (2003). Dynamic crack deflection and penetration at interfaces in homogeneous materials: experimental studies and model predictions. *Journal of the Mechanics and Physics of Solids* 51: 461–486.

8

Emerging Directions: Ceramics with Tailored Properties

8.1 Introduction

A new market report by Transparency Market Research shows that the global ceramics market reached 229.13 billion USD in 2018 and is projected to grow at a compound annual growth rate of 8.6% from 2019 to 2025 (Sergey 2020). Based on their inherent natural behaviors as well as their industrial demand, "ceramics and transparent materials" are technologically important materials. The dynamic behavior of materials continues to be an active research area due to sustained demand for lightweight, high hardness, and high strength materials for military, defense, and aerospace applications. Ceramics must also retain their strength and structure under the demanding conditions of high pressure, high rates of loading, and high temperature, which accompanies dynamic deformation. However, only a few materials currently exist that can address the growing demand of stringent criteria of material performance in today's age, e.g. light weight coupled with high hardness and strength at elevated temperatures, etc. While we have touched upon certain classes of ceramics in this monograph, yet, as demand for sophisticated performance grows, new materials continue to emerge with claims of superior performance or suitability for many applications. For example, 3D-printed materials are promising alternatives for conventional ceramics, but their behavior is yet unknown due to the nonconventional microstructure that evolves during this manufacturing process. Similarly, materials with dopants are suggested for further improvements in performance of icosahedral ceramics, but their dynamic constitutive response is yet to be unearthed.

In addition, the field of dynamic behavior encompasses a plethora of mechanisms and deformation modes including elastic, plastic, and shock-wave propagation, phase transformation, fracture, heat generation, coupled thermo-mechanical responses, and more. Their complex interplay during

Dynamic Response of Advanced Ceramics, First Edition.
Ghatu Subhash, Amnaya Awasthi, Dipankar Ghosh.

any given dynamic event is difficult to assess using mechanistic approaches alone. Statistical approaches coupled with new data-driven methods must be developed to gain in-depth insight for better design of materials with the desired resistance against the defined dynamic loads. Concurrently, developments in novel experimental and characterization methods as well as sensors and instrumentation that can capture rapid changes in the field are essential for fundamental advances in the field.

Ceramics are excellent contenders for being pursued through the materials informatics drive because of their vastly different crystal structures, material properties, varying static and dynamic behaviors, potential in a variety of applications for high performance and finally, numerous possibilities in research and development. From this perspective, we have highlighted few material systems that hold high promise for dynamic applications.

Broadly, "materials science" or "materials engineering" is a field that focuses on developing not only materials with improved properties but also with unique combination of multiple properties of interest. This interdisciplinary field utilizes information from both experimental and computational realms with the goal of achieving superior material behavior. Central to this philosophy of developing superior materials is a concept called the "Iterative Materials Tetrahedron" (Meyers and Chawla 2008), which calls for interdependence among components of materials development: "materials theory," "synthesis and processing technologies," "structural characterization," and "properties of interest." Within this broad perspective of materials development, new refinements have been proposed (Kalidindi and De Graef 2015; McMillan 2003; Liu et al. 2017, Zhao et al. 2016) due to the recent rise in computational power. Such efforts, introduced in the past decade, are based on materials knowledge gained from a variety of computational investigations. Collectively such works have been coined the general vision "Materials Data Science" (MDS). The branch of knowledge that focuses mainly on materials discovery using materials data, whether acquired computationally or through experimental means has recently been given the term "Materials Informatics" by Rajan (2005, 2013, 2015). The sheer large possibility of smart tools available to process the vast volume of data has geared into efforts known as "big-deep-smart-data" (Kalinin et al. 2015), with the goal of seeking novel functional materials. The last few decades have also witnessed establishment of new focused efforts in the United States, such as the Nanotechnology Initiative (NTI), the Integrated Computational Materials Engineering Initiative (ICMEI), the Materials Genomic Initiative (MGI), and the NSF's ten big ideas: "Harnessing Data for 21st Century Science and Engineering" (10 Big Ideas for Future NSF Investments 2019), which have earmarked considerable impetus to materials discovery.

8.2 Theoretical Materials Discovery

Ceramics are a unique class of materials possessing vast capability for extensive development. There are two governing features at the core of this concept: (i) wide range of properties exhibited by these materials, and (ii) their nanoscale structure, which allows infinite combinatorial possibilities. It is not surprising that these ceramic materials could be identified as good candidates for both bottom-up as well as top-down synthesis of advanced materials (Wakamatsu and Salomão 2011). In the following, we review a few advances in modeling nanoscale structures theoretically, which possess high potential of accelerating this field.

8.2.1 Advanced Algorithms for Predicting Crystal Structure

Perhaps the most difficult step in theoretical materials discovery is to predict the equilibrium structure of a compound from its stoichiometry (Maddox 1988). Despite increased research focus in this field, the problem of finding equilibrium structure, with just stoichiometry as the initial guess, is a tough challenge. Development of robust structure prediction tools, which take as input just the number and type of atoms and produce as output a prediction of exhaustive number of ways these atoms could bond, is perhaps the most important step that underscores theoretical materials discovery. The difficulty grows with number of atoms, and it may not be straightforward to find equilibrium configuration of thousands of random atoms. A few notable attempts that have advanced molecular/crystal structure prediction field are Universal Structure Predictor: Evolutionary Xtallography (USPEX) (Glass et al. 2006), XtalOpt (Lonie and Zurek 2011), EVO (Bahmann and Kortus 2013), GAtor (Curtis et al. 2018), and Crystal Structure Analysis by Particle Swarm Optimization (CALYPSO) (Wang et al. 2012), to name a few. Some of these tools, such as CALYPSO, are also equipped with additional features, wherein the software can perform equilibration, given external conditions as input (e.g. pressure). These tools could be effectively applied to study a variety of problems, including the effect of dopants, effect of defects, development of hybrid materials, etc.

Once the equilibrium structures are ascertained, obtaining theoretical estimates of stiffness and strength can easily be performed by existing density functional theory (DFT)/semi-empirical/molecular dynamics (MD) capability, and is relatively straightforward. Almost all currently available open-source and proprietary software possess capability to apply boundary conditions of deformation and stress and therefore studying material response over a given set of external conditions can be performed using in-built subroutines of atomistic packages. Such prediction of response to external conditions provide insight into complex atomic level changes that can occur under high rate deformation for screening materials and mechanisms. The struggle between computational cost and accuracy still remains, and accurate force fields for MD simulations are continuously sought.

8.2.2 Spectral Predictive Capability

Recalling the interconnectivity between high strain rate deformation and high-pressure response of ceramics (discussed in Chapter 6) and realizing the fact that experimental facilities to investigate high strain rate (shock) response are prohibitively expensive, it is but natural to look for validated computational approaches to gain insight into the behavior of ceramics. We have shown that computational Raman spectroscopy is an appropriate tool to predict high pressure deformation behavior of certain ceramics. A recent material development effort focuses on the correlation between material deformation and evolution of DFT-evaluated Raman spectra (Awasthi and Subhash 2019). It was found that Raman spectra predicted by DFT simulations were quite effective not only in identification of residual stresses but also polymorphic constituents (Kunka et al. 2017) in a stoichiometric mix of boron carbide. Such discoveries highlight the need to enhance synergy between experimental vibrational spectroscopy (Raman, IR, near-IR, UV, visible) and evaluation of vibration spectra using quantum computations. Even though prediction of spectra of single crystals is accurate, the quantum computational methods are limited in scalability due to the compute-intensive nature of these calculations. The community focused on developing Raman spectra of inorganic minerals, WURM (Caracas and Bobocioiu 2011), has seen a steady growth in the past years, tabulating over 450 DFT-evaluated Raman spectra of 325 minerals till date. Currently, the minimum size of laser used in most of Raman spectra equipment is capped at a diameter of 1 μm. A need has been felt (Awasthi and Subhash 2019) to increase resolution of areal coverage of Raman spectra to obtain accurate composition in sub-micron material domains, such as amorphized islands (see Figure 6.20a). Link between spectral signature and underlying molecular structure (Fisette et al. 2012) has given rise to interest in developing predictive capabilities for complex atomistic systems including biological molecules, especially geared toward protein structure determination (Fawzi et al. 2008; Fisette et al. 2010). This synergistic platform has found less application in research and development of ceramic materials than biological/pharmaceutical counterparts (Mazurek et al. 2020) and is hence an area of great potential for developing next generation of ceramics (for more details, see section 6.3.3 in Awasthi and Subhash (2020)).

8.2.3 Prediction of Hardness

Another development from the materials community aims at evaluating theoretical (intrinsic) hardness of material from its crystal structure. Four models exist currently, the electronegativity model (Li et al. 2008), the bond strength model (Šimůnek and Dušek 2017), the band gap model (Tian et al. 2012), and the thermodynamic model (Mukhanov et al. 2010). These models are primarily based on averaging strengths across all bonds in a crystal unit cell. These models provide a first guess on predicting true hardness of a material and their accuracy is continually being assessed against experimental hardness data for complex systems (Cheenady et al. 2020). However, there is no concurrence between the predictions of these models especially when the models are applied to broad classes of ceramics (e.g. icosahedral ceramics, transitional metal hexaborides, light weight meta oxides), and hence limit our ability to select suitable materials for specific applications or design of materials with proper selection and placement of atoms. Although this area is under active development, seeking more refined methodologies for predicting intrinsic hardness (Cheenady et al. 2020), yet, more detailed physics-based studies are needed that can accurately model the fundamental source of hardness at the atomic level, so that the methodology can be easily extended to new materials.

8.2.4 P-V Behavior from Atomistic Calculations

Another area of interest for proper selection of materials is to understand their behavior or response to extreme conditions of pressure, temperature, chemical environment, or radiation. For example, shock events occur at short time intervals, and experimental efforts to obtain shock behavior are prohibitively expensive, time-consuming, and most importantly, the data obtained from such experiments is limited to only shock velocity-particle velocity ($u_s - u_p$) relationship. Atomistic simulations can provide fundamental thermodynamic details such as temperature rise, diffusion rates, the bond angle between constituent atoms before and after shock, and the ability to probe and visualize operating mechanisms of deformation *in situ*. Therefore, emphasis should be placed on such simulations to gain better understanding of the deformation behavior of ceramics under such high-pressure states. Simulations of volumetric compression, as outlined by (Awasthi and Subhash 2019), for understanding thermodynamic states under shock loading in boron carbide, should be pursued further to decipher the influence of defects and propensities of polymorphs for high pressure loading. Recent advancements in MD shock simulations of boron carbide (DeVries et al. 2020), silicon carbide (Branicio et al. 2018, Li et al. 2017), and aluminum nitride (Branicio et al. 2006) using state-of-the-art intermolecular potentials (see Chapter 5) have not only demonstrated reliability of simulations in predicting experimental behavior but also alleviated the need for expensive and time-consuming experimentation.

8.2.5 Atomic-level Synthesis of Ceramics

The idea of mixing stoichiometries in preparing multi-polymorph/multi-polytype compounds has tremendous potential. This method has the capability to produce tailored mixtures for desired properties, especially in the dynamic regime. Previous computational studies on the icosahedral family of boron-rich ceramics (Kunka et al. 2016, 2017; Fanchini et al. 2006) has shown that they are good contenders for such efforts. The limitation still exists in experimental validation of those computational studies, which is complicated by complexities in evaluating the final atomic structure of fabricated mix of stoichiometric compounds. Concerted efforts are needed to connect the crystal structure of polymorph-mixes with overall properties of such compounds, along with development of capability to produce not just known stoichiometries but also known atomic-level crystal structure of such mixes. Such material mixing, including the effect of dopants can be studied for atomic-level structure through recently developed crystal-structure prediction methods, including the USPEX (Glass et al. 2006) and CALYPSO (Wang et al. 2010). Since all the ceramic materials and transparent glasses of interest are multi-element compounds, they are all good candidates for development of CALPHAD databases (Saunders and Miodownik 1998). These databases are compilations of thermodynamic properties of multicomponent materials, developed through computational methods, usually a combination of Gibbs free energy and equilibration algorithms.

8.3 Future Prospects and Challenges in Fabricating Advanced Ceramics for Dynamic Applications

On the experimental side, there are numerous new directions pertaining to the development of resilient high-performance ceramics. Such directions have come forth as the culmination of several years of research efforts. In addition to enumerating these directions, we also acknowledge challenges associated with each endeavor.

8.3.1 Boron–Aluminum–Magnesium (BAM) Materials

Some compounds have undergone renewed interest, the chiefs of which are B_6O and "Boron–Aluminum–Magnesium" (BAM) materials. Currently, no data exists on their response to dynamic loading including shock loading. BAM materials were first reported by Matkovich and Economy (1970), and though they have been shown to hold promise (Canter 2009; Grishin et al. 2015), these materials have been cursorily investigated (Zhukov et al. 2019) for their crystal structure, amorphous behavior, and hardness through very few studies (Ivashchenko et al. 2016). Hence, the challenge for icosahedral family of boron-based ceramics for investigation of their enhanced performance over a wide range of loading conditions including shock, and development of processing and fabrication techniques.

8.3.2 Nanograining

Another interesting direction with tremendous potential for improved performance of icosahedral ceramics is that of efforts related to tailoring of grain size, especially nanograining (DeVries et al. 2016), which, as discussed in Chapter 5, has shown to benefit, not only static but also dynamic properties of boron carbide. Nanograining has proven to reduce propensity of amorphization substantially in born carbide. However, nanograining has not been studied for other materials in the icosahedral family of boron-rich ceramics, particularly B_6O and "BAM" materials. There are very few facilities around the world that can synthesize nano powders of these ceramics and that limits the scalability of this research area. Challenges also exist in consolidation techniques such as spark plasma sintering (SPS) which can retain they nanograined structure during processing of large-scale ceramic components. There are very few facilities in the world with large capacity SPS machines to fabricate these materials at high temperatures and in large batches.

8.3.3 Nanotwinning

In recent years, nanotwinning, i.e. twins at nanometer spacing within the grains, has emerged as a promising microstructural feature to dramatically enhance macroscopic mechanical properties such as hardness, fracture strength, fatigue strength, yield strength, etc. of a vast variety of materials (Huang et al. 2014; Lu et al. 2009). In high hardness ceramics such as boron suboxide (B_6O), nanotwinning has been found to be a beneficial feature that enhances hardness and mitigates amorphization (Kunka et al. 2018a). However, the difference in energies for the formation of a nanotwinned structure and nontwinned structure is small (~3 meV) (Kunka et al. 2018b), and hence, it is difficult to obtain a uniform nanotwinned structure in the microstructure due to inevitable fluctuations that may occur in processing conditions. The benefit of nanotwinning should be probed by further improving control in processing parameter fluctuations for better microstructure design. Additionally, knowledge of how to introduce nanotwinned structure is in its infancy. Methods of artificially inducing nanotwinning, or fabricating material with tailored twinned nanostructure or having selectively twinned zones, are currently not available. There is a need for generating knowledge on the effect of twinning on static and dynamic properties across different length scales, likely through a mix of atomistic and mesoscale modeling. Such studies performed across the wide span of crystal structures of interest will be useful in determining the effectiveness of nanotwinning broadly. Thermodynamic drivers, including temperature and pressure that lead to nanotwinning in these ceramics is not known yet, and they are warranted to precisely predict and control the extent of nanotwinning in a given material.

8.3.4 Additive Manufacturing

Currently, the materials discussed in previous chapters are fabricated at high temperature using conventional processing methods such as hot pressing (HP), sintering, reaction bonding, or SPS. Additive manufacturing or 3D printing is rapidly evolving as a promising tool to fabricate high-temperature ceramics. Even though there have been tremendous developments in the utility of 3D printed ceramics for different applications (Scheithauer et al. 2015; Bose et al. 2013; Chia and Wu 2015), the knowledge of evolution of microstructure and properties in this method is however in its infancy (DeVries et al. 2018). Once optimal microstructures are developed through this process, concerted effort is warranted in connecting fabrication process to properties, and potential performance under static and dynamic loads. The vast differences in the microstructure between the conventionally manufactured and 3D printed materials is expected to yield vastly different dynamic responses and renewed efforts should be pursued to understand the response of these new generation of materials for continued progress and realization of their promise and potential. Finally, efforts toward theoretical prediction, validation, and optimization of properties for 3D printed ceramic materials is yet to be attempted.

8.3.5 Shock Studies

In situations where shock data is not available, confinement data in quasi-static conditions has been found to be equally effective in understanding material behavior and fragmentation behavior under high pressure loads. Recall that shock loading is primarily a high-pressure loading at high strain rate. Hence, new techniques that can investigate material response under multi-axial confinement are desired. On the theoretical side, while micromechanical models are useful in understanding the influence of various microstructural features, for e.g. crack-growth, effect of crack distributions, effect of inhomogeneities, etc. new models for ultra-high-pressure behavior are needed for effective use in computational simulations.

8.3.6 Effect of Distention

Few studies have investigated the effect of distention (porosity) on shock properties of ceramics (Grady et al. 2013, 2014). Additional work is warranted on the precise measurement of distention and its correlation with dynamic behavior of ceramics, especially in connection with their equations of state. Accurate theoretical models are needed to include the effect of distention on mechanical properties of ceramics. Once developed, validation schemes will also be needed to complement the modeling efforts.

8.4 Closing Thoughts

The next technological breakthrough in materials science will likely spur from materials discovery forced upon by current drivers of data science. Built upon decades of materials modeling and experimentation, the new era will provide capability to quickly create desired materials with capabilities precisely tailored to almost user-driven specifications, including extensive multifunctionality. Icosahedral boron-rich ceramics offer the needed test bed to create that reality. Based on research findings described in this book, icosahedral ceramics appear to be the most promising candidates for providing the desired strength characteristics for impact applications. This is due to the vast exploration space presented by their crystal structure and multifunctional properties spread over different time and length scales. We have only scratched the surface in understanding this class of compounds, and much more remains to be unveiled. Perhaps the biggest challenge facing MDS is the assimilation, usage, and creation of knowledge, specifically at a fast pace, based on inputs from different research groups from across the world. Our current pace of materials discovery is rather slow, and may have to be re-examined, so that time taken to perform research endeavors, including that for publishing, could be further reduced.

Authors hope that the breadth of knowledge presented in previous chapters will invigorate awareness about these materials and their potential in bringing technological revolution in material tailoring for specific dynamic applications. We also hope that such awareness will further incite the needed thrust in bringing together multi-skilled efforts for the goal of high-end materials discovery and development. As an example, recently, Awasthi and Subhash (2020) proposed a strategy for combinatorial-level material design through material discovery for icosahedral boron-rich ceramics. This calls for collective efforts in fields of computation as well as experimentation into synergistically driving materials informatics of icosahedral boron-rich ceramics (see Figure 8.1).

It is the dynamic performance of ceramics that makes a variety of human endeavors possible. Perhaps the most glaring of these applications is space

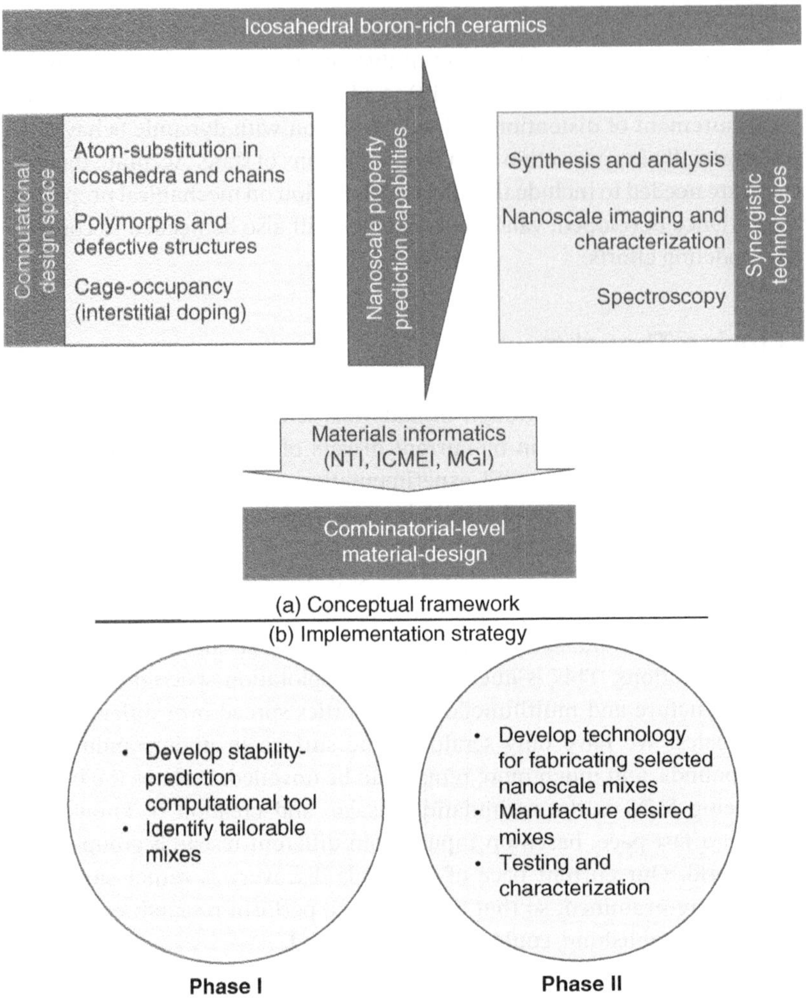

Figure 8.1 Strategy for material discovery development by combinatorial paradigm for icosahedral boron-rich ceramics. (a) Putting together computational design space inherent in these compounds with synergistic experimental technologies through materials informatics is key to the proposed strategy. (b) Proposed phases of development of the proposed strategy. Source: Reproduced with permission from Awasthi and Subhash (2020). © 2020 Elsevier.

exploration. Due to their resilience to compression, resistance to corrosion, and high melting points, ceramics are becoming more promising for the aerospace industry than ever before (Parker 2006). There are at least three hot-beds of application of high-performance ceramic materials in aerospace: (i) propulsion and exhaust washed structures, (ii) thermal protection systems, and (iii) hot primary structure. (Steyer 2013). There is significant thrust on developing ceramic matrix composites and ultra-high-temperature ceramics specifically geared for high-performance needs of the aerospace industry. Though their potential speaks-out clearly enough, industry-level production and use of advanced ceramics warrants development of rigorous theoretical analysis. The associated research areas currently underway for high-performance ceramics for aerospace applications include damage accumulation mechanisms/models, life predictions methodologies/modeling, nondestructive inspection techniques, and development of robust field and depot-level repair methods (Steyer 2013). Additive manufacturing could promise value in producing a host of aerospace components from advanced ceramics (Froes and Boyer 2019), and synergy between processing, characterization, modeling and data-sciences is needed to open new businesses targeted toward pushing development of high-performance ceramics for aerospace applications.

Perhaps the demand for highly versatile ceramics with a wide range of applicability is in the medical industry. Depending upon application, ceramic materials may even have to interface with human tissue and organs, e.g. in the orthopedic (Rieger 2001) and dental applications (Denry and Holloway 2010), and hence those ceramics are also termed "technical ceramics" or "tech-ceramics". Catchy jargon aside, ceramics are the only class of materials which have made these advancements possible. Obtaining tunable properties from ceramics, catering to both bioactive and bioinert applications, is one of the biggest challenges faced by the biomedical industry (Vallet-Regí and Salinas 2019). These applications and associated R&D-communities and industries currently face needs for materials with customized dynamic properties and behaviors, hence, rigorous advancements in material discovery is critical for such needs (Vallet-Regí 2001; Chevalier and Gremillard 2009).

Dynamic behavior of materials continues to be a vibrant field with numerous scientific issues yet to be unlocked. While the nature of the field is such that higher strain rates demand lower time scales, design of experiments to deliver such short time and high loads becomes prohibitively expensive. Atomistic modeling using MD simulations are an ideal avenue to unlock the mechanisms that prevail at such small spacio-temporal scales. However, one of the biggest drawbacks is availability of interatomic potentials for these materials. Hence, to effectively progress this field further, there is a greater need for dedicated research focus on both developing potentials for each class of materials and demonstrating usability of semi-empirical methods to treat such materials, for producing efficient computational models and boosting their overall efficiency. *In situ* measurement of field quantities is vital to understand the evolving mechanisms of deformation at high strain rates. These *in situ* diagnostic tools that measure temperature, particle velocity (or displacement) at the time and length scales of interest, and their development continues to be a challenge. It has also been well recognized in the community that experimental methods that bridge the gap between quasi-static deformation and dynamic deformation continue to be elusive. However, recent innovative modifications to Split-Hopkinson pressure bar (SHPB), called serpentine bars, have expanded the incident pulse durations to several tens of milli seconds thus allowing us to test materials at strain rates as low as $10\text{–}100\,s^{-1}$ (Whittington et al. 2015). Such advancements deem collaboration with theoretical and data scientists to create and validate material deformation models with higher levels of accuracy. Spectroscopic tools to measure differences in crystal structure and composition at small-length scales (e.g. amorphized regions) continues to be a challenge to fully decipher the mechanisms of deformation. Use of spectroscopic measurements in theoretical models, development of theoretical prediction of spectroscopic behavior owing to deformation and damage, and developing diagnostic devices informed with spectroscopic behavior of various modes of material deformation, are research areas far from maturity. Though the basic theory is well established at the atomic level, its multiscale application in building macroscale models and diagnostic devices is far from our current horizon.

For continued advancement in this field, developments in experimental, computational, and theoretical realms is essential along with training and education of next generation of scientists. This is where books of this nature are needed with compilation of results from recent literature summarizing the latest advancements in the field to make that knowledge accessible. Our sincere hope is that we have made a small step in the right direction. We hope that this book will inspire leaders and drive effective collaboration among different realms of the field of ceramics development into realizing their inherent value and performance for betterment of society.

References

Awasthi, A. and Subhash, G. (2019). High-pressure deformation and amorphization in boron carbide. *Journal of Applied Physics* 125: 215901.

Awasthi, A. and Subhash, G. (2020). Deformation behavior and amorphization in icosahedral boron-rich ceramics. *Progress in Materials Science* 112: 100664.

Bahmann, S. and Kortus, J. (2013). EVO–evolutionary algorithm for crystal structure prediction. *Computer Physics Communications* 184: 1618–1625.

Bose, S., Vahabzadeh, S., and Bandyopadhyay, A. (2013). Bone tissue engineering using 3D printing. *Materials Today* 16: 496–504.

Branicio, P., Kalia, R., Nakano, A., and Vashishta, P. (2006). Shock-induced structural phase transition, plasticity, and brittle cracks in aluminum nitride ceramics. *Physical Review Letters* 96: 065502.

Branicio, P., Zhang, J., Rino, J. et al. (2018). Plane shock loading on mono- and nano-crystalline silicon carbide. *Applied Physics Letters* 112: 111909.

Canter, N. (2009). Upgrading biodegradable lubricant performance. *Tribology & Lubrication Technology* 65: 14.

Caracas, R. and Bobocioiu, E. (2011). The WURM project–a freely available web-based repository of computed physical data for minerals. *American Mineralogist* 96: 437–443.

Cheenady, A.A., Awasthi, A., and Subhash, G. (2020). Intrinsic hardness of boron carbide: influence of polymorphism and stoichiometry. *Journal of the American Ceramic Society* 103: 7127.

Chevalier, J. and Gremillard, L. (2009). Ceramics for medical applications: a picture for the next 20 years. *Journal of the European Ceramic Society* 29: 1245–1255.

Chia, H.N. and Wu, B.M. (2015). Recent advances in 3D printing of biomaterials. *Journal of Biological Engineering* 9: 4.

Curtis, F., Li, X., Rose, T. et al. (2018). GAtor: a first-principles genetic algorithm for molecular crystal structure prediction. *Journal of Chemical Theory and Computation* 14: 2246–2264.

Denry, I. and Holloway, J.A. (2010). Ceramics for dental applications: a review. *Materials* 3: 351–368.

DeVries, M., Pittari, J., Subhash, G. et al. (2016). Rate-dependent mechanical behavior and amorphization of ultrafine-grained boron carbide. *Journal of the American Ceramic Society* 99: 3398–3405.

DeVries, M., Subhash, G., Mcghee, A. et al. (2018). Quasi-static and dynamic response of 3D-printed alumina. *Journal of the European Ceramic Society* 38: 3305–3316.

DeVries, M., Subhash, G., and Awasthi, A. (2020). Shocked ceramics melt: An atomistic analysis of thermodynamic behavior of boron carbide. *Physical Review B* 101: 144107.

Ektarawong, A., Simak, S.I., Hultman, L. et al. (2016). Effects of configurational disorder on the elastic properties of icosahedral boron-rich alloys based on B_6O, $B_{13}C_2$, and B_4C, and their mixing thermodynamics. *The Journal of Chemical Physics* 144: 134503.

Fanchini, G., McCauley, J., and Chhowalla, M. (2006). Behavior of disordered boron carbide under stress. *Physical Review Letters* 97: 035502.

Fawzi, N.L., Phillips, A.H., Ruscio, J.Z. et al. (2008). Structure and dynamics of the Aβ 21–30 peptide from the interplay of NMR experiments and molecular simulations. *Journal of the American Chemical Society* 130: 6145–6158.

Fisette, O., Morin, S., Savard, P.-Y. et al. (2010). TEM-1 backbone dynamics–insights from combined molecular dynamics and nuclear magnetic resonance. *Biophysical Journal* 98: 637–645.

Fisette, O., Lagüe, P., Gagné, S., and Morin, S. (2012). Synergistic applications of MD and NMR for the study of biological systems. *Journal of Biomedicine and Biotechnology* 2012: 1–12.

Froes, F.H. and Boyer, R. (2019). *Additive Manufacturing for the Aerospace Industry*. Elsevier.

Glass, C.W., Oganov, A.R., and Hansen, N. (2006). USPEX–evolutionary crystal structure prediction. *Computer Physics Communications* 175: 713–720.

Grady, D., Fenton, G., and Vogler, T. (2013). Equation of state and evidence of enhanced phase transformation for shock compression of distended compounds. *International Journal of Impact Engineering* 56: 19–26.

Grady, D., Fenton, G., and Vogler, T. (2014). Equation of state for shock compression of distended solids. *Journal of Physics: Conference Series* 500: 152007.

Grishin, A., Khartsev, S., Böhlmark, J., and Ahlgren, M. (2015). Ultra-hard $AlMgB_{14}$ coatings fabricated by RF magnetron sputtering from a stoichiometric target. *JETP Letters* 100: 680–687.

Huang, Q., Yu, D., Xu, B. et al. (2014). Nanotwinned diamond with unprecedented hardness and stability. *Nature* 510: 250–253.

Ivashchenko, V., Turchi, P.E., Veprek, S. et al. (2016). First-principles study of crystalline and amorphous $AlMgB_{14}$-based materials. *Journal of Applied Physics* 119: 205105.

Kalidindi, S.R. and De Graef, M. (2015). Materials data science: current status and future outlook. *Annual Review of Materials Research* 45: 171–193.

Kalinin, S.V., Sumpter, B.G., and Archibald, R.K. (2015). Big-deep-smart data in imaging for guiding materials design. *Nature Materials* 14: 973–980.

Kunka, C., Awasthi, A., and Subhash, G. (2016). Crystallographic and spectral equivalence of boron-carbide polymorphs. *Scripta Materialia* 122: 82–85.

Kunka, C., Awasthi, A., and Subhash, G. (2017). Evaluating boron-carbide constituents with simulated Raman spectra. *Scripta Materialia* 138: 32–34.

Kunka, C., An, Q., Rudawski, N. et al. (2018a). Nanotwinning and amorphization of boron suboxide. *Acta Materialia* 147: 195–202.

Kunka, C., Yang, X., An, Q., and Subhash, G. (2018b). Icosahedral superstrength at the nanoscale. *Physical Review Materials* 2: 063606.

Li, K., Wang, X., Zhang, F., and Xue, D. (2008). Electronegativity identification of novel superhard materials. *Physical Review Letters* 100: 235504.

Li, W., Yao, X., Branicio, P. et al. (2017). Shock-induced spall in single and nanocrystalline SiC. *Acta Materialia* 140: 274–289.

Liu, Y., Zhao, T., Ju, W., and Shi, S. (2017). Materials discovery and design using machine learning. *Journal of Materiomics* 3: 159–177.

Lonie, D.C. and Zurek, E. (2011). XtalOpt: an open-source evolutionary algorithm for crystal structure prediction. *Computer Physics Communications* 182: 372–387.

Lu, L., Chen, X., Huang, X., and Lu, K. (2009). Revealing the maximum strength in nanotwinned copper. *Science* 323: 607–610.

Maddox, J. (1988). Crystals from first principles. *Nature* 335: 201–201.

Matkovich, V. and Economy, J. (1970). Structure of $MgAlB_{14}$ and a brief critique of structural relationships in higher borides. *Acta Crystallographica Section B: Structural Crystallography and Crystal Chemistry* 26: 616–621.

Mazurek, A.H., Szeleszczuk, Ł., and Pisklak, D.M. (2020). Periodic DFT calculations–review of applications in the pharmaceutical sciences. *Pharmaceutics* 12: 415.

McMillan, P. (2003). New materials from high pressure experiments: challenges and opportunities. *High Pressure Research* 23: 7–22.

Meyers, M.A. and Chawla, K.K. (2008). *Mechanical Behavior of Materials*. Cambridge university press.

Mukhanov, V., Kurakevych, O., and Solozhenko, V. (2010). Thermodynamic model of hardness: particular case of boron-rich solids. *Journal of Superhard Materials* 32: 167–176.

NSF News. (2019). Special Report: 10 Big Ideas. www.nsf.gov/about/congress/reports/nsf_big_ideas.pdf.

Parker, K. (2006). Advanced ceramics soar to new heights: Advantageous physical properties help these inorganic, nonmetallic materials expand their usage into the manufacture of critical aerospace components. *Metal Finishing* 104: 16–18.

Rajan, K. (2005). Materials informatics. *Materials Today* 8: 38–45.

Rajan, K. (2013). *Informatics for Materials Science and Engineering: Data-Driven Discovery for Accelerated Experimentation and Application*. Butterworth-Heinemann.

Rajan, K. (2015). Materials informatics: the materials "Gene" and big data. *Annual Review of Materials Research* 45: 153–169.

Rieger, W. (2001). *World Tribology Forum in Arthroplasty*, 283–294. Suisse: Hans Huber Verlag Bern.

Saunders, N. and Miodownik, A.P. (1998). *CALPHAD (Calculation of Phase Diagrams): A Comprehensive guide*. Elsevier.

Scheithauer, U., Schwarzer, E., Richter, H.J., and Moritz, T. (2015). Thermoplastic 3D printing–an additive manufacturing method for producing dense ceramics. *International Journal of Applied Ceramic Technology* 12: 26–31.

Sergey, S. (2020). Market analysis. *Research and Reports on Metals* 3: 1–2.

Šimůnek, A. and Dušek, M. (2017). Generalized bond-strength model of vickers hardness: application to Cr_4B, CrB, CrB_2, CrB_4, MO_2B, MOB_2, OsB_2, ReB_2, WB_2, WB_3, and $Ti_{1.87}B_{50}$. *Mechanics of Materials* 112: 71–75.

Steyer, T.E. (2013). Shaping the future of ceramics for aerospace applications. *International Journal of Applied Ceramic Technology* 10: 389–394.

Tang, B., An, Q., and Goddard, W.A. III, (2015). Improved ductility of boron carbide by microalloying with boron suboxide. *The Journal of Physical Chemistry C* 119: 24649–24656.

Tian, Y., Xu, B., and Zhao, Z. (2012). Microscopic theory of hardness and design of novel superhard crystals. *International Journal of Refractory Metals and Hard Materials* 33: 93–106.

Vallet-Regí, M. (2001). Ceramics for medical applications. *Journal of the Chemical Society, Dalton Transactions*: 97–108.

Vallet-Regí, M. and Salinas, A.J. (2019). *Bone Repair Biomaterials*, 141–178. Elsevier.

Wakamatsu, M.H. and Salomão, R. (2011). (Unintentional) synthesis of ceramic nanoparticles. *Interceram - International Ceramic Review* 60: 364–369.

Wang, Y., Lv, J., Zhu, L., and Ma, Y. (2010). Crystal structure prediction via particle-swarm optimization. *Physical Review B* 82: 094116.

Wang, Y., Lv, J., Zhu, L., and Ma, Y. (2012). CALYPSO: a method for crystal structure prediction. *Computer Physics Communications* 183: 2063–2070.

Whittington, W., Oppedal, A., Francis, D., and Horstemeyer, M. (2015). A novel intermediate strain rate testing device: the serpentine transmitted bar. *International Journal of Impact Engineering* 81: 1–7.

Zhao, Z., Xu, B., and Tian, Y. (2016). Recent advances in superhard materials. *Annual Review of Materials Research* 46: 383–406.

Zhukov, I., Nikitin, P.Y., Grigoriev, M., and Vorozhtsov, A. (2019). Phase composition and hardness of $AlMgB_{14}$-based polycrystalline materials. *Russian Physics Journal* 62: 882–885.

Index

Dynamic Response of Advanced Ceramics, First Edition.
Ghatu Subhash, Amnaya Awasthi, Dipankar Ghosh.

Printed and bound by CPI Group (UK) Ltd, Croydon, CR0 4YY
19/09/2026
14965105-0001